OPTICAL THIN FILMS

USERS' HANDBOOK

McGRAW-HILL OPTICAL AND ELECTRO-OPTICAL ENGINEERING SERIES
Robert E. Fischer and Warren J. Smith, Series Editors

Published

Rancourt • *Optical Thin Films Users' Handbook*
Nishihara, Haruna, Suhara • *Optical Integrated Circuits*

Forthcoming Volumes

Fantone • *Optical Fabrication*
Fischer • *Optical Design*
Gilmore • *Expert Vision Systems*
Jacobson • *Thin Film Deposition*
Johnson • *Infrared System Design*
Ross • *Laser Communications*
Stover • *Optical Scattering*
Thelen • *Design of Optical Interference Coatings*
Vukobratovich • *Introduction to Optomechanical Design*

Other Published Books of Interest

Cselt • *Optical Fiber Communications*
Hecht • *The Laser Guidebook*
Hewlett-Packard • *Optoelectronics Applications Manual*
Kao • *Optical Fiber Systems*
Macleod • *Thin-Film Optical Filters*
Marshall • *Free-Electron Lasers*
Optical Society of America • *Handbook of Optics*
Smith • *Modern Optical Engineering*
Texas Instruments • *Optoelectronics*
Wyatt • *Radiometric System Design*

OPTICAL THIN FILMS

USERS' HANDBOOK

JAMES D. RANCOURT

Optical Coating Laboratory, Inc.
Santa Rosa, California

McGraw-Hill Publishing Company

New York St. Louis San Francisco Auckland
Bogotá Hamburg London Madrid Milan Mexico
Montreal New Delhi Panama Paris São Paulo
Singapore Sydney Tokyo Toronto

Printed in the United States of America

Printing: 2 3 4 5 6 7 8 Year: 8 9 5 4 3 2 1 0 9

Library of Congress Cataloging-in-Publication Data
Rancourt, James D.
Optical thin films users' handbook.

(The McGraw-Hill series in optical and electro-optical engineering)
Bibliography: p.
Includes index.
1. Optical films. 2. Light filters. 3. Thin films.
I. Title. II. Series.
TA1522.R36 1987 621.36 87-7693
ISBN 0-07-051199-3

For more information about other McGraw-Hill materials, call 1-800-2-MCGRAW in the United States. In other countries, call your nearest McGraw-Hill office.

Contents

Preface

This book is intended for the user of optical thin film products. Its primary aim is in the area of commercially available products: their preparation techniques and their characteristics. It is hoped that this will help users to specify filters, and to be aware of the limitations they might encounter in using them.

The reader interested in a design textbook is directed to a companion volume in this series written by Alfred Thelen entitled *Design of Optical Interference Coatings*.

The vast majority of the discussions in this text are independent of the method of deposition used—evaporation, sputtering, ion plating, epitaxial growth, or other scheme of depositing optical films. The emphasis is on pragmatic information that one needs when using, specifying, or designing into an optical system an optical interference filter.

A NOTE ON THE STYLE USED

In writing this book, the ubiquity of the computer—handheld, desktop, and mainframe—models, has led to the decision to minimize the number of tables of computed values in favor of curves and formulas. The curves hopefully give the reader a feeling of the phenomenon described by an equation. If precise values are required, they can be easily obtained from the equations.

ACKNOWLEDGMENTS

The support given to me by my employer, O.C.L.I., during the effort of writing this book, is gratefully acknowledged.

I want to acknowledge the help my colleagues and co-workers gave me in answering my many questions in a friendly and helpful way, and for the many suggestions they offered. In particular, I thank Shari Powell Fisher who wrote the BASIC program included in Appendix C, and Jacques Cote, who drew the figures and then cheerfully revised them as I changed my mind about them.

I owe a debt of gratitude and many thanks to Joseph Apfel for his very constructive critique of the manuscript.

Any errors, faults, or omissions in this book are not the fault of any of those mentioned above; they are solely those of the author.

Finally, I thank my family, and especially my wife, Jeannine, for all of her help, patience, and understanding during the writing of this book.

OPTICAL THIN FILMS

USERS' HANDBOOK

Chapter 1

Introduction to Optical Thin Films

In this introductory chapter, we discuss the origins of interference phenomena. We then consider the manner in which these effects are combined to produce practical interference filters. In this book, the terms *filters* and *coatings* are used interchangeably, and mean any type of optical coating deposited on a substrate. For example, an antireflection coating is a filter as is a narrow bandpass coating design. The coating may consist of one or more layers, the optical thickness of which is on the order of a wavelength of light.

We confine our discussion to those coatings that are used in optical devices and for optical purposes. We do not consider nonoptical uses for coatings such as decorative, protective, and electronic.

Assumptions that we make, unless specifically stated otherwise, are:

1. All materials are nonmagnetic
2. The plane of the coatings is infinite in extent so that diffraction may be ignored
3. Materials are isotropic and homogeneous

The term *performance* is frequently used throughout the text. This is intended as a generic term for any characteristic performance parameter that pertains to a thin film coating, such as reflectance, transmission, absorption, or phase.

1.1. INTERFERENCE OF LIGHT

The origins of interference lie in the wave nature of light and the superposition of coherent waves. The incident wave may interfere with one or more of the waves that are reflected from the interfaces formed by the thin film and the substrate. The phases and the amplitudes of these waves determine whether the resultant sum of these waves leads to constructive or destructive interference, and an increase or decrease in the reflectance or the transmittance of the incident light. The incident waves may vary in wavelength, and there is, in general, a dependence of the performance on the wavelength. Coatings with a strong wavelength dependence give rise to useful designs such as narrow bandpass filters. In other situations, one desires a wavelength insensitive performance such as may be found in a wide band antirereflection coating. These different performances are attained by varying the index of the layers, their thicknesses, the order of the layers, and the total number of layers in a stack. In addition, the indices of refraction of the incident medium and of the substrate are important.

Figure 1.1 shows schematically a layered structure of thin films deposited on a substrate. Each layer typically has an index that is different from the adjacent ones, although there are some instances in which they may be the same. The optical thicknesses; i.e., the product of the index of refraction and the physical thickness, of typical layers used in optical filters is of the order of one-quarter of the wavelength of light.

Light is reflected from, as well as transmitted through, each of the interfaces. These transmitted and reflected waves interfere with one another. They can also undergo multiple reflections, and these are taken into account in analytical treatments. The waves are usually all coherent because they originate from the same incident wave.

Figure 1.2 shows the effects of two simple cases of interference. In the first, the two waves add to zero, thus leading to complete destructive interference, whereas in the second illustration, constructive interference takes place.

The normal incidence amplitude reflectance at an interface between two semi-infinite, nonabsorbing materials of different indices of refraction (Fig. 1.3) is given by the Fresnel formula:

$$r = (n_0 - n_1)/(n_0 + n_1), \tag{1.1}$$

where r is the amplitude reflection coefficient, n_0 is the index of refraction of the incident medium, and n_1 is the index of refraction of the second medium.

The transmitted amplitude t is given by $1 - r$.

Note that n_1 is complex when the second medium is absorbing. The incident medium may not be absorbing because the incident wave is assumed to arrive from infinity with no intervening surfaces. Any absorption in the incident medium would attenuate the intensity of the incident wave to zero.

Equation 1.1 is also valid at the interface between two thin films if multiple reflectances are ignored or accounted for separately. In this case, both media may be absorbing, and the corresponding indices n_0 and n_1 are complex, as is the result,

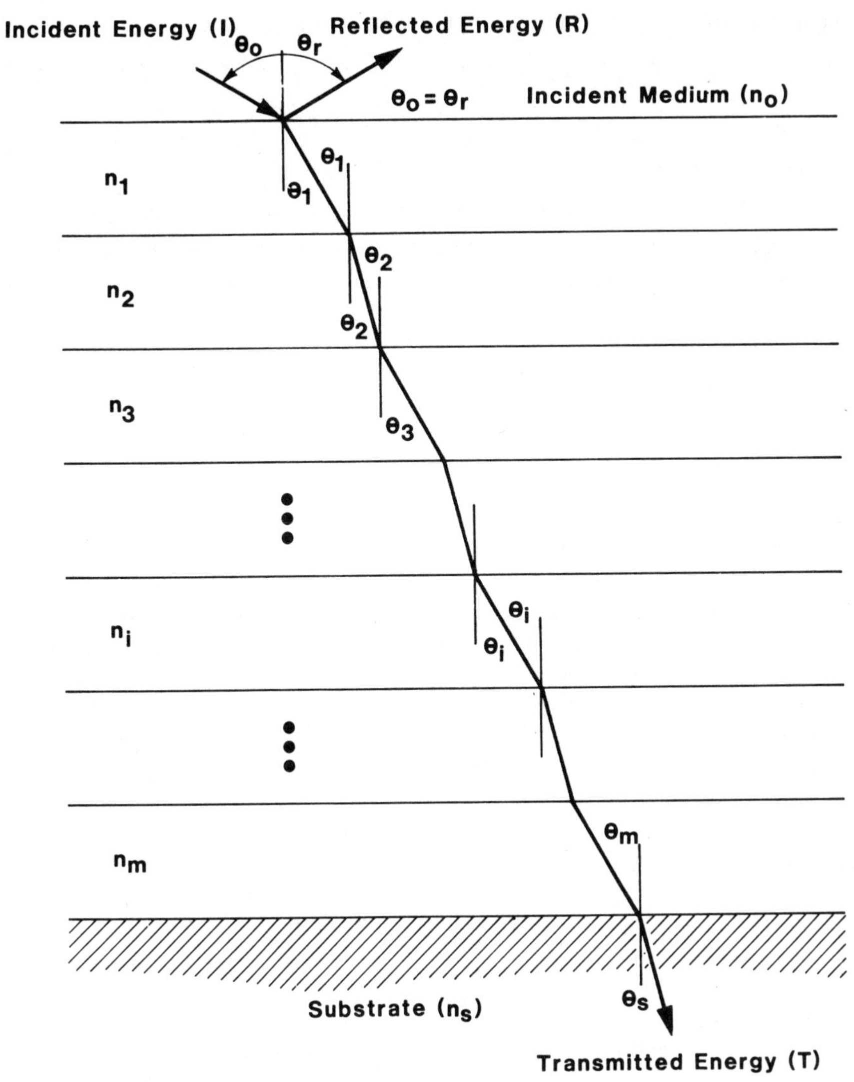

Figure 1.1. Structure of a thin film stack. The index of refraction is indicated by *n*, and the layers are numbered starting from the incident medium. Reflections occur at each interface, and the rays are refracted as they cross each interface between layers.

r. The implication of complex values for n_0 or n_1 is that there will be some absorption of the energy in the light beam.

The reflection coefficient generally refers to the ratio of the reflected intensity to the incident intensity and is given by the complex square of r: $R = r \cdot r^*$. For nonabsorbing substrates, this is

$$R = (n_0 - n_1)^2/(n_0 + n_1)^2,$$

whereas for absorbing substrates it takes on the form,

$$R = [(n_0 - n_1)^2 + k_1^2]/[(n_0 + n_1)^2 + k_1^2] \tag{1.2}$$

where the complex index of refraction of the substrate is given by

$$\hat{N} = n_1 - ik_1.$$

In the literature, the real part of the optical constant is almost universally represented by the letter "n." Some specialties of physics use the dielectric constant, represented by the Greek letter ε, instead of the refractive index. The relations between the two are

$$\begin{aligned}
\varepsilon &= \varepsilon_1 + i\,\varepsilon_2 = (n - ik)^2 \\
n &= \{\,[\varepsilon_1 + (\varepsilon_1^2 + \varepsilon_2^2)^{0.5}]/2\}^{0.5} \\
k &= -\,\{[-\varepsilon_1 + (\varepsilon_1^2 + \varepsilon_2^2)^{0.5}]/2\}^{0.5} \\
\varepsilon_1 &= n^2 - k^2 \\
\varepsilon_2 &= -2nk
\end{aligned}$$

The imaginary portion of the complex index of refraction (k) given above has a variety of names including *extinction index* (Stone, 1963) and *extinction coefficient* (Macleod, 1969). It also appears in different forms, including the product $n\,\kappa$, also called *extinction coefficient* or *attenuation index* (Born and Wolfe, 1970, p. 613). *In this work, only the form* $\hat{N} = n - ik$ *is used.* Care should be exercised when using tabulated data to be certain that it is clear which representation of k a particular author is using.

A value related to k is the absorption coefficient; it is sometimes represented by α and sometimes by β. It is defined as

$$\alpha = 4\,\pi\,k/\lambda$$

where λ is the vacuum wavelength.

It has units of reciprocal length, usually inverse centimeters, and is used to calculate the attenuation of a beam of light in an absorbing material:

$$I/I_0 = \exp(-\alpha d),$$

where I is the emergent intensity, I_0 is the incident intensity, and d is the thickness of the material.

The phase of the reflected wave for nonabsorbing substrate is

$$\phi = \begin{cases} 0 \text{ if } n_0 > n_1 \\ \pi \text{ if } n_0 < n_1. \end{cases}$$

A wave reflected at an air/glass interface undergoes a phase shift of π (180°), and a node in the electric field occurs at the interface. On the other side of the glass plate, at the glass/air interface, no phase shift occurs for the reflected wave.

For absorbing substrates, the situation is more complex; the phase is given by

$$\phi = \tan^{-1} \frac{2\,n_0 k_1}{n_0^2 - n_1^2 - k_1^2}$$

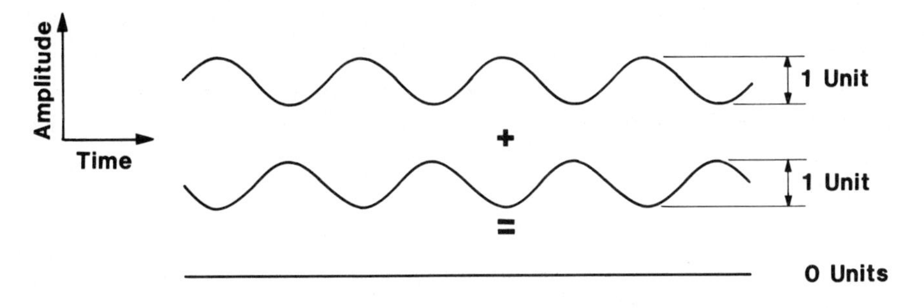

Destructive Interference

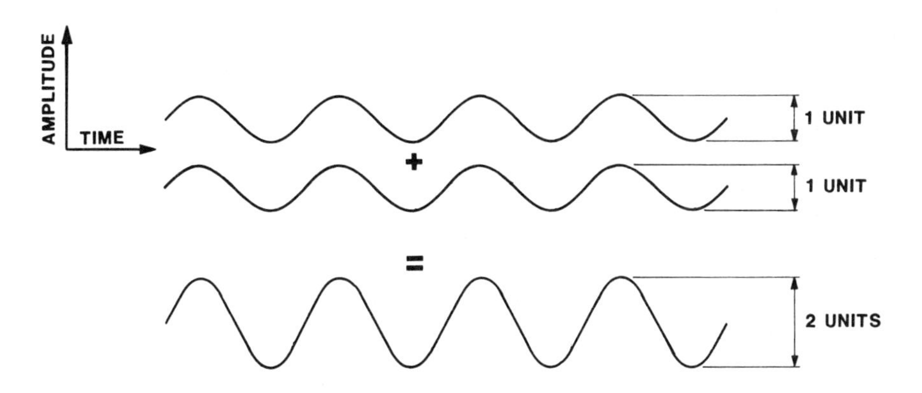

Constructive Interference

Figure 1.2. Constructive and destructive interference of light waves. The amplitudes of two waves with the same frequency (wavelength) add if the two waves are in phase, or cancel if they have the same amplitude and are exactly out of phase with one another. In other cases of unequal amplitudes or inexact phasing, the result is a wave with a different amplitude from the original two interfering waves. More than two waves may interfere.

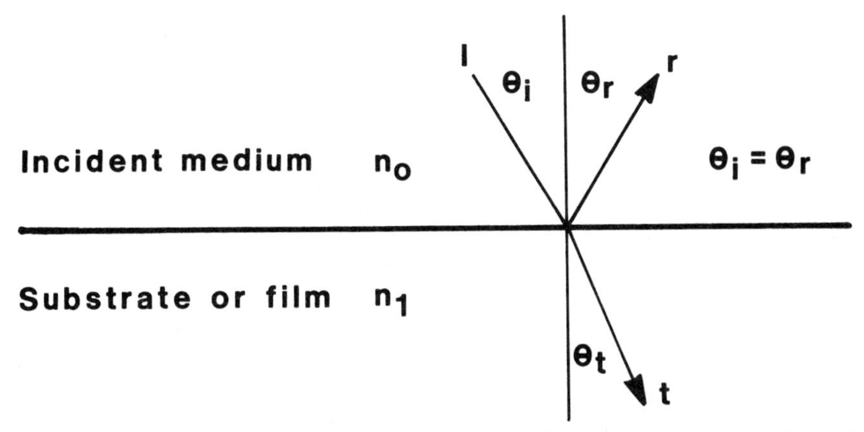

Figure 1.3. Schematic diagram of an interface. The angle θ_r of the reflected ray *r* is equal to angle θ_i of the incident ray. The deviation of the refracted ray is dependent on the indices of refraction of the media on either side of the interface and on the angle of incidence of the light ray.

In this case, there is no node at the boundary. In some situations, however, the indices of refraction can be such as to allow an approximation of a node to exist at the surface. Silver is one such material where the complex index of refraction in the infrared portion of the spectrum is approximately $(0.05 - 3.0i)$. The phase shift for these values is approximately 178 degrees, which is close to the 180 degrees required for a node. Table 1.1 contains some complex index values for this and other materials. This subject is covered in more detail in the section on electric field standing waves in Chap. 3.

It should be noted that the wavelength does not explicitly enter into the calculation of phase values at an interface, although it does determine which values of the refractive indices one should use.

It is straightforward to add layers to the initial interface. The Fresnel formula applies at each interface, and the phases of the individual waves must be included in the calculation (Fig. 1.4). For a single layer between two semi-infinite regions, the formula is

$$r = \frac{r_1 + r_2 \exp(-2i\beta)}{1 + r_1 r_2 \exp(-2i\beta)} \tag{1.3}$$

where r_i is the amplitude reflectance at the ith interface as given by Eq. 1.1 and β is the phase thickness of the film, and is given by

$$\beta = \frac{2\pi}{\lambda} n_f\, d_f,$$

λ = wavelength and $n_f\, d_f$ = optical thickness of the thin film.

The phase of the reflected wave is given by

$$\phi = \tan^{-1} \frac{r_2(r_1^2 - 1) \sin 2\beta}{r_1 (r_2^2 + 1) + r_2(r_1^2 + 1) \cos 2\beta}$$

When the film thickness goes to zero, Eq. 1.3 reduces to Eq. 1.1, the Fresnel reflection equation for the interface between the incident medium and the substrate.

Figure 1.5 and 1.6 illustrate the behavior of the reflectance as the thickness

Table 1.1 Typical metal indices and normal incidence phase shifts for normally incident light in air

Metal	*n*	*k*	Phase shift (degrees)	Comments
Silver	0.05	30	176	Infrared
Copper	12	60	178	10 μm
Aluminum	23	86	179	9.5 μm
Nickel	2.8	5	163	0.99 μm
Gold	7.6	72	178	10 μm
Gold	1.4	1.9	140	0.45 μm
Dielectrics	any	0	180	

Most indices are from Ordal et al., 1983.

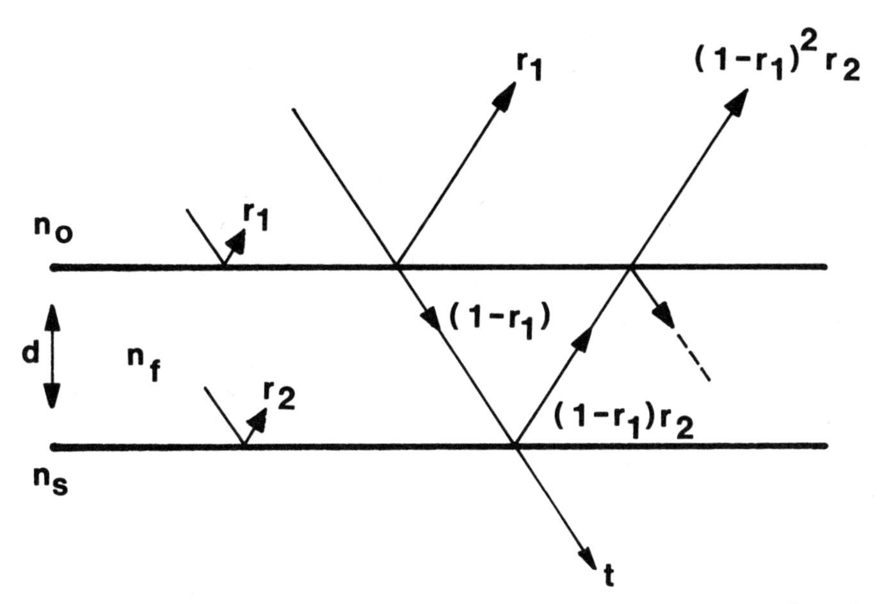

Figure 1.4. Schematic diagram of the summation of the partial reflectances to get the total reflectance from two interfaces. If the two interfaces are sufficiently separated so that the optical path between the surfaces exceeds the coherence length of the wave packets, then the partial reflected *intensities* are added to give the total reflectance. Otherwise, the *amplitude* reflectances add and the phase differences between all of the waves that are introduced by the thickness of the film must be taken into account (Eq. 1.2). The Fresnel reflection coefficients, r_i, are indicated at each interface. The indices of the incident medium, the film, and the substrate are indicated by n_0, n_f, and n_s, respectively. The refraction that occurs at each surface is not shown in this figure.

of a film increases. Two different presentations of this information are used. In Fig. 1.5, the complex reflectance amplitude is plotted on a polar coordinate representation of the complex reflectance plane. The tick marks indicate constant increments of thickness. Note that these are not equally spaced about the circle. In Fig. 1.6, the intensity reflectance is plotted against the film thickness. Note that the shape of the latter curve is not sinusoidal.

Several interesting effects can be gleaned from Eq. 1.3 and be seen in Figs. 1.5 and 1.6. One that has practical application is that the reflectance of the layer reaches an extremum when the argument of the exponential reaches $\pi/2$ or 90 degrees and neither the film nor the substrate is absorbing. The optical thickness that corresponds to this is called a *quarter wave,* and the layer is said to be a *quarter wave thick,* or one QWOT (quarter wave optical thickness). Whether the extremum at the quarter wave thick point is a minimum or a maximum depends on the relative values of the index of refraction of the incident medium, the film material, and the substrate. The extremum will be a minimum if either of the following two conditions is satisfied:

$$n_0 < n_f < n_s \quad \text{or} \quad n_s < n_f < n_0,$$

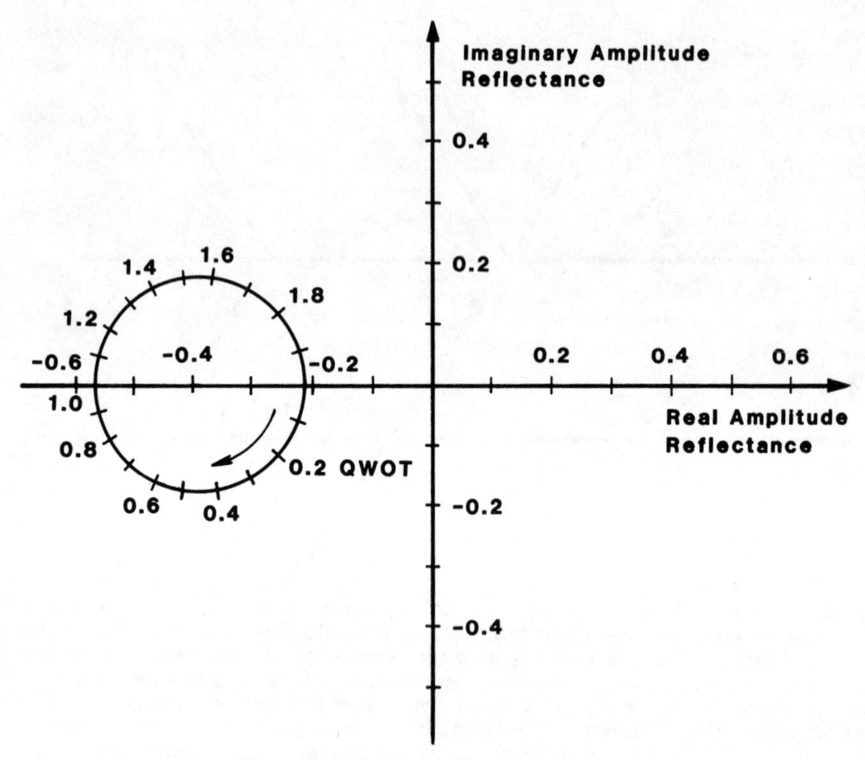

Figure 1.5. The amplitude reflectance of a film of index 2.35 of increasing thickness plotted on the complex reflectance plane. The tick marks on the circle are spaced at intervals of 0.1 quarter wave optical thickness. The arrow indicates the direction of increasing film thickness. The substrate has an index of 1.52. The amplitude of the reflectance is the distance from the origin to a point on the circle, and the angle formed by that vector is the phase angle of the reflected light. The intensity is the square of the amplitude value. Note that the spacing of the thickness intervals is not uniform.

where n_0, n_f, and n_s are the indices of the incident medium, the film material, and the nonabsorbing substrate, respectively. Otherwise, the reflectance reaches a maximum at the quarter wave point. When the index of the film is the geometric mean of the incident medium and the substrate, then the minimum is exactly zero:

$$n_f = \sqrt{n_0 \, n_s}.$$

If a given material has no dispersion in its index of refraction, then the same peak value of the reflection is obtained at all odd multiples of the quarter wave optical thickness.

If the substrate is absorbing, e.g., it consists of a polished metal surface, then the extremum does not take place at the point where the optical thickness of the film is exactly a quarter wave, as shown in Fig. 1.7. The nature of the reflectance for these cases is discussed in detail in in Chap. 3.

As the optical thickness of a nonabsorbing layer approaches twice a quarter wave, i.e., a half wave or a thickness of 180 degrees, the argument of the exponential term in Eq. 1.3 approaches 360 degrees, and the value of the exponential term becomes unity. The implication of this is that a layer that is one-half wave thick has no effect on the light passing through it. The complex exponential function is unity at integral multiples of a half wave, so that the film essentially disappears at those wavelengths also. At these integral multiples of a half wavelength thickness, the film is described as being *absentee,* and the reflectance returns to the initial value of the uncoated substrate.

The above effects are strictly true for homogeneous and nonabsorbing films. Films that are either absorbing, inhomogeneous, or both, will not be absentee when

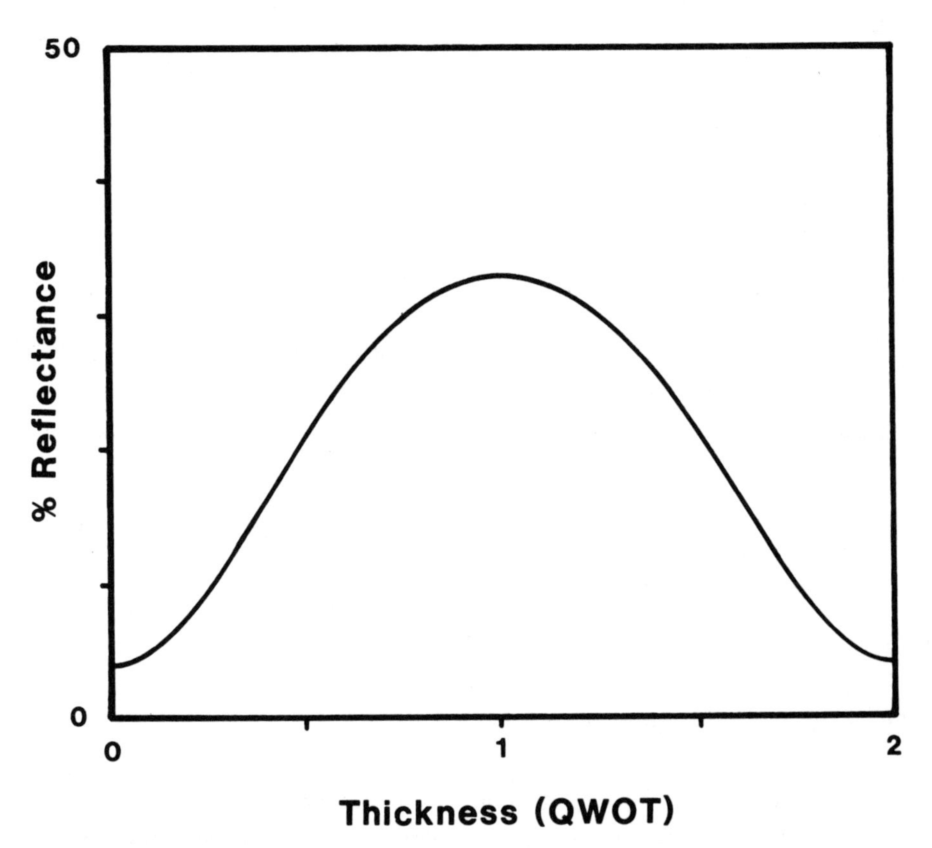

Figure 1.6. Reflectance (intensity) as a function of thickness for a nonabsorbing film The film index is approximately 2.35; the index of the substrate is 1.52. The peak reflectance occurs at the point where the film optical thickness is an odd multiple of a quarter wave. At even multiples of a quarter wave optical thickness, the film is a multiple of a half wave thick, and the reflectance returns to that of the substrate. The curve is not a simple sinusoidal function. Note that if the film index had been less than that of the substrate, the reflectance would have gone through a minimum at the quarter wave point. The phase content of the light is not recovered when only the intensity is measured.

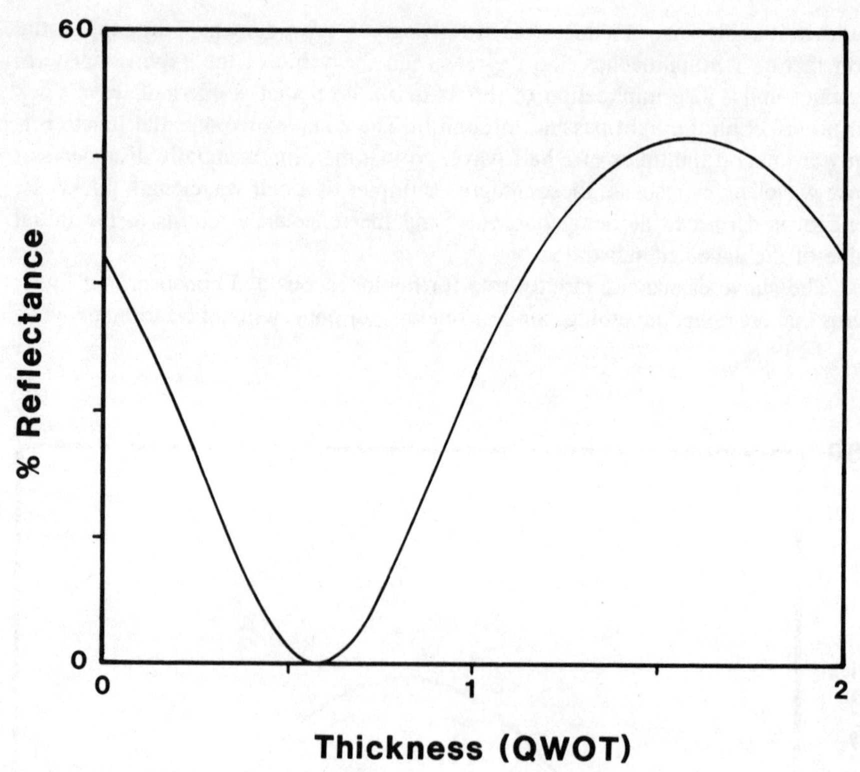

Figure 1.7. Reflectance of a dielectric film on a metal substrate. Note the initial drop in reflectance as the thickness of the thin film increases. The extrema do not occur at the quarter wave points. The substrate index is 2.0 − 2.0*i* and the film index is 2.35. At the half wave points, the reflectance returns to that of the bare substrate.

they are a half wave thick. This information can be useful in evaluating the properties of an unknown film.

The formulas given previously apply for normal incidence. Nonnormal incidence effects are discussed in section 3.6. A complete treatment of this subject can be found in Chap. 13 of Born and Wolf (1970).

1.1.1. Coherent Addition of Amplitudes

For interference to occur between waves, they must be coherent. This assumption was made implicitly in the previous discussion on the interference between the partial reflectances from the various interfaces of a thin film structure. Light waves in the useful range of a given filter typically have a coherence length that is larger than the total thickness of the filter, and the various partial waves can interfere with each other. This interference alters the reflective and transmissive properties of the original substrate.

In the discussion of interference, we added the wave from the first surface with that from the second, and we assumed that the two waves were coherent. We could do this because the coherence length of most light is several wave lengths long. For a continuous wave (CW) laser beam, the coherence length is several meters (a pulsed laser with a pulse length of 1 psec has a coherence length of 300 μm). With laser light sources, interference between waves can take place when the two waves travel paths that are quite different in length. White light from a thermal source on the other hand, does not have this coherence. In a typical thin film configuration, where the total thickness of the stack is usually a small number of waves thick, all of the waves reflected from the interfaces can be added coherently, i.e., the amplitudes add, and the phases of the waves have to be taken into account when adding these amplitudes.

The coherence length of light $\Delta 1$ is approximately

$$\Delta 1 = \lambda^2/\Delta\lambda,$$

where λ is the wavelength and $\Delta\lambda$ is the bandwidth.

In a monochromator that has a resolution of 1 nm, for example, this leads to a coherence length of approximately 0.25 mm for light with a wavelength of 500 nm. A transparent layer with a thickness on this order of magnitude, or thinner, will produce interference effects. If the wavelength is 5 μm, and the bandwidth of a monochromator is proportionately the same as the preceding case, then the coherence length will be 2.5 mm. This is on the order of magnitude of some substrate thicknesses. Under these circumstances, a substrate acts as an interference layer. The light waves reflected from the two surfaces interfere and produce a fringe pattern that is a function of wavelength. When the distance between the two reflecting surfaces is large, many interference maxima and minima can occur over a given wavelength range. This spectral pattern is called a *channel spectrum* (Jenkins and White, 1957, p. 284). This channel spectrum can be superimposed on the spectrum of a filter. This manifests itself as a high frequency ripple on the spectral curve of a filter. On a wavenumber scale, the ripple frequency will be constant over the entire spectrum. On a wavelength scale, however, the ripple frequency will increase as the wavelength decreases.

When the length differences between two paths are greater than the coherence length, the phases of the two waves are randomized and they cannot interfere with each other. In such a situation, the two waves behave as two incoherent sources. Such sources are combined by adding the intensities of the two waves, rather than their complex amplitudes [see, for example, Stone (1963) Chap. 13, for a derivation of the addition of incoherent sources].

1.1.2. Layer Thickness Specifications

The thicknesses of the layers in a thin film coating are specified in a number of ways. The first is the physical or metric thickness of the individual layers, i.e., the thickness that might be measured by an ultrasensitive micrometer. This thickness is found frequently in publications, and is practically always used when specifying absorbing layers.

A more useful way to represent the thickness of a layer is the optical thickness. The optical thickness is the product nd, where n is the index of refraction of the layer material and d is its physical thickness as described previously. It is this product that is frequently important, to first order, in optics. Thus, this is a common

A previously mentioned thickness specification that is commonly used in optical filter circles is called the quarter wave optical thickness, or QWOT for short. It is defined as

$$\text{QWOT} = 4\,nd/\lambda,$$

where nd is the optical thickness of the film and λ is the wavelength at which the *QWOT* condition occurs.

Note that this thickness specification implies a wavelength. Although a design may be designated as consisting of layers with a certain number of QWOTs, it is not fully specified until the wavelength at which those QWOTs are defined is given.

This measure is useful because many nonabsorbing thin film filters consist of layers that are one or more quarter waves thick. A very common design is a stack made up of layers of alternating indices, the thicknesses of which are all one quarter wave at some given wavelength. A design specified in these terms is instantly recognizable. If the coating design were given in terms of the physical thicknesses of the layers, it is not so readily identified, and its performance characteristics would become obvious only after some amount of computation.

It is important when looking at the design data of a thin film filter to understand which type of thickness representation is being used.

1.1.3. Nomenclature

The roots *reflect, transmit,* and *absorb* can have a number of suffixes such as *-ance, -ivity,* and *-tion*. The distinctions among the suffixes are not universally accepted, but the current common usages, and the one followed here, are

-ivity: e.g., transmissivity, reflectivity, absorptivity

The fundamental properties of a material, or theoretical or calculated values based on first principles.

-ance: e.g., transmittance, reflectance, and absorptance

Measured values.

-tion: e.g., transmission, reflection, and absorption

Generally refers to the phenomena rather than actual values. Frequently used interchangeably with the -ance version.

Because quarter wave layers are frequently used in thin film optics, a notation has evolved that represents such layers, as well as multiples and submultiples of them. An uppercase letter is used to designate a quarter wave layer with a particular index of refraction; e.g., H, represents perhaps, a QWOT of a high index material, and L represents a QWOT of low index material. A basic two-layer period consisting of a low and a high material would be designated as HL. If more than one period exists, an exponent is used: $(\text{HL})^{10}$ represents 10 periods of the basic stack (20 alternating layers). If a layer is less than a quarter wave thick, it may be designated as a fraction of the quarter wave; e.g., 0.5 H or H/2 is an eighth wave thick layer of the high index material.

1.1.4. Index of Refraction and Dispersion

The index of refraction discussed previously varies as a function of wavelength for most real materials. In some restricted wavelength ranges, this change may be negligible. The index of some materials can change as a function of a number of variables, such as time or UV (ultraviolet) light exposure. Therefore, the common nomenclature of *optical constants of a material* is somewhat of a misnomer.

In this text, the theoretical performance curves have no dispersion included in the calculation unless specified otherwise, so that the primary features in the performance curve can be seen, unobstructed by dispersion effects.

The change in index of refraction as a function of wavelength is referred to as dispersion. For most nonmetallic materials used in thin films in the visible portion of the spectrum, the index is typically higher in the blue end of the spectrum, and lower in the red end. Similarly, in the infrared, the index of most nonmetallic materials decreases as the wavelength increases. Near absorption bands, however, this trend may be reversed and the index can increase as the wavelength increases. Concurrently, the absorption coefficient, k, is generally nonzero.

1.1.5. Absorbing and Nonabsorbing Filters

Filters can be classified as either absorbing or nonabsorbing. In nonabsorbing filters, the energy conservation formula $R + T = 1$, where R and T are the intensity reflection and transmission ratios, respectively, applies exactly. Where there is some absorption, this equation must be modified to $R + T + A = 1$, where A is the absorption in the filter. In this discussion, we ignore scattering by assuming that it is insignificant.

A fundamental theorem of optical thin films is that the transmission of any coating is *always independent* of the directions of the arrival of the light [see Macleod (1969), p. 24 for a proof]. Thus, the transmission of light that is incident from the substrate side of the coating will be equal to that of light incident from the opposite side. A change in the index of either the incident medium or of the substrate, or both, will likely change the transmission value but it will remain independent of the direction of incidence. If there is no absorption in either the film materials (or the substrate), then by the law of energy conservation, the reflection must also be independent of the direction of the arrival of the light.

The reflection of a coating that contains absorbing materials does not follow the aforementioned rules. The reflection and the *reverse* reflection do not have to be the same. Because the transmission is *always* the same in both directions, and the reflection depends on the direction of the light's arrival, then in accordance with the law of conservation of energy, the absorption is also sensitive to the direction of the light's arrival. This effect is readily shown for a metal film with the complex index $2 - 2i$ and thickness 5.0 nm on a glass substrate ($n = 1.5$). The reflectances from the air side and the glass side are shown in Fig. 1.8. An

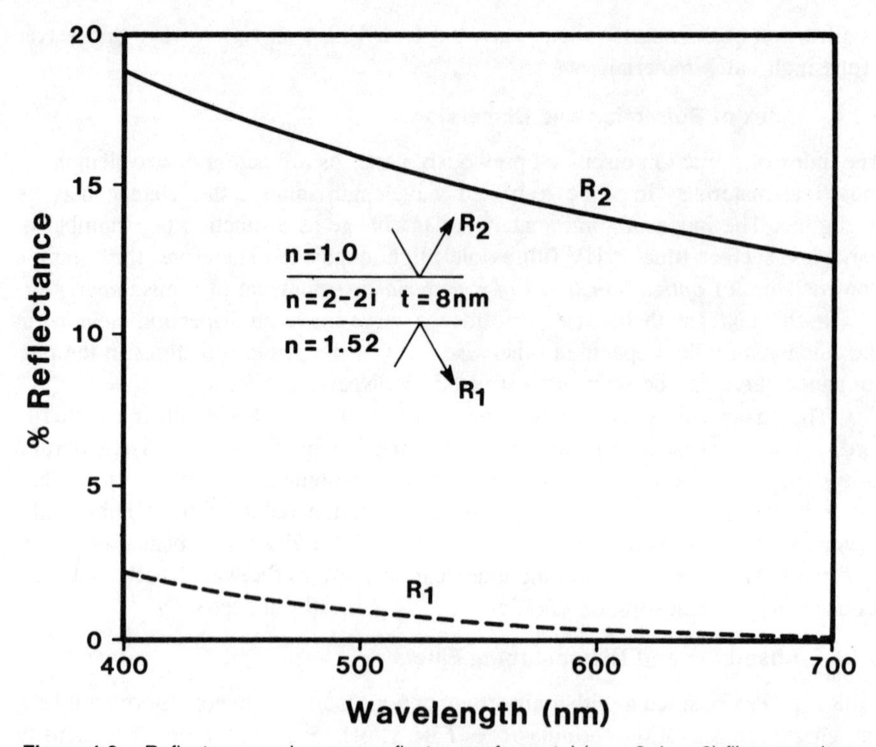

Figure 1.8. Reflectance and reverse reflectance of a metal ($n = 2$, $k = 2$) film on a glass substrate ($n = 1.5$). The reflectance (and absorptance) depends on the direction of incidence of light. There is no dependency in the transmission on the direction of arrival of the light.

almost perfect antireflection condition can be obtained when the interface is viewed from the glass side, yet the reflection is moderately high when the coating is viewed from the air side. Concurrently, absorption of the light arriving from the glass side is higher than that arriving from the air side since the transmission is the same, irrespective of the light's direction of arrival.

When a filter is measured in transmission, it is not important which side faces the light source and which, the detector. If the reflectance is being measured, however, it is important that the proper face of the sample be oriented toward the light beam if significant absorption exists in any of the layers in the coating in the spectral range of interest.

Along with absorption, scatter is another mechanism of energy loss from the specular beam. Scatter sites are generally microscopic volumes, the refractive indices of which are different from that of the surrounding medium. The light diffracted from these sites is distributed in all directions and is, therefore, lost from the specular beam. If the scatter sites are large enough, the reflection and refraction may also play a role and contribute to the overall light lost. Each site removes only a small percentage of the total energy in the beam, but as there can be many such sites, the cumulative effect can add up to a serious light loss.

In optical coatings, surface roughness and local index variations frequently give rise to scatter. The following are some of the situations that can occur:

1. The substrate can be rough. Because thin films for the most part do not smooth out a surface, some of the surface defects may replicate to the top surface.
2. The films can have inhomogeneous properties, such as inclusions or other defects that have a different index of refraction than the surrounding material.
3. The interfaces between films can be rough.
4. The surface of the coating may be rough, so that the air/film interface is the source of the scatter sites. The difference in index of refraction between air and the first film is large, so that this roughness can lead to a significant light loss.

Many thin film coating materials grow in a columnar structure as they condense on the substrate. The relatively limited lateral motion of the arriving atoms and molecules produce little smearing of the underlying structure. As a consequence, the original structure can be replicated through a large thickness.

1.2. SPECTRAL RANGES AND MATERIALS SELECTION

By far the most difficult problem in the field of thin film engineering is that of finding the proper materials for a particular application. It is a rather straightforward task to design a coating that meets just about any arbitrary specification if one does not have to worry about building it with the materials supplied to us by Mother Nature. The limitations imposed by real materials can be very severe.

In addition to optical problems, the designer also has to cope with coating process, mechanical, environmental, and sometimes chemical problems in converting a design from paper to reality. Some of the mechanical problems that are frequently encountered are stresses in the films that cause the substrate to warp or the films to delaminate through adhesive or cohesive failure. The process temperature, which will ensure an environmentally stable coating, may be too high for the substrates. The substrates may react with the coating materials in some ways that cause degradation. Some materials cannot survive outside the vacuum chamber because of oxidation or moisture reactions. Less severe cases of this may leave the film intact, but make it too fragile to be of any practical use in other than a controlled environment.

Some materials are useful only in certain narrow wavelength regions, whereas others have much broader utility. A material does not have to be transparent everywhere to be useful; for example, in the blocking of certain wavelengths, absorption is the property of interest, though usually, some region of transparency is also generally desired.

Some materials, such as magnesium fluoride, are transparent over very wide spectral ranges. Mechanical stresses, on the other hand, can severely limit the maximum thickness attainable without failure. The maximum quarter wave optical thickness of a magnesium fluoride film that can be typically obtained with a conventional evaporation process is on the order of 1 to 2 μm. Magnesium fluoride is an interesting example of a situation in which stress is not necessarily the only consideration. This material, when evaporated, forms films with high tensile stresses, yet it is a very usable material because its adhesive strength to glass is extremely high. The upper limit on the thickness of these films is not governed by adhesive failure to the substrate, but by cohesive failure in the film when the tensile strength limit of the film is reached. The failure point depends on the coating processes, the expansion coefficient of the substrate, and other factors. Magnesium fluoride is an example of what occurs to some extent in virtually all materials. These problems of material selection are discussed in more detail in Chap. 6.

1.3. ANGLE OF INCIDENCE EFFECTS

The performance of a filter that uses the absorption properties of materials to obtain the desired spectral shape does not change markedly as a function of the angle of incidence. This is readily apparent when a beam of white light is observed through a colored glass plate: its color does not shift spectrally when the filter is tilted. The optical density of the filter increases owing to the increase in the path length inside the absorber. The ratio of the internal transmission at some angle to that at normal incidence is given by

$$T_i = \exp[-\alpha d(\sec\theta_2 - 1)],$$

where α is the absorption coefficient, d is the thickness of the absorber, and θ_2 is the angle the ray makes with the normal inside the absorbing medium.

This equation is valid for relatively weak absorbers, such as most colored glasses, which are basically a glass matrix doped with specific ions or atoms. The complex part of the refractive index of the absorbing filter can be ignored in evaluating Eq. 1.2. It should be noted that this discussion ignores the effect of the surface reflectances: at high angles of incidence, these can be substantial.

1.3.1. Snell's Law

A light beam crossing an interface between semi-infinite media with different indices of refraction changes direction if the angle of incidence is nonnormal. The change in the direction of the light beam is given by Snell's law:

$$n_0 \sin\theta_0 = n_1 \sin\theta_1 \tag{1.4}$$

where n_i is the index of refraction in medium i and θ_i is the angle the beam makes with the normal to the interface plane.

The angle of incidence is the angle of the incident beam, whereas the angle of refraction is that for the refracted beam. The reflected and the refracted beams lie in the plane defined by the incident beam and the surface normal at the point where the beam strikes the interface.

In a thin film coating, the interfaces are essentially all parallel, and the product $n \sin\theta$ has a constant value in all layers (see section 1.3.2.). Equation 1.4 applies to both nonabsorbing as well as to absorbing materials. For complex indices, i.e., one or both media are absorbing, the angles and their sines are also complex.

1.3.2. Optical Invariant

In a nontilted optical system with plane parallel surfaces, the product $n \sin \theta$ is a constant throughout the system. This is a restatement of Snell's law. The same principle holds true at all points in a thin film filter stack. The constant is defined by the ray entering the coating and the index of the incident medium. The value is generally determined by the incident angle as measured in air, in which case the constant is simply $\sin \theta$ (Fig. 1.9).

For a plane parallel glass plate with a coating on the second surface (the surface away form the incident light beam), the coating is being used effectively immersed in an index of 1.52. The $n \sin \theta$ product is defined in air, because all of the surfaces are plane and parallel. Thus, the only effect of the immersion is that due to the change in the index ratios, and the angle of incidence effect is the same as it would have been if the coating were in air. Also, it should be noted that in such a situation the maximum angle of incidence on a coating on the second surface of a plane parallel glass plate is limited to $\sin^{-1}(1/n)$, where n is the index of the glass. For prisms and other nonparallel surfaces, the preceding statement does not hold, and the rays need to be traced individually to determine the appropriate angles.

1.3.3. Critical Angle

When the index of the incident medium is greater than that of the second medium and neither is absorbing, there exists an angle in the incident medium beyond which light will not be transmitted into the second medium. This angle is called the *critical*

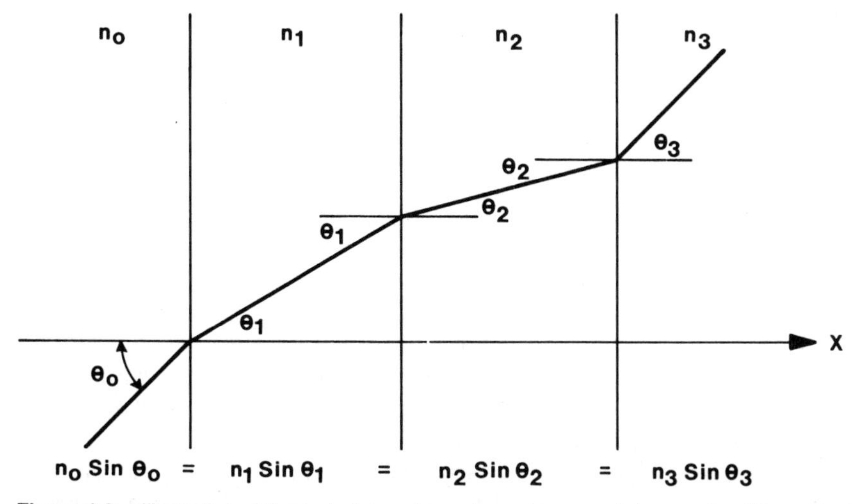

Figure 1.9. Illustration of the $n \sin \theta$ invariance. In a plane parallel geometry, this product is constant throughout the structure.

angle. At the critical angle, the angle of the refracted beam is 90 degrees; i.e., the refracted beam is travelling parallel to the interface. The critical angle θ_c is given by

$$\theta_c = \sin^{-1}(n_1/n_0)$$

where n_0 is the index of the incident medium, n_1 is the index of the second medium, and $n_1 < n_0$.

Note that there is no critical angle when the incident medium has an index that is less than that of the second medium.

Any increase in the angle of incidence beyond the critical angle results in total internal reflection (TIR), and no light energy enters the second medium. In practice, it is possible to obtain a reflectance value very close to the theoretical one of 100 percent when TIR is utilized. TIR is used in many prisms. The standard laws of reflection i.e., the angle of reflection is equal to the angle of incidence, apply to those cases of reflection beyond the critical angle.

Evanescent waves can penetrate the interface when the angle of incidence is greater than the critical angle. The amplitude of these waves decreases exponentially in the low index medium; their short attenuation length is on the order of a few wavelengths (Jackson, 1962, p. 221). With dielectric media, and in the absence of an absorbing layer at the interface, this attenuation of the amplitude of the wave does not remove any energy from a light beam, and the reflectance remains at 100 percent.

An absorbing medium can be located such that it interacts with the evanescent waves; for example, absorbing particles can be located on the reflecting face of a prism. This interaction is strong, and energy is removed from the beam that was undergoing TIR. This has been a fertile field for experimenters who are interested in the measurement of materials with very low absorption coefficients; [see Harrick, (1967)]. This technique of measuring absorption also is called *frustrated internal reflection* or *attenuated total internal reflection*. Figure 1.10 shows the relationship between the critical angle and the index of refraction ratio.

1.3.4. Polarization

The polarization of light is sometimes overlooked in specifying optical filters. In a complex optical system, some degree of polarization may come about inadvertently. For example, the nonnormal reflection coefficient for a dielectric coating or metallic mirror depends on the plane of polarization. An unpolarized incident beam may be partially polarized as it progresses through an optical system, and one polarization may emerge with a higher intensity than the other. This could be significant if there are many reflectors in a system. If such is the case, then it is prudent to do a careful analysis of the system to verify the nature and amount of the polarization introduced into the beam.

Laser light is frequently plane (linearly) polarized and this can make the analysis of a laser system simpler. It should be noted that coated mirrors can introduce differential phase shifts that can lead to circular or elliptical polarization (see section 1.3.7.). When a laser with *random* polarization is used, a system analysis needs

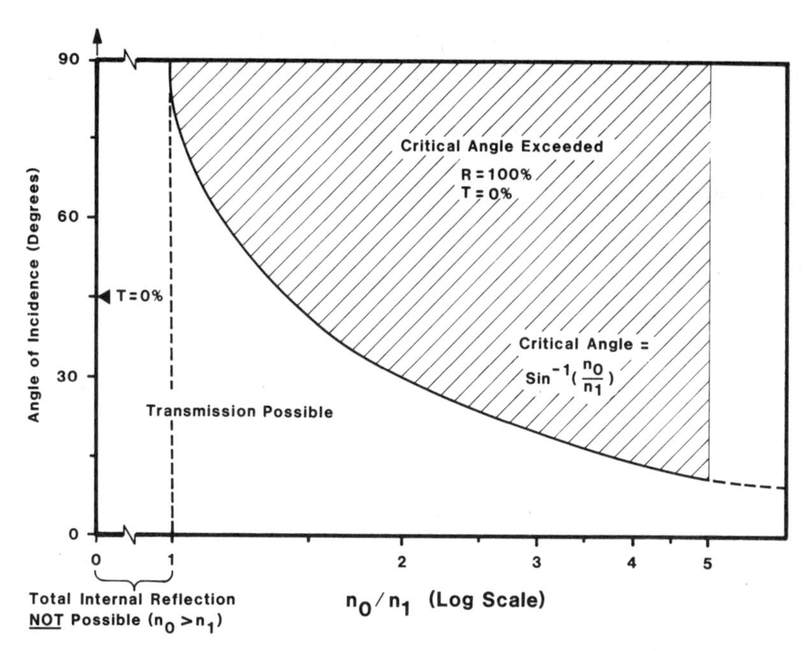

Figure 1.10. Plot of critical angle versus index ratio (n_0/n_1) for nonabsorbing materials. In the areas where transmission is possible, its actual value is determined by the reflection coefficient for the interface. The transmission is 100 percent for those cases where $n_0 = n_1$ (dashed vertical line), and the value of T decreases on either side of it. The transmission is zero along the vertical axis.

to be done separately for each plane of polarization. A beam of light can have two states of polarization. For plane polarized light, these states are specified with respect to the plane of incidence.

Alternative polarization states are sometimes used when the analysis lends itself more easily to these different forms. An example is the use of right and left hand circular polarization. Details on these topics may be found in many references, such as Shurcliff (1962) and Clarke and Grainger (1971).

The electric field of a plane polarized electromagnetic wave can be decomposed into two orthogonal components, commonly designated "*s*" and "*p*." In the former, the plane of oscillation is perpendicular to the plane of incidence (The "*s*" designation derives from the German word for perpendicular, *senkrecht*.). In the latter, the plane of oscillation is parallel to the plane of incidence ("*p*" state).

In a complex system, each needs to be examined as a beam that has an *s* polarization at one interface may, because of the system's geometry, be a *p* state at another interface.

The quality of a polarizer is frequently given in terms of its extinction ratio. This parameter ρ is defined by

$$\rho = T_{max}/T_{min},$$

where T_{max} is the maximum intensity of light which the filter transmits when its axis is aligned with the polarization of the probe beam and T_{min} is the minimum intensity of light that the filter transmits when its axis is rotated 90 degrees in the polarized beam (Driscoll and Vaughan, 1978, pp. 10–13).

1.3.5. Brewster's Angle

Brewster's angle at an interface between two nonabsorbing media is given by the angle of incidence whose tangent is the ratio of the two indices:

$$\tan \theta_B = n_1/n_0,$$

where θ_B is Brewster's angle in medium n_0, n_0 is the index of the incident medium, and n_1 is the index of the substrate.

At this angle, no light in the incident beam with polarization parallel to the plane of incidence (p plane) will be reflected. Conversely, the tangent of Brewster's angle in the second medium is given by n_0/n_1. By conservation of energy, all of the energy is transmitted into the second medium with 100 percent efficiency. An interesting aspect of this geometry is that the refracted ray and the reflected ray have an included angle of 90 degrees. Brewster's angle is the basis of some techniques for determining the index of refraction of films, such as Abeles' method (1950).

The design of high reflectors becomes difficult as the angle of incidence approaches Brewster's angle. The reason for this is apparent when one evaluates the effective indices for both planes of polarization in this angular regime.

1.3.6. Effective Indices

At nonnormal incidence, the Fresnel reflection equation is still applicable, provided that *effective* indices are used instead of the normal values. The effective indices depend on the plane of polarization and are given by

$$n_s = n \cos\theta \quad \text{and} \quad n_p = n/\cos\theta$$

where n_s and n_p are the effective indices for the s and p states of polarization, respectively, and θ is the angle of the beam of light in a medium with an index n.

Figure 1.11 shows the effective index for the two planes of polarization as the angle of incidence in that medium varies from 0 to 90 degrees. The effective index of two and four times the index of the medium in which the angle of incidence is measured is also shown. If the incident medium is air, the values of 2 and 4 approximate the indices of ZnS ($n = 2.2$) and germanium ($n = 4.0$), two common infrared coating materials. The effective index changes relatively little for the higher index materials, whereas it changes markedly for low index materials such as air. As a result, coatings that consist of higher index materials shift less as the angle of incidence increases than do coatings with lower average indices.

The curves for the s plane light do not intersect, whereas those for p plane light do. The angle at which the curves intersect correspond to the Brewster angle

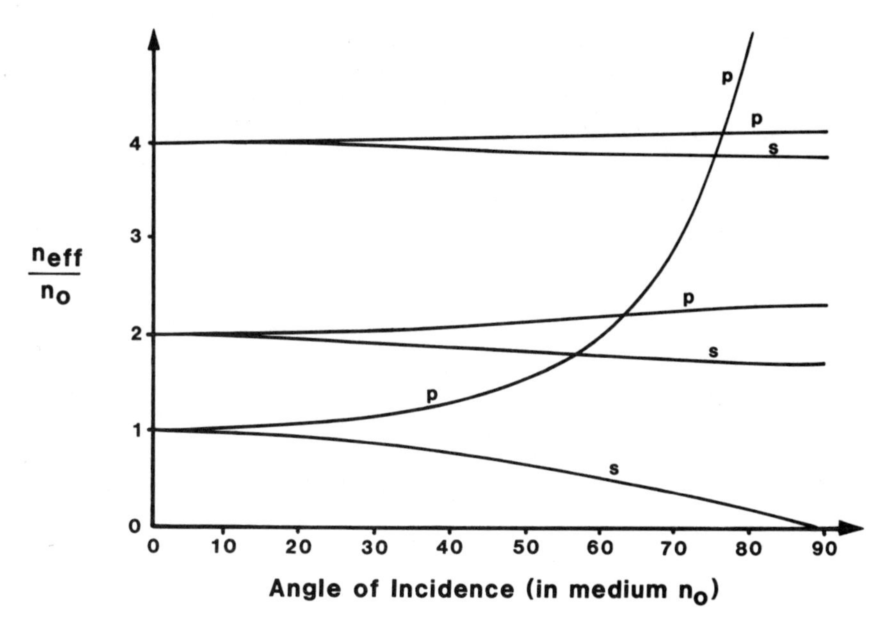

Figure 1.11. Effective index as a function of angle of incidence. The effective index is normalized to the index of the medium in which the angle of incidence is defined (n_0). The *s* and *p* plane values of the effective index are marked. The intersection of the *p* plane curves occurs at Brewster's angle for that pair of indices. At index values below n_0, the curves have the shape of the unity curves, except for a scaling factor along the abscissa; the effective index of the *s* plane component then goes to zero at the critical angle.

for that pair of indices: since the effective indices are equal, the interface disappears at that angle and there can be no reflectance from it.

Figure 1.12 shows variation in the reflectance as a function of the angle of incidence at air interfaces with glass ($n = 1.5$), a poorly reflecting metal ($n = 2.0$, $k = 2.0$), and a highly reflecting metal ($n = 0.5$, $k = 3.0$). The reflectance of light with *p* polarization is zero at Brewster's angle. At all angles (except normal, of course), bare interfaces act as polarizers to some extent.

1.3.7. Birefringence and Differential Phase Shifting Coatings

Birefringence is a phenomenon related to the optical properties of materials. It occurs when a crystal or other molecular system is anisotropic so that it has more than one optical axis. Materials that are ordinarily isotropic can exhibit birefringence when placed under stress. The strain can break the symmetry of the crystal or molecular structure of an otherwise symmetric lattice and the strained system will be birefringent. Thin films are known to be birefringent (Johnston and Jacobs, 1986), but as they are so thin, the effect only rarely manifests itself in an observable way.

The effects of materials' birefringence and those of the phase shift differences introduced by thin film coatings can lead to similar polarization changes. Thin film designs with such properties have been called birefringent. Strictly speaking, thin film filters are not birefringent because this term refers to a material property. Ellipsometric techniques are typically used to measure the differential phase shift.

When the angle of incidence is not 0 degrees, the phase shift for the two planes of polarization is not equal. In some instances, this is a useful property. By proper design, a coating can change the relative phase of the two components such that incident linearly polarized light will be reflected as circularly polarized. In effect, this type of system functions as a reflective quarter wave plate. These are useful in spectral regions where transmissive birefringent quarter wave plates are difficult to obtain. Figure 1.13 shows the performance of a typical design in the infrared portion of the spectrum. These, and similar designs, are available commercially. By using more than one surface to achieve the desired phase shift, it is much easier to fabricate and use these devices, as they are quite wavelength sensitive when the entire shift is attempted on one reflection. The shifts are additive, so a cascade system is effective. Several papers have been published on this subject recently (Apfel, 1982, 1984; Southwell, 1980).

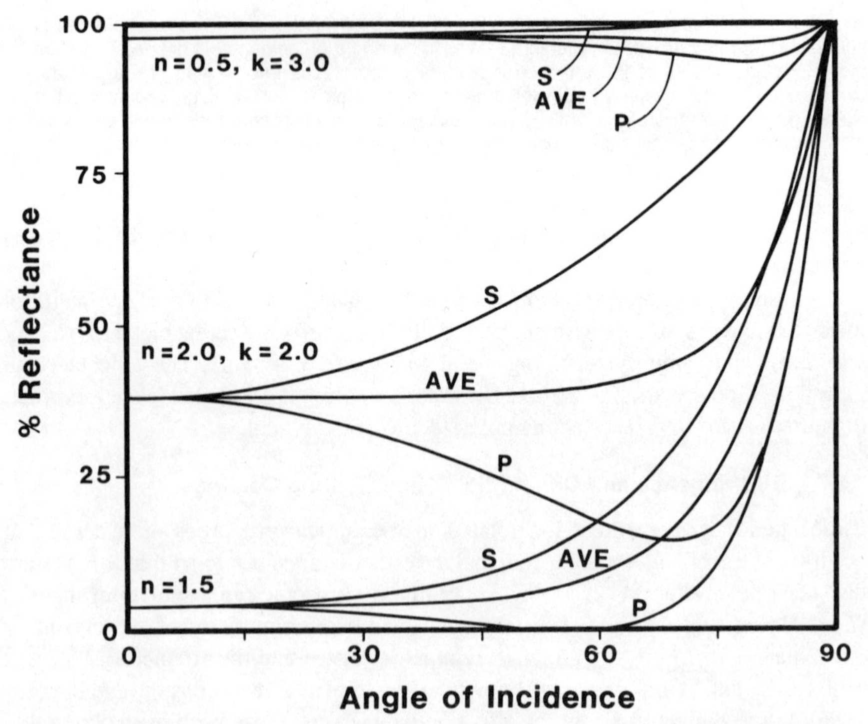

Figure 1.12. Reflectance at various air/bare substrate interfaces as a function of angle of incidence. At Brewster's angle for glass, the *p* component of the light is not reflected. The reflectances of the *p* component for metals also have a minima.

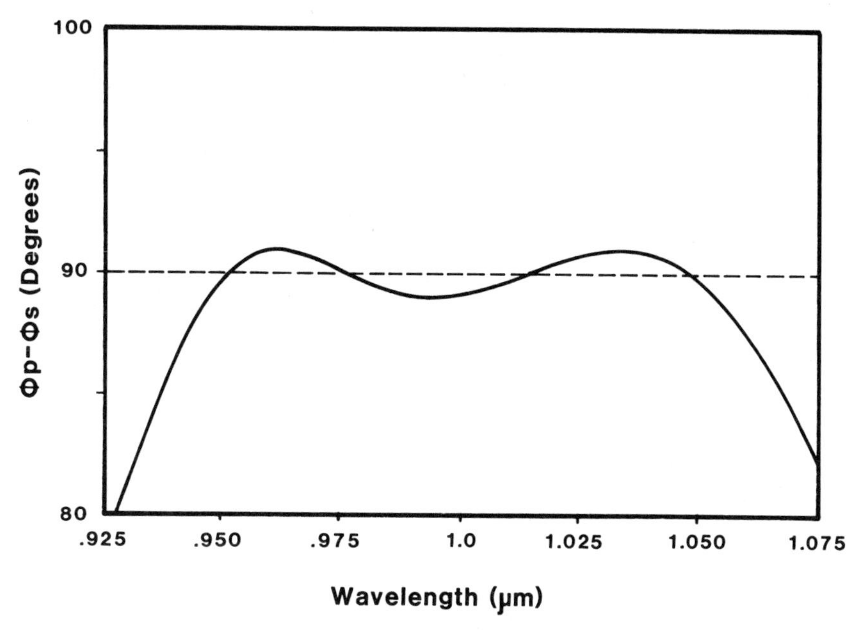

Figure 1.13. Phase difference ($\phi_p - \phi_s$) upon reflectance from a mirror as a function of wavelength. This design has a nominal phase difference of 90 degrees.

1.3.8. Steep Surfaces

In some diffraction limited optical systems, the angle of incidence of a ray on a lens or mirrors surface can be steep (angles of incidence may be in excess of 30 to 45 degrees). In these severe situations, it is necessary, when doing an optical systems analysis, to take into account the angular effects, such as polarization, for each of the different ray angles that strike these elements. A correct analysis uses a ray tracing system that takes all of the surface polarizations into account, and then combines the rays according to the rules of diffraction optics in the image plane. Such an analysis should also take into account the angle shifts of the performance of thin film designs, as well as their contributions to polarization phase shift differences.

Chapter 2

Deposition Technology

Thin film optical filters can be prepared using a number of different coating technologies. A few of these methods have found their way into commercial production processes. In this section we discuss some of the methods currently used for commercial filter production. Although the subject matter is extensive, this section is necessarily brief. For more detailed information on this subject, the reader is referred to the text by Maissel and Glang (1970).

2.1. COATING PROCESSES

2.1.1. Vacuum Evaporation

The first thin film interference coatings for optical applications were made by resistively heated thermal evaporation, and this process is still in use today. A charge of the material to be evaporated is placed in a crucible of a refractory material. An electrical resistance heater heats this charge (either by conduction or by radiation) after the chamber has been evacuated to about 10^{-6} torr, or one billionth (10^{-9}) of an atmosphere (Table 2.1 contains factors to convert between some common units used to express pressures and vacuums). As the temperature of the charge rises, its vapor pressure rises and a significant evaporation rate develops. The vapor

Table 2.1 Conversion Factors Between Various Pressure Units for Gases and Vacuum

1 Pascal	= 10^{-5} bar = 7.5×10^{-3} torr
1 bar	= 0.987 atmosphere = 10^{5} Pa
	= 14.5 pounds per square inch
1 micron (of Hg)	= 10^{-3} torr = 1.3×10^{-6} atmosphere = 0.13 Pa
1 torr	= 1 mm of Hg = 133 Pa

Note: The Pascal (Pa) is currently the official metric unit (Systéme Internationale, SI) for designating pressures.

pressure-temperature relationship is the primary determinant of the evaporation rate. The evaporant then condenses on the cooler substrate. This process of indirectly heating the charge is still used today, though perhaps with some refinements.

The crucible material may be either a refractory metal or a ceramic container. In indirect heating, the heating element does not come into direct contact with the charge. This can be accomplished radiantly by placing a tungsten filament over the charge or by conduction with a filament or other element that is in contact with the crucible.

Direct heating takes place when the charge is directly heated by the hot element. In one method, which is widely used, the charge is placed in a depression on a refractory metal strip. In some materials available in wire form, i.e., aluminum and gold, short pieces of the wire can be wrapped on a stranded tungsten heater wire. A current through the refractory metal heats it and the charge. Once the material in the charge reaches its boiling or sublimation temperature in the vacuum (lower than the normal atmospheric boiling point), atoms and molecules of the material leave the charge and condense on the cooler surfaces in the chamber. Thus, a film builds up on the substrate (as well as the rest of the chamber).

The equipment used in the pioneering days of thin films coatings suffered from a lack of sufficient vacuum pumping capacity as the best high vacuum pumps available were small mercury diffusion units. As a result, the lowest or base vacuum that could be attained in a typical coating chamber was not very good by today's standards. Furthermore, as the evaporation process progressed, the evaporation source radiantly warmed the chamber, causing the walls to release some of their adsorbed gases and water vapor (at the worse possible time, i.e., during the actual deposition of the coating). The coatings thus produced contained entrapped residual gases from the vacuum system. These soft, porous films had relatively low durability. On the other hand, this approach is relatively gentle, and can be used for compounds that have a tendency to decompose.

The technology of vacuum coating improved with time. The development of high temperature, low vapor pressure silicone oils led to the construction of larger diffusion pumps with much larger pumping capacity. The elimination of the mercury also made them safer.

Diffusion pumps operate efficiently only at low pressures at both the inlet and outlet ports. Mechanical pumps are used to evacuate the chamber to a pressure within the operating range of a diffusion pump. This is called the *rough* vacuum. Once the chamber has been *roughed,* the mechanical pump is used to keep the outlet port of the diffusion pump at a low pressure, called the *backing pressure*. Reliable mechanical pumps of the rotary vane type for roughing and backing of the diffusion pumps, and of rotary ("Rootes") blowers for increasing the mechanical pump throughput set the stage for modern vacuum systems and coating technology.

In time, more efficient means of heating the evaporant charge were developed, including the electron beam and the laser beam. This latter method of heating a charge in a vacuum evaporation chamber was first demonstrated by Smith and Turner (1960) with a pulsed ruby laser. Hass and Ramsey (1969) used a carbon dioxide laser to evaporate dielectric and semiconductor materials. Sankur and Hall (1985) used the laser heating method to produce films with little ionization of the evaporant. In an electron beam heating system, electrons from a high voltage (typically 5 to 15 kV), high current (up to several amperes) electron gun are focused on the surface of the charge that is contained in a water-cooled metal, typically copper, crucible. This heating process is a very efficient way of depositing energy into a material in a vacuum. As compared to a resistively heated charge, relatively little radiant heat, except from the surface of the molten charge, is produced by the evaporation source in an electron beam heated system. A resistively heated source with a comparable coating rate radiates from a number of surfaces and this heats the coating chamber much more. Another major advantage of using the electron beam heating technique is that it makes it practical to evaporate refractory materials, the melting points of which can exceed that of tungsten metal. These materials cannot be evaporated using resistively heated boats because they cannot be heated to sufficiently high temperatures to obtain good coating rates. In particular, the oxides of zirconium, tantalum, silicon, titanium, and aluminum became practical coating materials with the advent of the electron beam heating system. Another drawback of the resistance method of heating a charge is the fact that some materials may react chemically or alloy with the hot crucible and cause it to fail.

At room temperature, oxides are geneally good electrical insulators. Under intense bombardment from a sufficiently energetic electron beam, however, all insulators and other refractory materials can be heated sufficiently to cause them to evaporate (unless they first decompose).

2.1.2. Sputtering

Sputtering is another vacuum coating process that is used extensively for depositing optical thin films. This process is most easily characterized as a momentum transfer process in which argon or other ions and atoms from a plasma bombard a *target* made of the material to be coated. The area of a sputtering target may be an order of magnitude larger than an equivalent evaporation source. The collision of the argon atoms and ions with the surface atoms and molecules of the target knocks off (sputters) the target material, which then forms a deposit on the substrate. Flowing water keeps the target cool, so thermal evaporation of the target material does not occur. Figure 2.1 shows the relative scale of a sputtering system's geometry as compared with that typically used in an evaporation system.

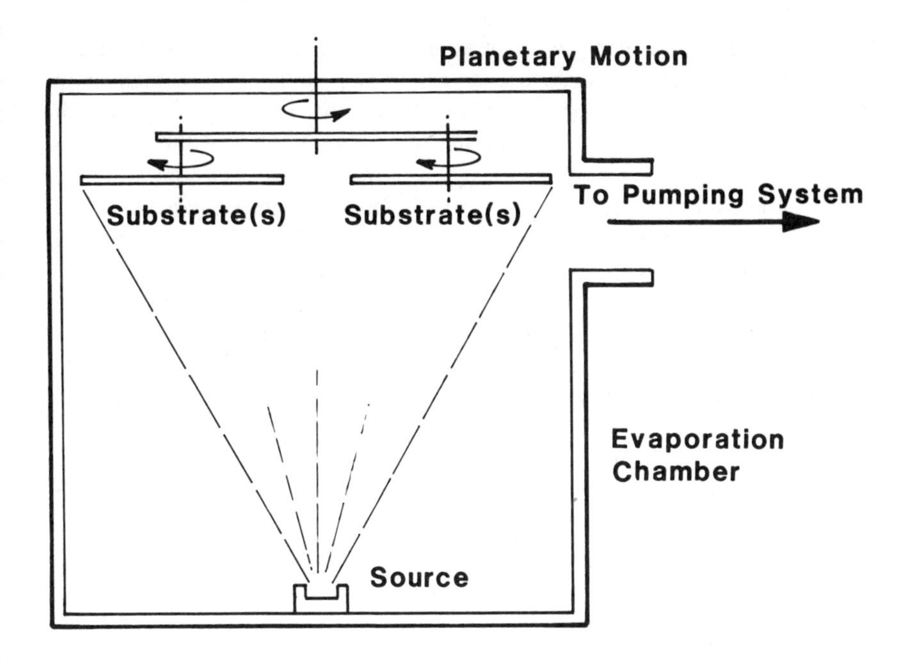

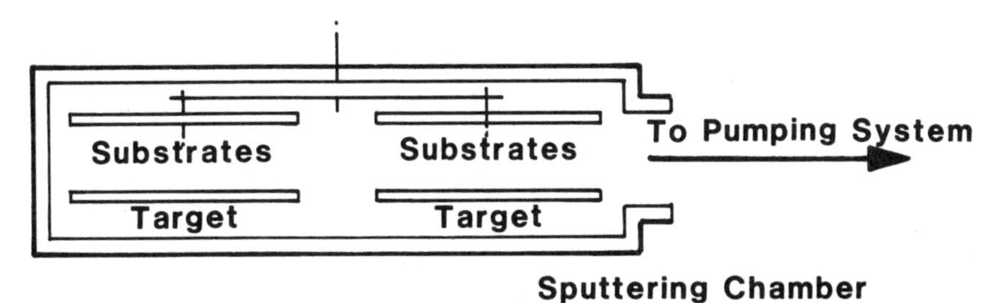

Figure 2.1. Diagram of sputtering and evaporation chambers on approximately the same scale. The substrate size capability of these chambers is the same. The evaporation source is approximately a point source, whereas the sputtering targets are a large area source. Note that the sputtering system can be made much smaller than a comparable evaporation system.

The target to substrate distance in typical sputtering systems is short as compared with that found in evaporation systems; this is a consequence of the nature of the sputtering process. A number of advantages result from this and are outlined later.

Because the sputtering process does not rely on keeping the target surface at a high temperature as in an evaporation type of process, it is possible to use relatively large area targets and to place them close to the substrate without having them heat the substrate radiatively. This is an important consideration as sputtering processes yield far fewer atoms and molecules per area of target per second than does an evaporation source. In particular, oxide targets frequently result in low deposition rates, and large area targets are required to obtain practical rates.

A heavy, chemically inert gas, such as argon, is frequently used as the major constituent of a sputtering gas. A heavier gas is more efficient in a momentum transfer process such as sputtering. Few target materials sputter well when light gases, such as nitrogen and oxygen, are the major components of the sputtering gas.

The gas pressure in a sputtering system is relatively high compared with that in an evaporation system (10^{-3} mbar vs. 10^{-6} mbar); therefore, the mean free path of a sputtered particle is three orders of magnitude less than a particle in an evaporation system (Fig. 2.2). The sputtered target atoms and molecules are energetic as a result of the momentum transferred when they were struck by the argon atoms and ions. For the target atoms to carry this energy to the substrate and not lose much of it through collisions, the distance between the target and the substrate must be as small as possible. This is another reason why the substrate must be placed closer to the target in a sputtering system than it is in an evaporation system.

Because the target is a large area source, as opposed to the approximate point

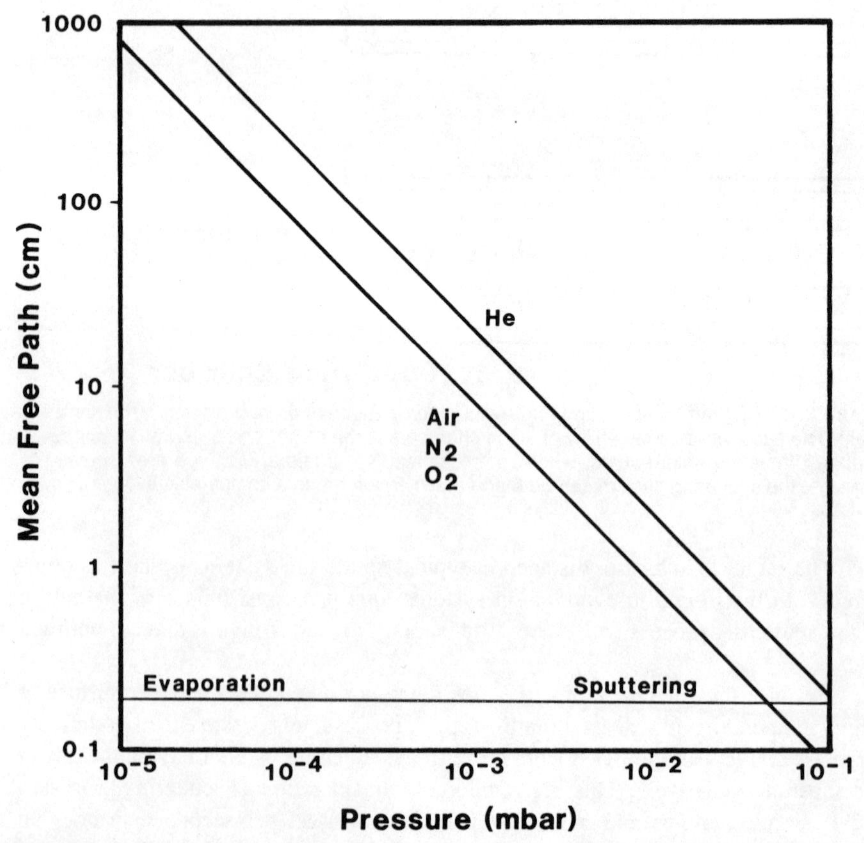

Figure 2.2. Mean free path versus pressure. The operating ranges of evaporation and sputtering processes are indicated.

source characteristics of an evaporation source, there is relatively less of a problem with nonuniformity in the thickness of the coating with a sputtering system. The size of an evaporation system is indicated by the need to be far enough away from the source that the distribution of the evaporant flux is within certain predefined tolerance limits. This is not the case with a sputtering system.

A short target to substrate distance is advantageous from the standpoint that there is a better coating material utilization factor in sputtering processes than in evaporative systems.

There are several variations of the sputtering process. If a conducting target, such as a metal, is used, then a DC voltage can be applied to it to maintain the plasma discharge that accelerates the argon ions. This is referred to as a *diode* arrangement. Sometimes, a grid is included in the system to help stabilize the discharge, and these are called DC *triode* systems. If the target is not conductive, then it is not possible to use it to carry the direct current. One then uses an alternating current and couples it capacitively to the plasma. A radio frequency of approximately 13 MHz is used for this purpose. Power levels may range from a few hundred watts to a few kilowatts.

Reactive sputtering can be used in situations where a nonconductive material is to be deposited. If a metal target is used, then a DC configuration can be used. Oxygen is mixed with the argon that is supplied to the plasma, and both gases become ionized. This highly active oxygen then can oxidize a surface layer of the target, which then gets sputtered, or it can oxidize the growing film on the substrate. The exact mechanism that is operative is determined by the geometry of the system, as well as by the gas mixture and oxidation chemistry. If the oxidation process is too vigorous, an oxide layer can build up on the target, effectively shutting off the direct current that excites the plasma, with the consequence that the plasma is extinguished.

In both the DC and the AC sputtering systems, there are many free electrons in the plasma. These electrons are highly mobile and are accelerated by the electric fields present in both types of discharges. The low mass of the electrons allow them to follow an alternating electric field easily. These electrons bombard the target and cause it to get hot, thus water cooling lines are included in the target designs to carry away this excess heat. The substrates are also close to the plasma, and they too get bombarded by the electrons and become heated. Substrate heating is one of the limitations of this type of coating system.

Magnetron sputtering is one technique that was developed to reduce the substrate heating by electron bombardment. The concept is that the free electrons can be trapped by magnetic fields and controlled. The magnetic fields are weak enough so that the paths of the heavier argon ions and charged sputtered species are not significantly affected. Thus, the electrons with their small mass to charge ratio as compared with the ions are confined to the vicinity of the target, and do not bombard, and thereby heat, the substrate. An extra benefit can be realized from this arrangement: as the electrons are trapped in orbits away from solid surfaces, they can ionize the gases in the plasma zone more efficiently. Consequently a lower gas pressure can be used to maintain the plasma. This means that there are fewer collisions between the sputtered atoms and the background gas and plasma as the sputtered molecule makes its way to the substrate. Magnetron types of systems are becoming very common today, and are commercially available in AC or DC versions. Their coating rates are far superior in most cases to what one can achieve with conventional sputtering.

A recent development in sputtering technology is the ion beam sputtering technique. In this approach to sputtering, a beam of particles is accelerated in an ion gun and directed at the target. These particles knock atoms from the surface, which then coat the substrate. Such a system can operate at the same vacuum levels as evaporation processes, thus combining the benefits of both evaporation and sputtering. Unfortunately, this coating process currently has deposition rates that are far slower than that obtained with standard sputtering techniques. Thus they are used at present only in research environments and when specialty films are required.

2.1.3. Flash Evaporation

Flash evaporation is a technique used to deposit multicomponent materials that (1) are incompatible in bulk form at high temperatures and, therefore, cannot be heated to the evaporation point together or (2) interact with crucibles or other containers at the evaporation temperature.

In flash evaporation processes, the source material, generally in a powdered form, is metered slowly onto a surface that is heated to a temperature well above the evaporation temperature of the most refractory component in the mixture. All components of the source material evaporate essentially instantly and completely. Because no liquid pool forms, there is no chance for a chemical reaction to occur. And as all of the material is evaporated before more arrives, fractionation doesn't occur and the coating is homogeneous. Thus, the deposit is a homogeneous mixture of the components that were dropped onto the hot surface. This process finds relatively limited use as the deposit is not as durable as those produced by conventional evaporation or sputtering.

2.1.4. Nonvacuum Processes

Some nonvacuum techniques are used to produce optical thin film coatings. Two major categories of these are listed here, though their applications are limited at present.

2.1.4.1. Organometallic Solutions. A nonvacuum technique for producing thin optical films is the use of organometallic solutions that are applied to a substrate's surface either by dipping or by spraying. The substrate is then dried and baked in a furnace to convert the film to a pure oxide while the organic constituents become volatile oxides. Thickness control is achieved by adjusting the temperature, viscosity, and other properties of the solution. This technique has been used successfully to coat high power laser optical components with antireflection designs that have extremely high damage thresholds. It is also used to apply multilayer antireflection coatings to glass substrates. An etch treatment after the baking step can produce a coating with a graded index that can result in a surface with extremely low reflectance over a broad wavelength region (Minot, 1976).

2.1.4.2. Electroless and Electroplating Solutions. These chemical processes are used to form opaque metal deposits. Both processes are generally performed at atmospheric pressure, therefore large steel vacuum vessels are not required. Silver films from electroless processes are commonly used to coat large astronomical mirrors that will not fit into readily available coating chambers. Many of the chemicals that are used in these processes are environmentally hazardous, and they must be disposed of carefully. In a production environment, where the quantity of chemicals used may be large, the costs associated with their disposal can become prohibitive, and therefore, the electroless process is not very widespread commercially. On the other hand, the use of this process on an occasional basis on a mountain top astronomical observatory could be quite attractive economically in comparison with the alternatives of acquiring a vacuum chamber, or moving a large mirror to another location for coating. A fundamental disadvantage of electroless coatings is that they are soft and, therefore, not very abrasion resistant.

The thickness of an electroplated or electroless deposit is not measurable in real time during the coating process; however, because the coating is typically opaque, this is not a significant problem.

Metal electroplating is a technique that has been known since the first days of electrochemical cells. The optical quality as well as the durability of some commercially available gold plate is excellent.*

Soft copper can be plated to a thickness of a few tens of micrometers and then turned with a single point diamond tool. The plated copper is an excellent material for diamond turning. As a result of its softness, the turned surface should not be touched by any solid object lest the surface be damaged.

As far as this author is aware, there are no ultraviolet, visible, or infrared transparent materials deposited by these techniques.

2.1.5. Process Conditions

The quality of a nonmetallic film is frequently improved when the substrate temperature is elevated above the ambient temperature before and during the vacuum coating process. This can be achieved with chamber heaters, or, as in the case of sputtering, from the process conditions themselves. The increase in surface temperature allows the arriving species to move about on the surface prior to their being incorporated into the growing film. The increased surface mobility increases the chances that a low energy lattice site will be encountered, which in turn will lead to a denser, more durable film. In some instances, high substrate temperatures are required to obtain the proper crystallographic form of a material (Macleod, 1983, pp. 201, 204; Maissel and Glang, 1970, pp. 17–14)

*Epner Technology, Inc., of Brooklyn, N.Y., 11222, is one plating company that actively supplies gold electroplated parts to optical industry types of specifications.

In practice, temperatures from 50°C to several hundred degrees are used. The upper limit on the temperature used is limited by a number of factors. In some coating equipment, the amount of heating that can be done is limited by the available heat sinking or cooling, as all of the heat that enters into a system must eventually be removed. In other situations, the substrates, such as plastic materials, may not be able to tolerate elevated temperatures (see Chap. 6). Finally, the time it takes to heat and cool a substrate in a vacuum may be economically prohibitive.

It should be noted that radiation is the primary means of heating and cooling a substrate in a vacuum chamber. In some special applications, conductive types of heating techniques can be employed, but these are generally complicated. In radiative environments, the emissivity of the heater and the substrate are the key elements. If the substrate is highly reflective in the wavelength range in which the heaters emit their energy, there will be poor coupling between the two and the substrate temperature will rise only very slowly. Similarly, the coupling between the substrate on the chamber walls is important when cooling takes place. A hot substrate with a infrared reflective coating will cool slowly.

2.1.6. Coating Time

The time required to coat a filter is dependent on a number of parameters, all of which can significantly alter the process time, and all of which depend on the actual equipment that is used. The pumping speed, the size of the vacuum chamber, and the state of cleanliness of the chamber, all affect the time it takes to reach base vacuum, that point from which all subsequent actions are referred. The following qualitative equation includes most of the important factors:

Coating Time Per Part = (Loading
+ Pump down
+ Heating
+ In situ cleaning
+ Coating
+ Cooling
+ Venting
+ Unloading
+ Machine clean up) / Parts in one run

The coating rate, layer thickness, number of layers in the design, and any special coating operations enter into the coating time term in the equation. Because an infrared layer may be 10 to 20 times the thickness of a similar layer in a visible coating simply due to the difference in wavelength, a sizable difference in coating times can be expected for comparable designs in the two portions of the spectrum.

As previously indicated, the type of coating process, whether evaporative, sputtering, or other process, significantly affects the coating time. In particular, if it is necessary to vary the temperature during the coating process, the processing time increases dramatically as such processes are inherently slow (see previous section on heating and cooling in a vacuum). Finally, because cooling rates can have a significant effect on yields (cooling glass is much more likely to cause breakage, and broken parts cannot be sold as profitably as whole filters), the coating temperature and substrate type are additional cost considerations. In summary, it is not possible to judge the amount of time that may be required to coat a substrate without intimate knowledge of the actual process that is used.

2.2. COATING PARAMETERS

2.2.1. Source Characteristics

The choice of coating technology depends to a great extent of the behavior of the coating material. Some materials may decompose when evaporated, whereas others, such as magnesium fluoride, decompose in a typical sputtering system. Thus, the selection of the appropriate coating system is dependent on materials related characteristics.

2.2.1.1. Compositional Variations. If the charge in an evaporation system consists of more than one component, it is unlikely that the two components will evaporate at the same rate at a given temperature. In that event, one component will evaporate faster than the other until the concentration of the two components is such that they evaporate at the same rate. This composition is called a congruently evaporating one, and the composition of evaporant stream is stable over time until one of the components is exhausted or the charge is changed. The mixture can consist of a single generic compound, which has several stable phases, e.g., the titanium oxide system (Samsonov, 1978), or of two materials that form a nonreacting mixture, e.g., a mixture of zirconium and tantalum oxides.

A coating source, the composition of which changes with time, leads to a deposited layer that is inhomogeneous. Such a coating will not perform according to theoretical calculations based on uniform, homogeneous layers.

2.2.1.2. Stoichiometry. Many oxide materials lose a certain amount of their oxygen during deposition, and this is especially true in systems in which the charge is heated with an electron beam. This is not surprising based on the fact that the chemical binding energy is of the order of a few electron volts, whereas the electron beam that is supplying the evaporation energy is accelerated by a potential of 5 to 10 kV. It is common practice to admit some oxygen to the coating chamber during the deposition of these materials to ensure that the deposit approaches stoichiometry. The oxidation reaction occurs at the substrate surface rather than in the gas phase. This can be readily understood since the mean free path in a vacuum chamber is on the order of a meter. Thus it is quite unlikely that a sufficient number of the evaporant atoms and molecules will strike an oxygen atom before hitting the substrate surface.

For certain materials it is important to obtain good stoichiometry, whereas for others it is much less important. For example, a slight loss of oxygen in silicon dioxide is generally not a serious problem, because the next lower stable oxide, SiO, is also rather transparent over a wide wavelength range. On the other hand, only a very slight oxygen loss from TiO_2 results in an absorbing coating. Thus, in this latter case, it is necessary to pay particular attention to be certain that the titania is completely oxidized on the substrate.

2.2.1.3. Coating Distribution. The distribution of the evaporant cloud is more important in an evaporation system than it is in a sputtering system. The proximity of the target to the substrate averages out most deviations in the distribution in the sputtering process. The large source to substrate distances in evaporation systems, however, make the distribution a concern to the operator of such a system. Factors that can cause the distribution to change as a function of time include the level to which the source has been loaded, the gas pressure in the chamber, and the way in which the evaporation charge is consumed. A change in distribution can affect the uniformity of the deposit on the substrate, and this in turn will affect the performance of the resulting filter.

2.2.2. Deposition Environment

The properties of a coating are strongly affected by the temperature of the substrate during deposition. The crystallinity, intrinsic stress, and transparency of a film are all functions of the coating temperature. Dielectric films are generally more durable when coated at higher temperatures.

The addition of a small amount of oxygen into the coating chamber during the deposition of oxide materials helps to ensure that they are fully oxidized, given the other process parameters. Suboxides are frequently absorbing, especially near their short wavelength absorption edges, which frequently occur in the visible portion of the spectrum. Oxidation reactions typically proceed more readily and completely at elevated temperatures, therefore, raising the coating temperature reduces the level of oxygen partial pressure needed to drive the oxidation reaction to completion. This is desirable as a lower background pressure yields a more durable deposit.

When the substrate is a common optical glass or a material that can tolerate high temperatures, there is generally no problem with coating them at elevated temperatures on the order of or greater than 100°C. Some substrates, however, cannot be heated very much: plastic substrates can decompose, melt or soften, or outgas some of their constituents and change their properties. Tempered metal or glass substrates the surface of which has been accurately figured to a diffraction limit can warp out of tolerance when heated. For such substrates, only limited heating can be used, and the accompanying loss in durability is a trade-off that has to be addressed by the system designer.

Outgasing, the release of gases into a coating chamber either before or during the coating process, can occur whether or not the substrate is heated. Substrates can have cavities that trap air, which then slowly leaks out. If the leakage occurs during the coating process, the gases can disrupt the coating process and cause the deposited film to be soft, scattering, or otherwise degraded. A blind screw hole is one common source of outgasing. This type of gas source is often called a *virtual leak*. There are standard design techniques that can alleviate these problems, but there are no definitive texts currently in print on the subject, though O'Hanlon's (1980) text and Maissel and Glang (1970) do give some guidelines.

2.2.3. Nucleation and Growth

During the deposition process, films do not always grow uniformly, monolayer by monolayer, especially in their earliest growth stages. Under some deposition conditions, metallic deposits can form tiny islands in their initial stages of growth rather than a very thin continuous layer. The atoms arriving at the surface can be mobile enough to move about on the surface until they encounter a suitable nucleation site. Such a site may be either directly on the substrate or on a part of the growing film. Eventually, the individual islands become large enough to touch one another and the layer becomes continuous.

The result of islands forming and then coalescing is an index of refraction that is dependent on thickness (in this thickness range). Gold and silver are two metals with well-known thickness dependencies of their complex index of refraction when they are in the 10 to 200 Å thick range.

A similar phenomenon of thickness dependent refractive index occurs with some dielectrics, but the effect in these materials occurs at larger thicknesses. For instance, it has been recently reported (Klinger and Carniglia, 1985) that the crystalline structure of an evaporated zirconium dioxide film changes from cubic to monoclinic as it grows beyond 70 nm in physical thickness. The amount of cubic phase does not change indicating that only the newly arriving material contributes to the increasing monoclinic phase, whereas the original cubic phase remains stable. They show that this accounts for the inhomogeneity observed in films of this material.

2.2.4. Rate and Thickness Monitoring

In a coating process, both the rate of deposition as well as the total thickness of the coating need to be monitored and controlled. There are several types of monitoring techniques that do this: (1) A quartz crystal oscillator measures the mass of a deposit as a function of time. (2) An optical monitor measures changes in the transmission or reflection of a part or witness exposed to the coating flux. (3) Atomic emission spectroscopy measures the intensity of the light emitted by excited evaporant atoms.

2.2.4.1. Crystal Oscillator. The most common monitoring technique that measures both rate and thickness simultaneously is the quartz crystal oscillator. In this device, an oscillating quartz crystal in a water-cooled holder is exposed to the flux of material that is coating the substrate. The mass of material that deposits on the crystal changes its resonant frequency. This change is sensed by an electronics package and is converted to a display of deposition rate and coating thickness. It is important to note that this technique produces an output that is a function of the mass of the deposit, but it provides no optical thickness information.

If the deposit on the crystal has excessive stress, the calibration factors of the crystal can change and cause an error in the output value. The deposit can crack or delaminate as a result of this stress and an abrupt change in crystal frequency occurs, leading to an indeterminate result. Modern electronics packages that control the oscillator can detect an abrupt change in oscillator frequency. Several means have been developed to get around these problems when the deposition process is underway and cannot be interrupted:

1. A second crystal can be employed; it is positioned adjacent to the first and shielded from the coating material by a shutter. At the first indication of trouble, the electronics in the control package activate a shutter that uncovers the second crystal.
2. A way of extending the life of a crystal is to reduce the rate of buildup of thick deposits on the crystal by using a shutter or screen in front of the crystal. Only a known fraction of the full flux is sensed, thus reducing the amount of material deposited on the crystal. A shutter that is open intermittently can consist of a rotating slotted wheel or a blade shutter that opens momentarily from time to time to sample the rate. It is assumed in these sampling schemes that the source is sufficiently stable that sampling gives an accurate picture of the total amount of material deposited and the rate at which this takes place.

Crystal oscillator systems are generally used to monitor metal deposition processes. There are several reasons for this: (1) Metals are highly absorbing and most metal films are opaque, therefore, standard optical techniques are not useful except for very thin films. (2) Metal films are typically not as stressy as dielectric films, therefore, greater total thicknesses can be deposited on the crystal before it fails.

A quartz crystal can be used equally well with a sputtering or an evaporation system. In fact, a quartz crystal system is ideal for a sputtering system that has a large target and close spacing of the substrate to the target. Special crystal holders are available that are optimized for each type of deposition system.

Sputtering is a very well-controlled process, therefore, in a production environment, the thickness of a deposit can be determined or controlled on the basis of the target power input and the time of deposition. This is fortunate, as it is rather difficult to monitor optically the growth of a sputtered film owing to the geometrical constraints of the close proximity of the target and the substrate. If the coating is done intermittently by moving either the substrate or the target, then it is possible to monitor the transmission or reflection between the coating passes.

Standard quartz crystal oscillator systems are available commercially from a number of suppliers.

2.2.4.2. Optical Techniques. Optical techniques are generally preferred when coating dielectric films for precision optical designs. The most common situation where such techniques are useful is in the construction of high reflector coatings that consists of a stack of layers of alternating index, each layer's being one quarter wave thick. If one monitors the reflectance from the substrate at the wavelength at which the highest reflectance is desired and terminates the coating of a layer when the reflectance reaches a peak value, then an error made in one layer will be automatically compensated in the next layer. This applies to small to moderate errors, and, of course, is not a substitute for good coating techniques.

Reflection or transmission optical monitoring is possible, and commercial units are available off the shelf to make these types of measurements. Figure 2.3 shows the optical layout of such systems. These systems are very useful when one is coating a stationary part. If the part(s) are in single rotation in the chamber, then it is still possible to monitor the transmission or the reflection as a given part follows a relatively simple circular path. This is especially convenient if the part is large and the geometry is such that a stationary beam is always reflected off the coated surface as the part(s) rotates. If double rotation or planetary systems are used to simulate a random motion of the substrate with respect to the sources, then it is

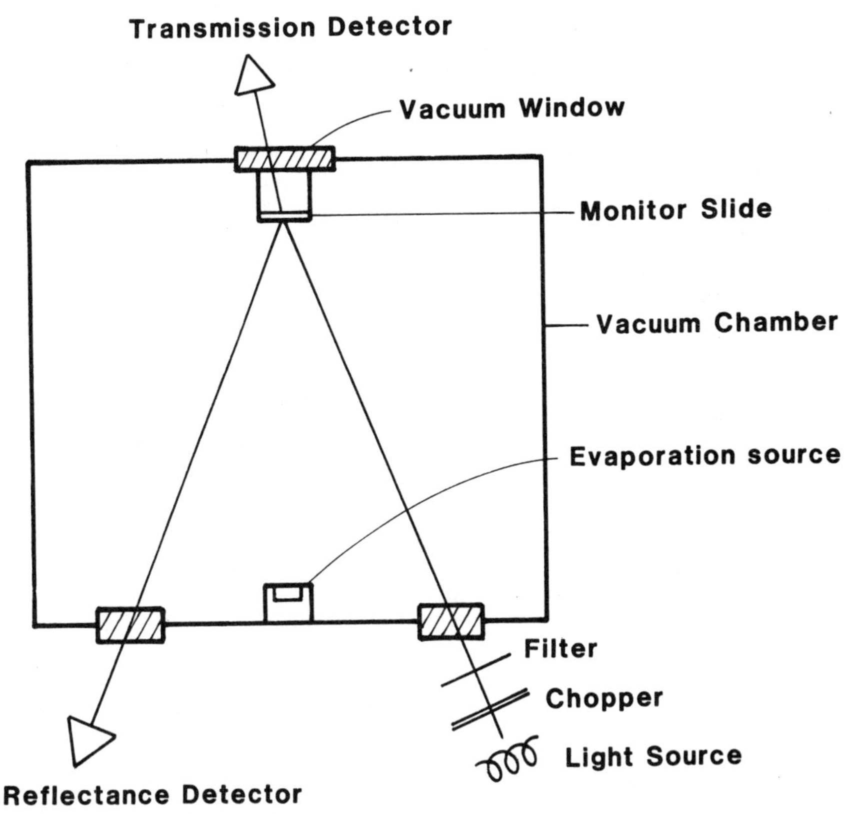

Figure 2.3. Optical schematic diagram of reflection and transmission monitors for vacuum evaporation. A number of variations on this basic arrangement are possible if the coating machine geometry does not lend itself to this arrangement.

difficult to measure the parts directly. In such cases, the optical properties of a stationary witness part must be employed.

2.2.4.3. Atomic Emission Techniques. A system is commercially available that monitors the rate in a *noncontacting* way, and therefore, can run indefinitely. It is based on ionizing a portion to the evaporating flux of material with a beam of electrons. The electrons are directed perpendicularly to the direction of the flux, and at right angles to both of these, a detector evaluates the light emitted by the excited atoms and molecules in the flux. The emitted light has the characteristic spectral lines of all of the materials that are being coated. A filter selects the line that represents the material of interest. The process is excellent for metals, but can be problematical when compounds, especially oxides, are being deposited in a gas reaction process. The band spectra obtained from the excited gas molecules obscure the elemental lines, thus reducing the sensitivity to too low a level to be practical. Further development will probably improve this technique further.

2.2.5. Sticking Coefficient

The sticking coefficient is defined as the fraction of the incident molecules from the source that actually adhere to the substrate. This coefficient is very dependent on the material, the temperature of the substrate, the nature of the surface, and so on. Some materials actually inhibit almost all molecules from sticking to a surface. It is interesting to speculate what are the dynamic parameters, e.g., kinetic energy, or velocity vector, after an atom or molecule strikes a surface but does not adhere to it. For example, is the atom or molecule that strikes the wall once thermalized?

One such material that makes surfaces nonadherent is available commercially as "Hi Torr"™. Such materials are useful in situations where the coating material is very valuable, such as gold, and one desires that it coat only on certain areas. This aids in prolonging the time between vacuum chamber cleanups, as well as reducing coating material costs. The commercially available material is a fluorinated hydrocarbon. Unfortunately it is not perfected to the point where it can be used in chambers where the coating is done at elevated temperatures.

For some materials, the sticking coefficient can be strongly dependent on the temperature of the substrate. For example, the sticking coefficient of ZnS, when the substrate is in the range of 150 to 200°C, varies markedly. Small variations in the substrate temperature produces large variations in coating thickness. At temperatures well below 150°C, the sticking coefficient is close to unity, whereas at temperatures well above 200°C, it is nearly zero. Temperatures near 200°C yield different crystalline forms of zinc sulfide, thus there is interest in coating in this temperature regime, but very careful attention has to be given to temperature control and the uniformity of the temperature over the entire surface to be coated. Zinc selenide is another material that shows this sort of behavior. Undoubtedly, other materials exhibit similar characteristics in some temperature ranges; however, most oxides and fluorides, the most common of the coating materials, are quite well behaved in typical coating temperature ranges.

2.3. COATING EQUIPMENT

Vacuum coaters use a variety of coating equipment to manufacture thin film filters. The common types of equipment are presented here from the filter user's standpoint. In particular, we describe the implications of the size of the equipment on filter performance and cost, and the effect of the substrate support structures in the vacuum chamber on useful filter aperture.

2.3.1. Size of Coating Chamber

A major factor in the unit cost of producing optical filters is the quantity of parts in a coating lot. This is no different than many other commercially produced items. When a large quantity of one item is desired, then the cost may be reduced, given appropriate capital investment. Thus, the equipment size and the order size need to be matched. Figure 2.4 shows the unit cost of a standard product produced over a number of years and in different volumes (Seddon, 1977). Note that a mismatch in equipment size to order quantity results in an increase in unit cost.

The size of commercially available vacuum chambers varies from 30-cm di-

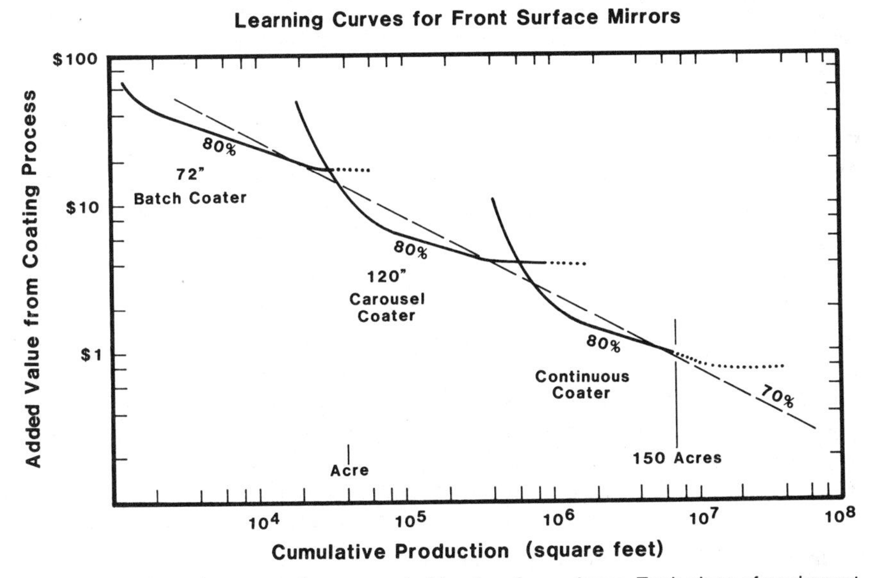

Figure 2.4. *Learning curves* for evaporated front surface mirrors. Each piece of equipment has a slope of about 80 percent, whereas the overall curve has a 70 percent slope. This means the the costs drops by 20 percent each time the volume doubles. Note that it is important to use properly sized equipment for a given size lot—both too small or too large a chamber will increase costs over the optimum machine. (After Seddon, 1977).

ameter glass bell jars used in research facilities and teaching laboratories, to behemoth chambers that are connected in tandem with large glass-making machines and are used to apply architectural coatings continuously to large glass panes. In between these extremes, one finds that production chambers are typically in the 0.5- to 1.5-m diameter range. These larger evaporation chambers are frequently made of stainless steel and are customized for specific applications. Their geometry may be horizontal or vertical cylinders, or boxlike with flat sides. Sputtering chambers may have two of their dimensions similar to that of a comparable evaporation unit, but the third dimension corresponding to the target to substrate dimension may be much smaller.

Generally, one does not build a production coating chamber to handle both evaporative processes and sputtering at the same time. The reason for this is that the geometry of the two techniques is not very compatible.

Nonvacuum processes (see section 2.1.4.) use tanks or vats that correspond to the size of substrate being coated. The capital cost for this equipment is far less than that of comparably sized vacuum systems.

2.3.2. Chamber Geometry and Process Considerations

The size of a vacuum coating chamber can have an important effect on several coating parameters. The surface to volume ratio can affect the residual gas load and composition and the temperature reached during coating. The distribution of coating on the part, and the fraction of evaporated material that actually coats the substrate is related to the size of the chamber with respect to the size of the substrate or the coating area.

2.3.2.1. Surface to Volume Ratio. The quality of the coating is influenced by the residual gases in the chamber during the coating process. The source of these background gases may be intentional, e.g., oxygen is metered into the chamber to control the stoichiometry of oxides, or it may be unintentional such as leaks or internal sources. The latter sources of residual gases include desorption of gases on the chamber walls and other surfaces. The wall area increases as the square of the average dimension of the chamber, but the volume increases as the cube (an r^2 vs. an r^3 relationship). If we assume a constant gas evolution per unit area, we find that the gas load increases with the size of the chamber. The volume that contains these gases, however, increases even faster yet with the chamber size. Thus, based on the very simplified model of volume and surface area considerations, larger chambers are less prone to problems with wall outgasing. The actual chamber pressure is determined by a number of other parameters including time and the pumping capacity on the vacuum system.

The amount of residual gases in a chamber can be reduced by heating the chamber and its contents during the pumpdown before a coating run to drive off the adsorbed gases. This process is called a *bake*.

Considering the upper limit of coater size, one would like an infinitely large chamber to coat a small part. On the other hand, it would take an infinite amount of time to pump out such a volume. The worst case is a chamber just large enough to hold a part of a given size. Then, the pumping time to base vacuum is very short; however, unless the chamber were well baked out prior to the start of deposition, the outgasing from the walls and the substrate as the coating process heats up the chamber may be quite severe.

The preceding remarks do not apply to continuous coating systems, either evaporation or sputtering, where the coating chamber is not exposed to atmospheric pressure over the course of days or longer.

2.3.2.2. Thickness Uniformity and Coating Efficiency. The coating uniformity obtainable with an evaporation system is dependent on a number of geometric considerations, including the ratio of the size of the area to be coated to the distance between the source and the substrate. The distribution of the material leaving an evaporation source can be approximated as a point source with an angular dependence that closely follows a cosine function with an exponent in the 2 to 4 range, the exact value of which depends on the source type, the background gas pressure, the source power level, and other parameters. Figure 2.5 shows the distribution for three values of the power. A small substrate area subtends a smaller part of this distribution. Preferably, the part is located near the peak of the distribution where the slope is the smallest and there is the least variability in thickness over the part.

The variation of the coating flux in an evaporator can frequently be modeled in a manner similar to that used for light intensity calculations in an optical system: the drop-off in thickness is proportional to $\cos^4 \theta$: one cosine factor comes from the source distribution (Lambertian), two come from the $1/r^2$ increase in distance between the source and the substrate, and the last factor comes from the change in the subtended surface area at the substrate. Tilting the substrate so as to make the surface normal to the flux arrival direction eliminates one of the cosine factors. The thickness uniformity can be further improved by introducing relative motion between the substrate and the coating source.

A trade-off must be made between the active coating surface, i.e., the average substrate area being coated and the coating distance. As a larger fraction of the distribution, shown in Fig. 2.5, is intercepted by the substrate, the better the utilization of the coating material. Much of the material evaporated is deposited on the walls of the chamber. This material is not only wasted, but it contributes to an increase in the frequency of cleanups of the chamber. The material that coats the walls is generally porous because the walls are cool and the angles of arrival of the evaporant are poor. Invariably, the wall deposits are hygroscopic because of the large surface area and capillaries afforded by the many pores. Thus it serves as a source of moisture during the pump down and coating procedures. This is exacerbated in that the walls generally warm during coating, and this tends to drive out moisture at the worse possible time. A bake before coating, if allowed by equipment and substrate constraints, helps reduce this problem. This tends to stabilize the process by driving out the adsorbed gases before the process starts. The chamber may be cooled, if necessary, before the actual deposition starts.

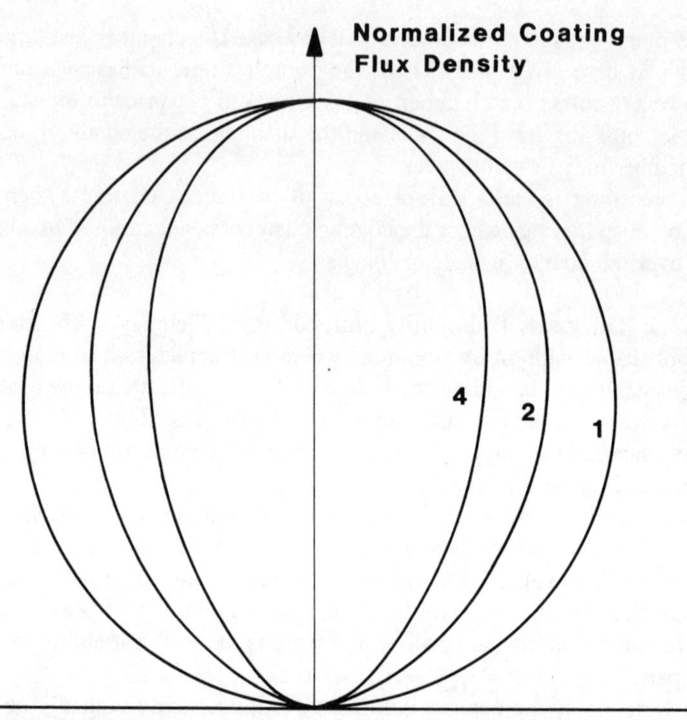

Figure 2.5. Polar plots of three normalized distribution patterns of a vapor source. The exponent of the cosine function is indicated on the plot. As the exponent increases, the source becomes more directional, and a larger fraction of the material deposits on the substrate. Conversely, the area at the substrate plane over which the uniformity is at a given level decreases as the exponent increases.

Sometimes, the only practical way to coat certain geometries is with a sputtering process, e.g., coating the inside of a glass tube with an infrared reflective coating (Harding et al., 1979). Sputtering targets, by virtue of the fact that they are solid, can be arranged to coat in any direction: up, down, or sideways. This flexibility is not usually found in evaporative systems as many of these sources contain molten or granular material and are restricted to coating upward. This is not to say that evaporation sources cannot be designed to coat in directions other than vertical, but significant source design trade-offs have to be made to obtain this feature. With sputtering targets, none of the drawbacks apply; therefore, sputtering in the downward direction is as easy as sputtering in the upward direction. One reason that downward sputtering is not practiced too commonly in high quality thin film manufacturing is that arcs occasionally develop between the target and the plasma, the substrate, or some other conductor or source of electrical charge. Though the electronic power supplies are designed to deal with this and extinguish the arc very quickly, there still may be a shower of fine particles resulting from the arc. If the substrates are lying face up, they will then be contaminated and the coating degraded by these particles.

2.3.3. Substrate Rotation for Coating Uniformity

Historically, the first thin films were made in chambers in which all parts were stationary. Next, single rotation systems were developed to enable the coating of larger substrates. Finally, more complex motion systems evolved as economic production of precision coatings became the dominant factor in thin film technology.

Substrates in a coating chamber generally have a support system that provides them with a motion with respect to the source. This is done to achieve a more uniform deposit over the surface of the substrate. It is conceivable that one could move the source with respect to a stationary substrate. Even when large substrates are involved, however, it is simpler to move the substrate rather than deal with the high current, high voltage, or high power electrical feedthroughs and the cooling water couplings that are necessary to operate movable evaporation or sputtering sources.

The relative motion may simply be linear, as in coating glass sheets for architectural purposes or the coating of webs of flexible material in a roll coating machine. In precision coating applications, the substrates are generally rotated in a planetary fixture. In such a geometry, the parts are mounted in racks that rotate about one axis (planet), and this axis in turn orbits around a central axis usually located at the center of the coating machine (sun). The ratio of the gearing that generates the orbiting motions is chosen such that many orbits are required before a point on the substrate retraces its path. As long as many orbits are completed during each layer's deposition, the motion will randomize thickness variations. The result is a much more uniform deposit than would be obtained without the complex motion (Macleod, 1969, Chap. 9). Because the substrate passes over the source many times during the deposition of any given layer, it is in fact built up of a succession of very thin deposits. This intermittent coating process has some desirable characteristics in addition to its improving the uniformity of the deposit. For example, if the coating is an oxide and there is a residual gas of oxygen in the coating chamber, then the intermittent coating process gives the deposited material a chance to react before another layer of evaporant molecules arrives.

2.3.4. Jigging and Substrate Support

Periodically, someone will offer an off-the-cuff suggestion that a suction cup be used to hold the substrates during coating. Such a system is flexible and avoids the need for customized tooling. Unfortunately, in a vacuum chamber there is an even better vacuum on the front side of the substrate to pull it off the suction cup! Thus another means of holding parts is needed.

With a few small parts, it is straightforward to make adapters that mate to permanent tooling inside a coating chamber. If the parts have an irregular shape, then it can be difficult to hold them so that the coating is applied uniformly over the surface. If there are many parts to be coated, the making of the adapters could become a significant fraction of the total cost of a coating task. If the quantity of parts is large enough so that many coating runs are required to complete the task, then special purpose tooling becomes very attractive from an economic standpoint.

The addition of parts to a load usually will not change the cost of producing the run, therefore, it is an incentive when large quantities have to be coated to spend more resources on designing tooling that will maximize the load size.

Most part holders in coating systems that coat upward leave a small uncoated area near the edge of the part. It is here that a small lip in the tooling actually supports the substrate. The part of the substrate that is hidden does not get coated, therefore it is desirable to make this zone as small as possible so that it does not intrude into the active aperture of the optical element. There should be some means of constraining the substrate from moving laterally in the tooling fixture to prevent scratching of the front surface by the lip. The tolerance between the side of the substrate and the holder must include provision for thermal expansion coefficient differences between the tooling rack material and the substrate.

Substrates with a mass of several hundred kilograms have been coated in double rotation systems. When the substrates become this large, then the tooling problem becomes one for mechanical engineers. The vacuum tooling designer is faced with a number of specialized problems. A major one centers on the tendency for two similar metals to weld when they come in contact in a good vacuum and some pressure is applied to them. This makes the design of bearings difficult. Lubricants have to be selected carefully lest they contaminate the vacuum environment.

It is important to avoid deposits on surfaces that are not intended to be coated. For example, some coating processes can deposit materials on surfaces that are not in a line of sight to the source. This happens for a number of reasons, among which are

1. The gas pressure during the coating process is relatively high and the coating material is scattered around to the back of the substrate
2. There is decomposition of the source material to gases that diffuse to the back of the part where they recombine and form a deposit
3. The source material may not stick well on the walls of the chamber, and thus bounce off and eventually find its way back to the substrate

These unwanted coatings are typically soft, nondurable, and uncontrolled. They must be avoided, especially if a surface that is susceptible to such coatings has any value; e.g., it will be subsequently, or has already been, coated in another coating operation.

2.3.5. Temperature-Related Substrate Degradation

The temperature at which a coating is applied is generally above ambient. This temperature can have a detrimental effect on certain substrates. In precision optics, the surfaces of which have been polished so that deviations in them away from the desired shape are less than a fraction of a wavelength, heating can cause optical distortions greater than the allowed tolerance. Organic substrates, i.e., plastics, pose severe coatings problems. As previously mentioned (section 2.2.2.), relatively low coating temperatures can cause the substrates to melt or otherwise deform (Agranoff, 1985). In addition, the heat of condensation of the evaporant or the radiant heat from the source in an evaporation process or the plasma in sputtering systems, can easily cause the substrate temperature to rise. This leads to warpage or outgasing.

A large number of different kinds of plastics are used as optical devices today. In addition, many of the compositions of these plastics vary from one manufacturer to another. Each supplier adds his own proprietary combinations of ingredients, including volatile components such as plasticizers and fillers. The volatile components can interfere with the coating process as they diffuse out of the plastic. It can be expected that the properties of the plastic will also be different after the coating process and the plastic has lost its volatile components. This limits the number of materials that can be coated on plastics and the temperature at which they are laid down. As a result, it is rare to find coatings that are as hard on plastic substrates as they are on glass and other materials where the coating temperature can be elevated to over 100°C.

2.4. STRESS EFFECTS

Although process heating during coating can have an effect on a substrate's optical figure, stress is an even more significant parameter when the substrate is thin. This section considers the sources of stress, and the effects of stress in the coating on both the coating itself and on the substrate.

Stresses in a thin film can have several components. *Intrinsic* stresses develop as the film is being formed. The growing crystal lattice is not perfect and stresses develop as it relaxes. These stresses can be affected to some extent by the nucleation sites available to the first few monolayers deposited, but after a few layers of atoms, the film structure tends to some dominant crystallographic form independent of the initial conditions. Intrinsic stresses are strong functions of the coating temperature, coating rate, and other coating conditions. Note that no temperature *change* is required to obtain intrinsic stresses. Postcoating annealing steps can sometimes reduce the intrinsic stress in a coating.

A second major source of stress is due to the differences in the coefficient of thermal expansion of the film and that of the substrate or adjacent layers. The stress develops when the temperature changes from the deposition temperature. It is difficult to avoid this type of stress when films are deposited at temperatures above ambient. Even when films are deposited nominally at room temperature, some heating of the substrate can occur during the deposition and condensation process. Finally, temperature variations during use may lead to changes in the stress level. The sign of the thermally induced stress can change from tensile to compressive, or vice versa, as it is a function of the differences in the thermal expansion coefficients of the film and substrate materials. For example, if the substrate expansion coefficient is greater than that of a film that is compressively stressed at some given temperature, there exists theoretically a higher temperature at which the stress in the film becomes tensile.

Stress by itself is not necessarily bad as far as the films themselves are concerned (warpage of the substrate is another matter altogether, however). For years, magnesium fluoride films, which have extremely high stress levels (see Table 2.2), have been used very successfully in forming visible quarter wave thick antireflection coatings on glass surfaces. The reason for this success is that the *adhesion* of the film to the substrate is excellent and the substrates are strong in compression. As a result, enough of the tensile stress in the film is transferred to a compressive stress in the surface of the massive substrate that the cohesive strength of the film is not exceeded. If a magnesium fluoride film is made thick enough, however, the stresses in the film become high enough to overcome the film's cohesive strength. A number of fine cracks forms to relieve the tensile stresses; a film with these fine cracks is said to be *crazed*. This happens with magnesium fluoride at quarter wave optical thicknesses greater than about 1.5 μm, thus precluding its use as an infrared coating material.

In some instances, a film's cohesive strength and its adhesive strength to a substrate can be strong enough to exceed the cohesive strength of the bulk substrate material. For example, the stresses in a tantala-silica stack on fused silica can be made large enough so that as the film fails in tension, it tears off small particles of the silica surface, leaving behind a rough, pitted surface.

The failure mechanism of films which are under compressive stress is delamination of the film from the substructure. The adhesive strength of the film-substrate bond is the dominant parameter in preventing failure of a film in compression; the cohesive strength of the film is not a factor.

To first order, the contribution of a material to the total stress in a coating is proportional to the product of the total thickness and the stress per unit thickness of the material. Coatings that consist of stacks of periodically repeating assemblies of layers (periods) have a total stress level proportional to the number of periods and also to the net stress in a typical period. The net stress of a period depends on the sum of the stresses (tensile and compressive) in the individual layers.

In a period with a pair of layers, one of which has tensile stress and the other, compressive stress, there will be only one thickness ratio that will yield a net stress of zero. The balancing of stresses is only one of the considerations a thin film designer must take into account in arriving at a producible final design. Although optical considerations may dictate a particular thickness ratio, stress considerations may suggest a different ratio of material thicknesses.

The majority of dielectric materials have tensile stress when deposited as a thin film in conventional processes (simple evaporation and sputtering). The stress levels of these films have of late been altered by new techniques such as ion beam bombardment of an evaporated film or ion beam sputtering (see Chap. 7). Fortunately for the thin film engineer, some materials have compressive stresses also. These materials can be used to compensate for tensile stresses in an adjacent layer. A typical material pair that has a good net balance between stresses when both layers have approximately equal optical thicknesses is ZnS and ThF_4 (Ennos, 1966). Another useful material with compressive stress is silicon dioxide. It is the lowest index oxide material commonly used for durable coatings, and is used to compensate for the tensile stresses found in the high index materials that are frequently used in conjunction with it, such as titanium dioxide and zirconium dioxide. There seems to be no relation between the stress type (compressive or tensile) and the index of refraction.

The cohesive and adhesive strengths of thin film materials are not well known quantitatively. Apart from the difficulty of measuring these quantities for thin film specimens, the values of these parameters can vary with time as moisture penetrates the films and as the internal forces relax. Also, as previously mentioned, the internal stresses and strengths are strong functions of deposition conditions. What is needed is a complete study of this problem so that it can be reduced to a more firm engineering practice.

Ennos (1966) took a first step in this direction by publishing a study he made of in situ stresses. He observed the deflections of a thin substrate while various films were being deposited in an evaporator. He was able to deduce the stress levels in films of various materials from his data, and these are summarized in Table 2.2.

In a few instances, a coating with crazing is acceptable. The majority of these cases involve optical systems that are only collecting energy and are not image-forming systems. In an energy-collecting system, for example, the key requirement is to gather energy, either reflected or transmitted and to deposit it on some receiver. A coating that is crazed loses a small fraction of the incident energy to scattering, but this lost energy is sometimes insignificant in the overall system performance, especially when the benefits of the coating are taken into account. The total surface area occupied by the cracks is very small in comparison to the total surface area covered by the coating. Diffraction at the edges of the cracks leads to scattering. An example of a coating in which the crazing is tolerable is in the energy conserving coating applied to the fused silica envelopes of tungsten-halogen lamps (Rancourt and Martin, 1986). The scattered light is still useful in lighting applications, and the coating performs its task of reflecting the infrared radiation back onto the filament. Other similar applications exist in solar energy collection systems.

Crazed films cannot be used in some designs at all; for example, they could not be used in an antireflection (AR) coating. An AR coating is intended to reduce

Table 2.2 Measured Stresses in Coated Films

Material	Thickness (angstroms)	Stress		
		(kg/cm²)	(lb/inch²)	(Pa)
Single Layers				
ZnS	675	+1800	+25,600	$+1.77 \times 10^8$
Cryolite	1170	−250	−3,500	-2.41×10^7
$ThOF_2$	1090	−1600	−22,800	-1.57×10^8
PbF_2	910	+51	+725	$+5.00 \times 10^6$
MgF_2	2500	−4500	−64,000	-4.43×10^8
Multilayers				
ZnS/Cryolite	—	+900	+12,800	$+8.83 \times 10^7$
$ZnS/ThOF_2$	—	+100	+1,400	$+9.66 \times 10^6$
PbF_2/Cryolite	—	−800	−11,400	-7.86×10^7

Note: Positive values indicate compressive stress. Negative values indicate tensile stress.
From: Ennos, 1966

the reflected light from a surface; a scattering coating will tend to do just the opposite by increasing the amount of light coming from it.

Stress relaxation by deformation of the substrate occurs when the substrate is thin enough or flexible enough to bend under the net force produced by the coating. If the substrate is a thin polymer film, for example, this can make the substrate curl to the point of rendering it useless. If the substrate is a solid, such as a thin glass or silicon wafer, the warpage may lead to breakage, or if an optical element, to unacceptable distortions. Assuming that the coating is thin with respect to the substrate thickness, the deflection of a flat plate substrate with parallel surfaces can be approximated by the following formula (Glang et al., 1965):

$$\delta = 3[(1-\nu)/E]\ (t_f/t_s^2)\ \rho^2\ \sigma$$

where δ = the deflection at a distance ρ, ρ = the distance from the center of a circular disk, ν = Poisson's ratio, E = Young's modulus, t_f = thickness of the film, t_s = thickness of the substrate and, σ = film stress with dimensions of force per unit length.

This formula can be used to estimate the stress in a coating by measuring the deflection of a substrate of a known material, such as silicon or silicon dioxide. Once the stress in a film is known, the deflection of other substrates can be estimated.

The sense of the curvature of a substrate can be used to determine whether a coating has tensile or compressive stress as shown in Fig. 2.6. Stress failures are easily classified as being caused by tensile or compressive stresses with a microscopic examination of the coating (see section 5.4.7).

2.5. SUBSTRATE CLEANING

Cleaning is one of the most difficult problems in the manufacture of low reflectance coatings, and it is a prime problem with other coating designs as well. One of the reasons that cleaning is so difficult is that it is dependent on all of the prior history of the part: at all stages of handling more dirt, grime, and organics get added to the surface. If cleaning is not considered from the beginning of the fabrication of

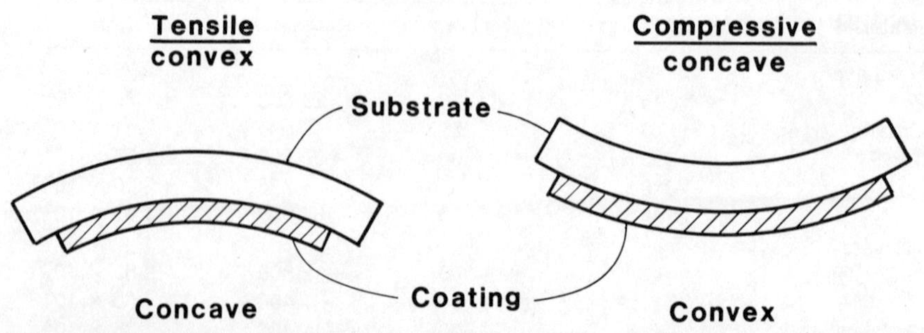

Figure 2.6. Curvature induced by compressive and tensile coatings. If the substrate is thin or flexible enough, the direction of bowing can be determined with the unaided eye. It is not always easy, however, to determine which side is coated.

a part, either on an optical polishing machine, a diamond-turning lathe, or an injection molding machine for plastics, then it may never be possible to clean the surfaces properly at a later time. A good example of where this has been found to be highly important is in the fabrication of diamond-turned metal optics. In this machining operation, the temperature must be precisely controlled to obtain the high level of accuracy desired, therefore, an oil shower is used to bathe all of the work and the tool in a constant temperature fluid. If this oil is not cleaned off the substrate within minutes of the completion of the turning operation, the oil somehow combines chemically with the fresh metal surface, and it then becomes impossible to remove (Buckmelter et al., 1975). Because the diamond-turned surface is extremely fragile, contact methods cannot be used to scrub the contaminant off the optics.

Some optical glasses are rather prone to staining as indicated in the standard catalogs of optical glasses (e.g., Schott Glass catalog). The cleaning of the glasses classified by the Schott Glass Company as "Group 5" can be problematical. These glasses are stained by an exposure of less than 12 minutes to a test solution buffered at pH 5.6 (only slightly acidic).

The common vigorous cleaning techniques can be divided into several classes:

1. Contacting: Physical contact is made with the surface, including scrubbing with or without abrasive materials, either manually or automatically.
2. Ultrasonic: The substrate is placed in a cleaning solution that is agitated with ultrasonic waves.
3. High pressure spraying: A stream of high pressure water or other solvent is directed at the substrate to dislodge particulates.
4. In situ: Vacuum processes, such as glow discharges or sputter etching, are used before starting the actual deposition of the coating.

2.5.1. Contacting Cleaning

Scrubbing is the oldest cleaning method. In this technique, a solid object actually touches the surface that is to be coated. The "solid" object may be as soft as a cotton ball or as hard as an abrasive particle. When cleaning a diamond-turned metal surface, even a wet cotton ball is "hard" and can damage the surface. Thus, "hard" is a relative matter. The more common means of contact cleaning is with brushes, as are found in automated scrubbers that find use in washing flat glass sheets. Sometimes an abrasive, such as rouge, is used to enhance the action of the brushes. The choice of these methods depends on:

1. The state of cleanliness of the incoming substrates
2. The fragility of the substrate surface
3. The strength of the substrate

Sometimes, a technique known as *tissue drag* is used to clean a surface prior to coating. The technique is a cross between the contacting methods and the fluid ones in that the contact between the surface and the cleaning agent is for the most part a fluid one. The process consists of soaking a lens tissue in acetone or other purified solvent and then dragging the tissue over the surface of the part (Stowers and Patton, 1978).

2.5.2. Ultrasonic Cleaning

Ultrasonic cleaners do a good job of removing most particulate matter from surfaces as long as they are not too tightly bound to it. A danger exists, however, with this approach in that if the substrates are left in the active cleaning tanks too long, the substrate can suffer surface damage in the form of micropits. These defects both reduce the specular transmission as well as increase the stray light. The solutions used in ultrasonic cleaning usually contain surfactants to accelerate the process and to keep the particulates in suspension in the solution. A good review of this technique has recently been presented by Guenther and Enssle (1986).

2.5.3. High Pressure Sprays

High pressure sprays were originally developed to clean silicon wafers for integrated circuit production. The Lawrence Livermore National Laboratory applied the technique to the cleaning of metal and glass optical surfaces (Stowers, 1978). The technique appears to be ideal for fragile surfaces that cannot be otherwise cleaned. The cleaning fluid can be a fluorinated hydrocarbon, water, or other suitable solvent. The pressures used are of the order of 7×10^6 Pa (70 atmospheres or 1,000 lb/inch^2). Particles with diameters down to 5 μm can be removed with this technique. Such a process is rather complicated and is used only when less complex means do not work. It should be noted that the fluid remaining on the surface after the cleaning step needs to be removed without its being allowed to dry on the surface.

2.5.4. Drying

In all cleaning processes where solvents or water are used as the cleaning fluid, the most important step is likely to be the one where the remaining liquid is removed from the surface after the cleaning process is complete. Unless the liquid is removed before it can evaporate, stains known as *water marks* form where a pool of liquid remained and evaporated. These cleaning-induced stains are most difficult to remove, and likely will not be removed by additional cleaning with the same technique that put them there in the first place. A high pressure jet of a clean gas, such as oil-free nitrogen or air, blows the fluid off the surface rapidly; this is a commonly used technique. The use of a contacting method, such as wiping with a paper or cloth towel, is not recommended in that the wiping process tends to move dirt from one area to another.

A vapor degreaser is often used to dry parts. Figure 2.7 shows a schematic diagram of the equipment used for this. The solvent in a closed (but not sealed) container boils and fills the container with its vapors. The substrate to be dried is placed in these vapors, which then condense on it as the substrate initially is cooler than the boiling point of the solvent. The condensate gives the surface a final rinse with a pure solvent that has just been distilled. Once the substrate has warmed to the temperature of the vapors, no more condensation can occur, the part dries, and

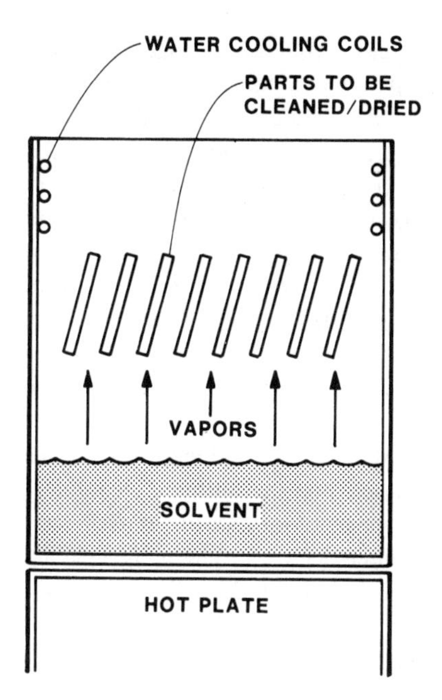

Figure 2.7. Vapor degreaser schematic. As cool parts are placed in the hot vapors, freshly vaporized solvent condenses on the parts. This gives the parts a final rinse. The solvent may be either an organic or a fluorinated hydrocarbon. If a flammable or toxic material is used, provisions must be made for adequate venting of escaping vapors.

can then be removed. Solvents typically used are trichloroethane, fluorinated hydrocarbons, and alcohols. Naturally, some of these solvents are flammable, especially when hot, and/or toxic. Care must be exercised when using them. Adequate ventilation is a must to prevent the accumulation of fumes.

2.5.5. In Situ Processes

There are several in situ techniques to clean surfaces. These are usually used regardless of what technique was used to clean the substrates before they were loaded into the vacuum chamber.

In sputtering systems, one frequently has the capability to *sputter etch* the surface of the substrate. In essence, this is a reverse of the normal sputtering process, where the substrate is made to be the target and a small amount of the surface is removed (on the order of a few atomic layers). This leaves an atomically clean surface to which the coating can adhere.

In evaporation systems, the common in situ means of cleaning substrates is the glow discharge. After the chamber has been evacuated to as good a vacuum as possible in a reasonable amount of time (called the *base vacuum*), a gas is introduced into the chamber to raise the pressure in it to about 10^{-3} mbar. A high voltage, ranging from about 0.5 to 5 kV, is applied to electrodes in the chamber. The voltage causes the gas to break down and form a glow discharge. This results in a flood of excited atoms and ions in the chamber. The gas frequently used to form the discharge is oxygen because it will remove organic contaminants on the surface of the substrate by oxidizing them. When substrates can be damaged by the oxygen plasma, another gas, such as argon, is used. The argon atoms bombard the surface and can remove some material in a manner analogous to sputtering, except that in a glow discharge cleaning step the material removal process is much less efficient than a sputter etch one. A glow discharge can heat a coating chamber (Maissel and Glang, 1970, pp. 6–41; Macleod, 1969, p. 229). If extensive cleaning is done with this technique, cooling periods may need to be included in the process, lest the substrate overheat. The more heat sensitive are the substrates, the more care that needs to be exercised in this type of cleaning process, because the heating can be nonuniform.

The above-mentioned techniques are frequently used in combination with one another. The choice depends on the initial cleanliness, on the ultimate cleanliness required, and on the characteristics of the substrate material and finish. The substrates remain clean for only a short while. For example, a glow discharge is typically repeated if the coating process does not start within a few minutes (Cox and Hass, 1958). Guenther and Enssle (1986) report that a *clean* surface remains that way for less than 3 hours, regardless of how well it is stored.

Chapter 3

Performance of Optical Filters

3.1. REFLECTANCE, TRANSMITTANCE, AND ABSORPTANCE

Conservation of energy requires an accounting of all of the energy in the incident light beam. Generally, a consideration of the reflectance and transmission intensities and of the absorption accounts for most of the incident light. In some instances, scattering also plays a role; this is not considered here, but is covered in the next section. If there is no absorption in the spectral range of interest, then reflectance and transmittance must equal unity, or 100 percent if the values are given as percentages. In this text, we deal with the ratio form of these values in equations, unless otherwise noted, and the percentage form is used in the narrative because of its clearer meaning.

The mathematical derivations of the equations governing the behavior of thin film interference filters indicate that the transmission of a coating is independent of the direction of the arrival of the light (Heavens, 1965, p. 77). This is well established in practice, and it simplifies the measurement of transmission, as the side of incidence of the filter is unimportant.

There may be reasons, however, for selecting one side of the substrate over the other on which to locate a nonabsorbing coating. The most common reason for selecting an inner surface of an optical system is to afford the coating some protection from environmental hazards. There may be optical reasons in special circumstances that dictate which surface is most desirable for a coating. For example, where high standing wave electric fields may be encountered, i.e., a high-power laser system, the system designer would prefer to have the electric fields distributed to minimize the chance for dielectric breakdown. A high-power pulse arriving from one direction produces damage at a different threshold than a pulse arriving from the other direction. The actual fields need to be considered in detail to arrive at a valid conclusion to the question of which side is better (see section 3.5). Such nonlinear effects do not ordinarily arise with most systems but do need to be considered when high-power lasers are used.

Nonabsorbing filters are sometimes called dielectric filters to distinguish them from *metallic* filters, which are absorbing. It should be pointed out that the term *nonabsorbing* is applicable to a restricted spectral region, as all materials are absorbing at some wavelength. Thus, the term nonabsorbing implies, either implicitly or explicitly, a wavelength range over which the material is completely transparent.

Nonabsorbing is also a relative term. In high-power laser systems, absorption on the order of a few tens or hundreds of parts per million may be significant, whereas in ordinary systems, absorption levels several orders of magnitude higher than this would barely be measurable, and the level of significance could be another order of magnitude higher than this. Thus, we must be careful in defining our terms to be certain that the reader or the listener is calibrated to the same context and significance level as the originator of the information.

In filter designs with absorbing layers, the reflectance can be asymmetric if the design is asymmetric. In other words, the reflectance from one side is not equal to that where the light is incident from the opposite side of the filter. A simple example of an asymmetric coating is a semitransparent metal layer with a quarter wave thick dielectric layer on one side, and that is immersed in a medium with an index of 1.5. The reflectances from the two sides are different as shown in Fig. 3.1. If the coating design is symmetric, and if the incident medium and the substrate do not have equal refractive indices, then the reflectance will not be the same from both sides of the filter. A single metal film on a glass substrate with an air incident medium will have different reflectances from the substrate side than it will from the air side, as shown in Fig. 1.8. One characteristic of any filter, either absorbing or nonabsorbing, is that the transmission is independent of the direction of the incident light. Because the sum of the reflectance, the transmittance, and the absorptance must be 100 percent, and as the transmittance is independent of the direction of incidence of the measurement beam, we conclude that the absorption also depends on the direction of incidence.

3.2. ABSORPTION AND SCATTERING

It is not possible from specular measurements of transmission and reflection with a spectrophotometer to determine whether a particular sample is losing light energy through a mechanism of absorption or one of scattering. Of course, techniques are available that directly measure any of these parameters and thus one can obtain a measurement on each one individually.

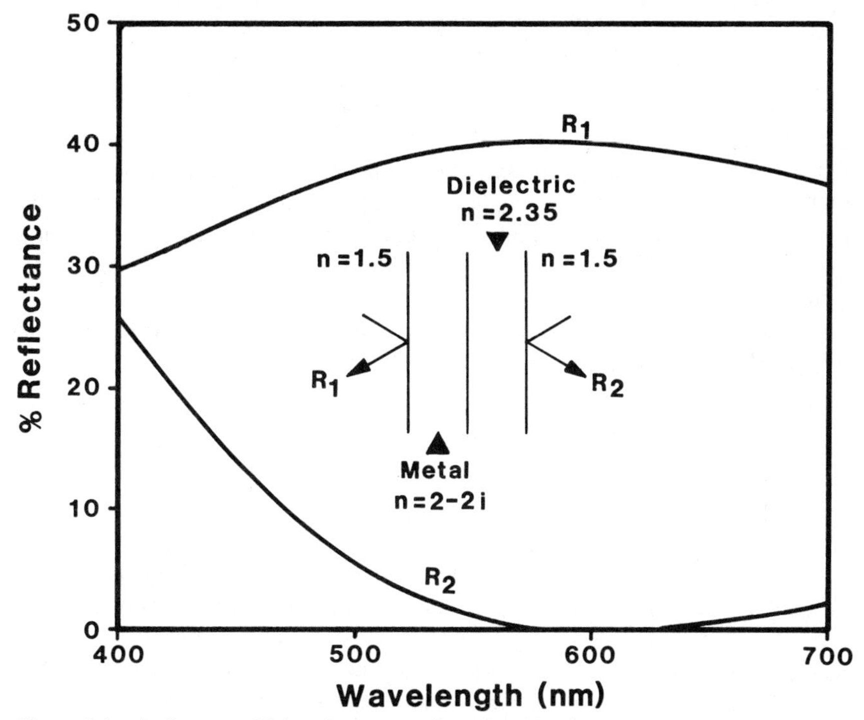

Figure 3.1. Reflectance (R_1) and reverse reflectance (R_2) for a metal film with a dielectric film on it. The dielectric film (n = 2.35) thickness is a quarter wave at 400 nm, whereas the metal layer ($A = 2 - 2i$) is 40 nm thick. The incident medium and substrate indices are both equal to 1.5. The reflectance from the one side can be made much higher than that from the other. The transmittance for this design is approximately 20 percent.

A calorimeter measures an increase in temperature and indicates the absorption in the coating. The temperature of a thin sample substrate increases as it absorbs energy from an intense laser beam. The temperature rise can be related to the absorption coefficient; a certain amount of analysis is required to calibrate these instruments to take into account the specific heat of the substrate, the heat lost by radiation, and so forth. Further discussion of this topic is included in section 5.1.11.

A scattermeter is used to measure the scattered light, usually using a laser as source of the incident light beam. An integrating sphere collects both the specular and the scattered light. Alternately, detectors can scan over 4 π steradians to get the complete scattering function.

The bidirectional reflectance distribution function (BRDF) describes the full four-dimensional scatter distribution function (two angles to describe the incoming beam, and two more to describe the scattered beam). Bartell et al. (1980) discuss the details of converting from the differential form of the mathematical function of BRDF to the measured quantities of an experiment made with finite-sized detectors.

Total integrated scatter (TIS) is a simple number that describes the amount of scattered light integrated over an entire hemisphere (Bennett, 1978). This value can be related to the average size of surface microfeatures:

$$\mathrm{TIS} = 1 - \exp[-(4\ \pi\ \delta/\lambda)^2]$$
$$\cong (4\ \pi\ \delta/\lambda)^2$$

where δ is the rms height of surface microirregularities and λ is the wavelength of the light.

The assumption is made in deriving this relation that the distribution of the irregularities is Gaussian. It appears that this is appropriate in practice. In addition, Bennett shows that the angle of the scattered light can be related to the spatial frequency of the irregularities of the surface, and the intensity can be related to the amplitude.

Scatter sites can exist in a number of locations, and each can lead to somewhat different results (Carniglia, 1979). These sites include:

Substrate bulk (Bennett, 1978)
Substrate surface (insufficient polish)
Film interfaces
Film *bulk*

The correlation between the sites also is an important parameter.

Scattering can play a surprising role when a coating is used at a wavelength not originally intended. A case in point is a coating that was designed for use at 3.8 μm. Unbeknownst to the manufacturer, the system in which this filter was used was aligned with the help of a 0.633-μm laser. In one batch of filters, the random errors normally present in a filter resulted in a filter with low scatter at the alignment wavelength. In another batch of filters, the random errors were such that the electric fields in the coating that existed for 0.633-μm light were peaked at some of the internal interfaces. The result was a filter that was totally opalescent at 0.633 μm, while being perfectly satisfactory at the specification wavelength of 3.8 μm.

3.3. BLOCKING RANGES

A filter is said to *block* when the transmission over a given spectral range is less than a specified value (Fig. 3.2). Blocking can be obtained from the filter itself, from the substrate, or from other filters. The transmission characteristics of the substrate are frequently used to block wavelengths that are far removed from the spectral zone of interest. It is more common to find materials that will block shorter wavelengths while passing the longer ones. The electronic band structures of most materials used in the visible portion of the spectrum have a forbidden energy band gap that corresponds to a photon energy located in the near ultraviolet (UV) or violet portion of the spectrum. Short wavelength light with an energy greater than that of the band gap can cause electronic transitions across the gap. This results in energy absorption of light with wavelengths shorter than this energy. Most solids and liquids have this characteristic for some photon energy. Table 3.1 lists the gap energy for several useful optical materials. Silicon and germanium are two semiconductor materials with forbidden gaps in the red or near infrared (IR); they can block the visible portion of the spectrum while transmitting IR wavelengths (see section 6.5. on material blocking properties for further details).

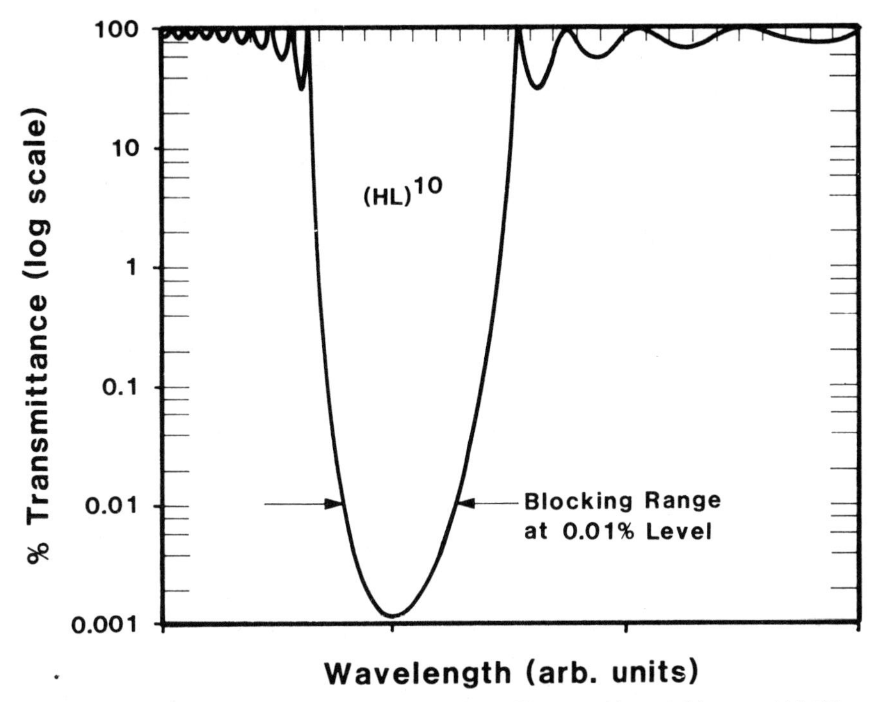

Figure 3.2. Plot of transmission of a blocking filter. The range for a 0.01 percent blocking level is shown. The shape of this low transmission band is typical of many blocking filters, and is slightly asymmetric when plotted on a wavelength scale.

Similarly, some protective covers for silicon solar cells used in space have cerium oxide doping to block wavelengths shorter than 350 nm. This keeps the UV light from damaging the cement used to affix the cover to the cell. The absorbed energy is converted into heat. In some cases, such as low-light level situations, this absorption is not a concern. In other cases, such as the space satellite usage just mentioned, the absorption increases the heat load on the satellite and can reduce the efficiency of the cell and the overall spacecraft.

To get long wavelength blocking, the characteristic (reststrahlen) oscillation of the crystal lattice can be used. In the mid- to far-infrared, the wavelengths excite collective oscillations of the lattice atoms. The vibrations absorb energy from the incident beam and make the substrate opaque to light of that wavelength. Table 3.2 lists representative data for the long wavelength edge of some common optical materials.

The materials in a coating can also have absorption characteristics that are useful for blocking purposes. The absorption edges of the various coating materials are well established. One function of a thin film designer is to make judicious choices in this regard.

The relatively fixed absorption characteristics of natural materials limits their use as blocking filters. In some cases, the shape of the transition from transmitting to blocking as available naturally is inappropriate for a particular application. To

Table 3.1 Values of the Energy Gap Between the Valence and Conduction Bands of Selected Materials at Room Temperature

Crystal	Energy gap (μm)
InSb	5.39
PbTe	4.13
InAs	3.76
PbS	3.54
Ge	1.85
Si	1.09
GaAs	0.89
CdTe	0.86
CdSe	0.71
CdS	0.51
ZnSe	0.48
TiO_2	0.41
SiC	0.41
ZnS	0.34
MgF_2	0.25[a]
C (diamond)	0.23
Al_2O_3	0.2
SiO_2	0.15[a]

[a]These values are approximate and were obtained from a number of sources.
Adapted; from Kittel (1966), chap. 10, table 1, with permission.

get a precise definition or positioning of the transmission to blocking transition wavelength, we can use an interference stack and position one of its edges where the blocking is desired. Figure 3.3 shows the reflectance versus wavelength performance of a simple quarter wave stack that consists of layers with alternating materials. The optical thickness of each of the layers is one quarter wave thick at a given wavelength. The width of the high reflectance zone, commonly called the *stop band* is determined by the index ratio of the materials that make up the stack. The total width in percent of the center wavelength of the high reflectance band is given approximately by

$$\%\ \text{total width} = \frac{400}{\pi}\ \text{Arcsin}\left|\frac{n_b - n_a}{n_b + n_a}\right| = \frac{400}{\pi}\ \text{Arcsin}\left|\frac{1 - n_r}{1 + n_r}\right| \tag{3.1}$$

where n_a and n_b are the refractive indices of the two materials used in the quarter wave stack, $n_r = n_b/n_a$, the vertical bars indicate the absolute value is to be used, and Arcsin is the principal value in radians of the arcisine function, i.e., $0 < \text{Arcsin}\ |\ x\ | < \pi/2$.

The behavior of this function is shown in Fig. 3.4. This is the maximum width

Table 3.2 Long Wavelength Limit for some Common Optical Materials

Material	Limit (μm)
Diamond	80
KRS5	40
Ge	23
CdTe	16
PbF_2	16
CdS	16
NaF	15
BaF_2	15
Si	15
GaAs	15
CaF_2	12
MgF_2	7.5
Al_2O_3	6.5
TiO_2	6.2
Spinel	6.0
SiO_2	4.5

(From Department of Defense, 1962, Mil-HNDBK-141, Fig. 8–13)

obtainable with the given index ratio. Table 3.3 lists the approximate values that can be obtained with several common coating material pairs.

The spectral location of the stop band is a function of the optical thicknesses of the individual layers that are periodically repeated in the stack. Finally, the level of the reflectance achieved is determined by the accuracy of the individual layer thicknesses and by the number of periods used in the design. With all of the parameters at hand, it is a relatively straightforward matter to design a filter that will block over the range specified. The matter of building it may be entirely different as, for example, the thicknesses may preclude its construction owing to stress and other effects. It is at this point that the use of materials properties become critically important. Frequently, it is possible to make a blocking filter less expensively with a combination of interference stacks and the use of absorbing materials, rather than relying on dielectric stacks to do all of the blocking. Another advantage of absorptive materials is that the eventual disposition of the energy is known precisely, and can be accommodated accordingly. This is especially important in systems where a large amount of energy is being blocked, such as sunlight or laser light, and the light energy may warm the absorber and distort the system. If the energy were reflected, it is possible that it may be focused on some nearby object that could be damaged.

An auxiliary filter can block the light; the spectral performance of this filter overlaps that of the main filter, so that the pair (or more) of them limit the transmitted

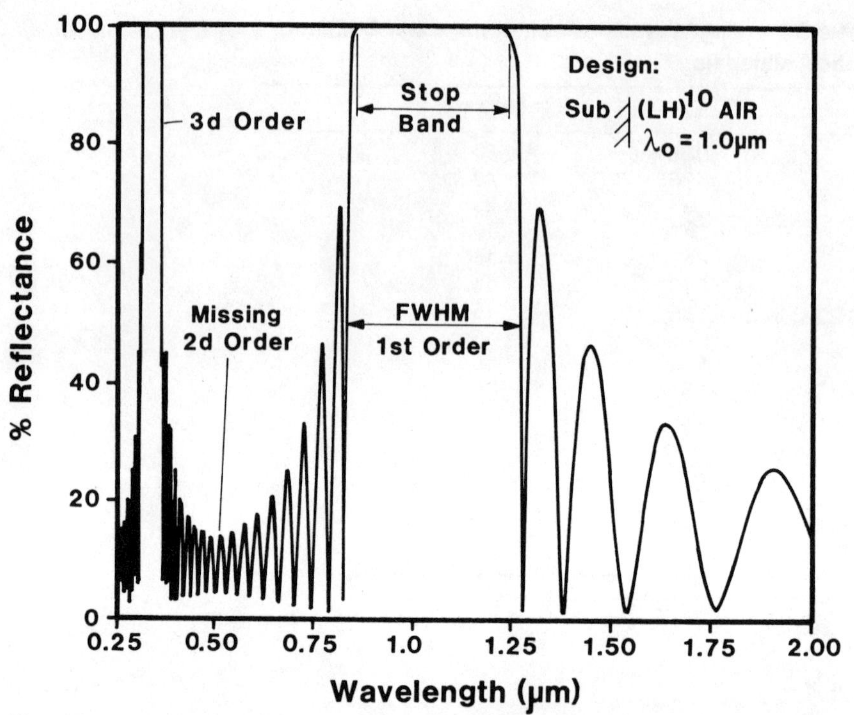

Figure 3.3. Reflectance on a linear scale of the design in Fig. 3.2. The figure shows the stop bands associated with the first and third order harmonics of the design. In this design, the second order high reflectance band is suppressed. Note that the full width at the half maximum point (FWHM) are not the same for the first and third orders. The ratio of width to wavelength is constant, however.

light energy. A series of stop bands can be cascaded to provide the necessary blocking width. In some instances, the energy blocked may extend from the far ultraviolet to the radio spectrum. Once again, the auxiliary filter may consist of either absorbing or nonabsorbing materials, or both. The auxiliary filter can be placed on the second surface of the substrate that holds the primary filter. This is commonly done, both for economic as well as for stress reasons.

3.4 FILTERS IN SERIES: INCOHERENT ADDITION OF INTENSITIES

Care must be exercised when two optical filters are placed in series to block light. The blocking properties of the filters do not necessarily combine linearly. If there is some reflection as well as transmission through them, the light that gets reflected by the second filter can be reflected back by the first and contribute to the transmitted intensity.

The total reflectance (R_{tot}) from a pair of parallel mirrors is

$$R_{tot} = \frac{R_1 + T_1^2 R_2}{1 - R_1 R_2} \tag{3.1}$$

where R_1 is the reflectance of the first mirror, R_2 is the reflectance from the second mirror, and T_1 is the transmission through the first mirror.

This formula assumes that there is no absorption in the space between the two mirrors, though the individual mirrors can have some absorption. If the mirrors do not have any absorption, the total reflection can be expressed in terms of the reflectances only:

$$R_{\text{tot}} = \frac{R_1 - 2\,R_1\,R_2 + R_2}{1 - R_1\,R_2}.$$

The total transmission is given by a similar equation:

$$T_{\text{tot}} = \frac{T_1\,T_2}{1 - R_1\,R_2}.$$

Note that when the reflectances R_1 and R_2 are low, the effect of the denominator is minimal. The denominator becomes more significant as the reflectances both increase. At high reflectance values, the error in using only the numerator is also small. When $T_1 = T_2 = 1 - R_1 = 1 - R_2$, the maximum error in using only the numerator occurs at $R = 0.618$ ($T = 0.382$). The magnitude of the maximum error in the total transmission is approximately 0.09. Because the denominator is

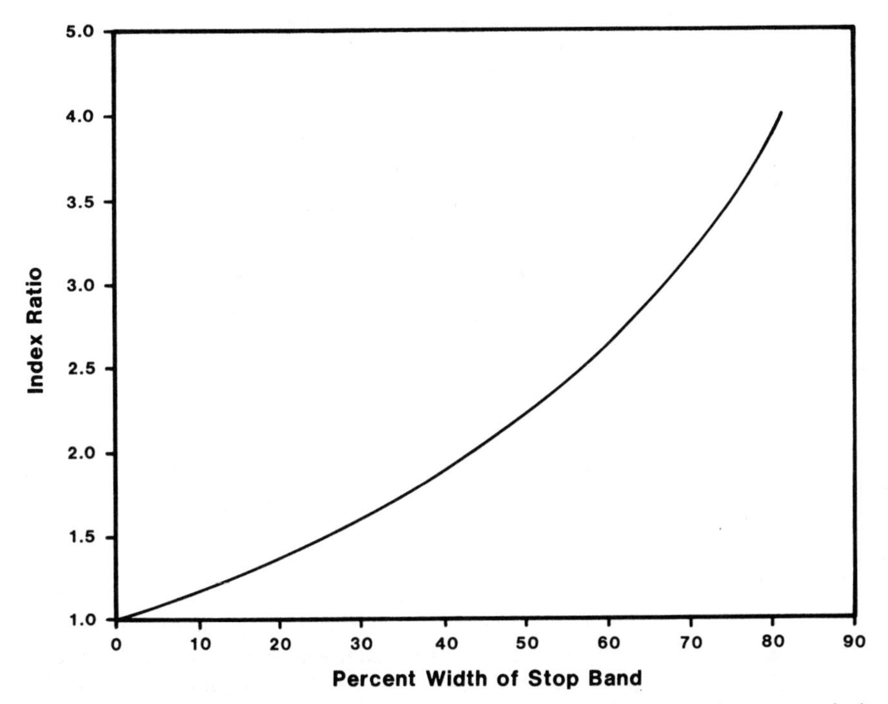

Figure 3.4. Plot of stop band width for a stack of layers, all of which have an optical thickness of a quarter wave. The stack consists of alternating layers with high and low indices of refraction. The ordinate of the plot is the ratio of the high index of refraction to the low one.

Table 3.3 Stop Band Width for Some Common Material Pairs

Material	Pair	Index ratio	Percent stop band width
TiO_2	SiO_2	1.62	9.7
ZnS	ThF_4	1.47	7.8
Si	SiO	1.84	12.2
Ge	ZnS	1.82	12.0

always less than unity, the error introduced by using only the numerator will always be negative, i.e., the estimated transmission will be smaller than the actual value.

In Eq. 3.1, the location of two mirrors does make a difference. To maximize the total reflection, the mirror with the higher reflection should be placed nearer to the light source.

A technique that can be used with filters in series is to tilt each filter so that no surface is parallel to any other reflecting surface. Light that is reflected from one surface makes its way out of the cavity between the filters and it does not contribute to the transmission. The tilt must not be so large as to introduce aberrations in the optical system or to shift the performance of the interference filter, yet it should be enough to avoid multiple reflections.

A pass band is the inverse of a stop band. In a bandpass filter, the transmission is low everywhere except in a relatively narrow region called the *pass band*. High-reflectance regions (rejection regions) exist on either side of the pass band. With metallic reflectors, there may be some absorption. The classical Fabry–Perot type of filter is one such example. A means of identifying the particular stop band is to number them, starting with the one that occurs at the longest wavelength. The number corresponds to the harmonics of the design, and is called the *order*.

The rejection regions on either side of the pass band are often narrower than desired, and therefore, additional coatings are required to obtain the desired width. A pass band can be considered as the region between two stop bands. The width of the pass bands in the lower orders is rather wide, but as the order number increases, the relative width decreases. This is the inverse of the situation of the stop band in a quarter wave stack.

A wavenumber scale is commonly used to show the performance of some filters over large spectral ranges. This scale is also common in infrared work, and is frequently used by solid state physicists and others as it is a linear photon energy scale. A common unit for this scale is reciprocal centimeters. The conversion factor between micrometers and reciprocal centimeters is 10^4. Because the dimensions of these two units are the reciprocal of each other, it is a matter of dividing into 10,000 the value one has to get the other. One way of visualizing the unit of reciprocal centimeter is that it represents the number of vacuum wavelengths of light that can fit into a distance of 1 cm. Table 3.4 lists some convenient points that relate the wavelength and the wavenumber scales.

A wavelength scale emphasizes the infrared portion of the spectrum by stretching it with respect to the blue portion. The wavenumber presentation does the opposite by squeezing the infrared and expanding the ultraviolet.

Figure 3.5 shows the performance on a wavelength number scale of the same design as that used for Fig. 3.3. The periodicity of the performance of a quarter wave stack on this scale is apparent in this illustration.

The blocking level of a filter is frequently shown on a logarithmic (base 10) plot. The attenuation level of some of these filters can reduce the transmitted intensity six orders of magnitude or more. A measure of the blocking level of a filter is *optical density* or *O.D.* This unit is defined as

$$\text{Optical Density} = -\log(T)$$

where T is the transmission of a filter expressed as a ratio (not %).

3.5. STANDING WAVE ELECTRIC FIELDS

Two coherent waves, i.e., having the same frequency or wavelength, traveling in opposite directions in the same region of space, generate standing waves. In a thin film stack, the waves reflected at each interface are coherent with the incoming wave, and therefore, standing waves are set up in the coating. In instances where high-power lasers are incident on the coating, the existence of these fields can play a vital role in determining whether the coating will survive the high-power levels or not.

The absorption of energy in an electromagnetic field is proportional to the time average of the square of the electric field, represented by $\langle E^2 \rangle$, times the product of the index of refraction (n) and the absorption coefficient (k) at that point:

Table 3.4 Convenient Reference Points Relating Wavelength and Wavenumber Scales

Wavelength (μm)	Wavenumber (cm^{-1})
0.25	40,000
0.40	25,000
0.50	20,000
0.633	15,803
0.70	14,286
1.00	10,000
1.06	9,434
2.00	5,000
2.8	3,571
3.8	2,631
5.0	2,000
10.0	1,000
10.6	943
15.0	667
20.0	500

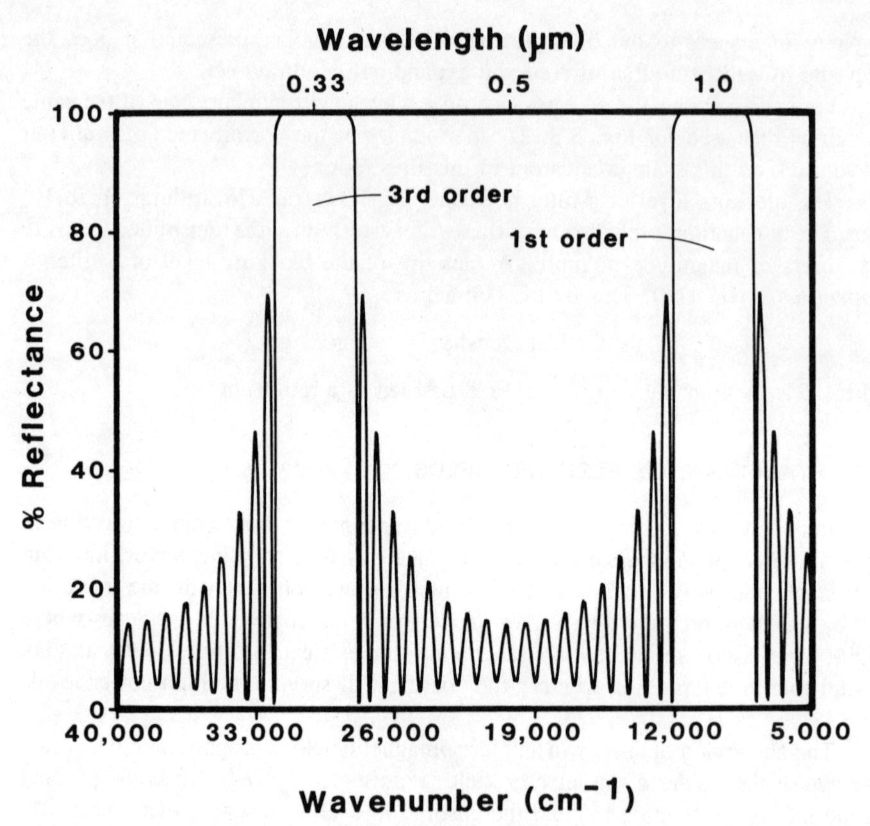

Figure 3.5. Reflectance of a quarter wave stack plotted on a wavenumber scale. The design is the same as that of Fig. 3.2, therefore this figure may be directly compared to Fig. 3.3.

$$A = C \cdot n \cdot k \cdot \langle E^2 \rangle \tag{3.2}$$

where C is a proportionality constant.

Generally, one considers the value of k constant through the volume of a film, but recent studies tend to indicate that the absorption is greater at interfaces than it is in the "bulk" of the film (Donovan et al., 1980). Consequently, more heat is deposited at the film interfaces. This is just about the worst place to deposit this energy as the interfaces are also the weakest part of the coating, thus are the most likely to fail. It is preferable to reduce the fields at these interfaces. Figure 3.6 shows the standing wave field for light incident from both sides of an interface with a quarter wave antireflection coating. The fields are sensitive to the direction from which the light is incident.

The boundary conditions that apply to Maxwell's equations require that a node or an approximation of one exist at a metallic high reflector/air interface. Figure 3.7 shows the standing wave in front of a silver reflector for both the IR and visible wavelengths. Relatively little heat is deposited in the metallic mirror as the product

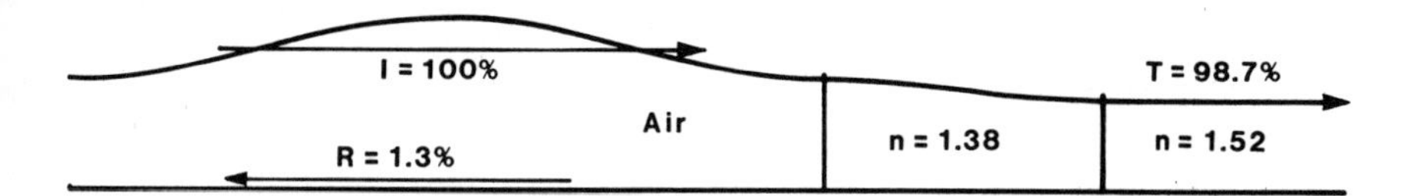

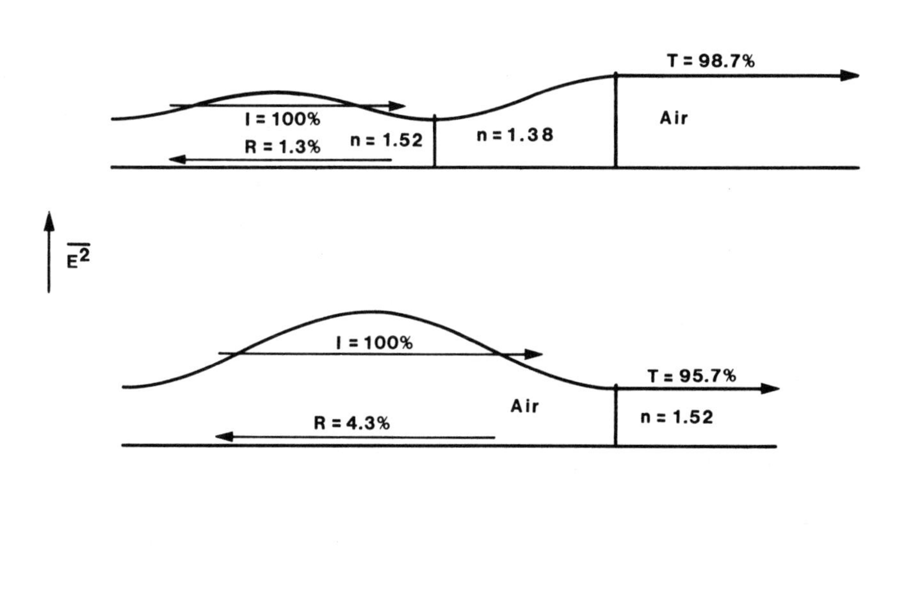

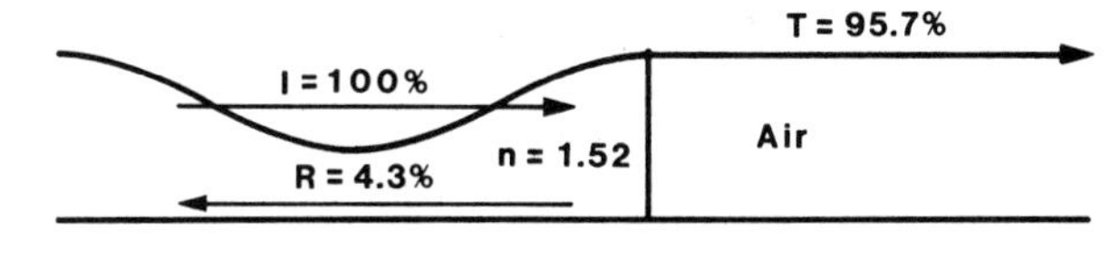

Figure 3.6. Standing wave $\langle E^2 \rangle$ patterns for light incident from both sides of glass/air interfaces with and without a quarter wave of magnesium fluoride antireflection coating. The arrows associated with *I*, *R*, and *T* indicate the incident, reflected, and transmitted energy, respectively.

of the field strength and *nk* is virtually zero everywhere. If the surface consists of a metal, such as nickel, which is a relatively poor reflector in the visible, there is significantly more penetration of the field into the metal surface, and the absorption is higher (Fig. 3.7c).

In a high-reflectance dielectric mirror, the electric field within the coating has maxima and minima, as shown in Fig. 3.8. The thin film structure gives us some design flexibility to adjust the electric field profile to optimize the overall filter

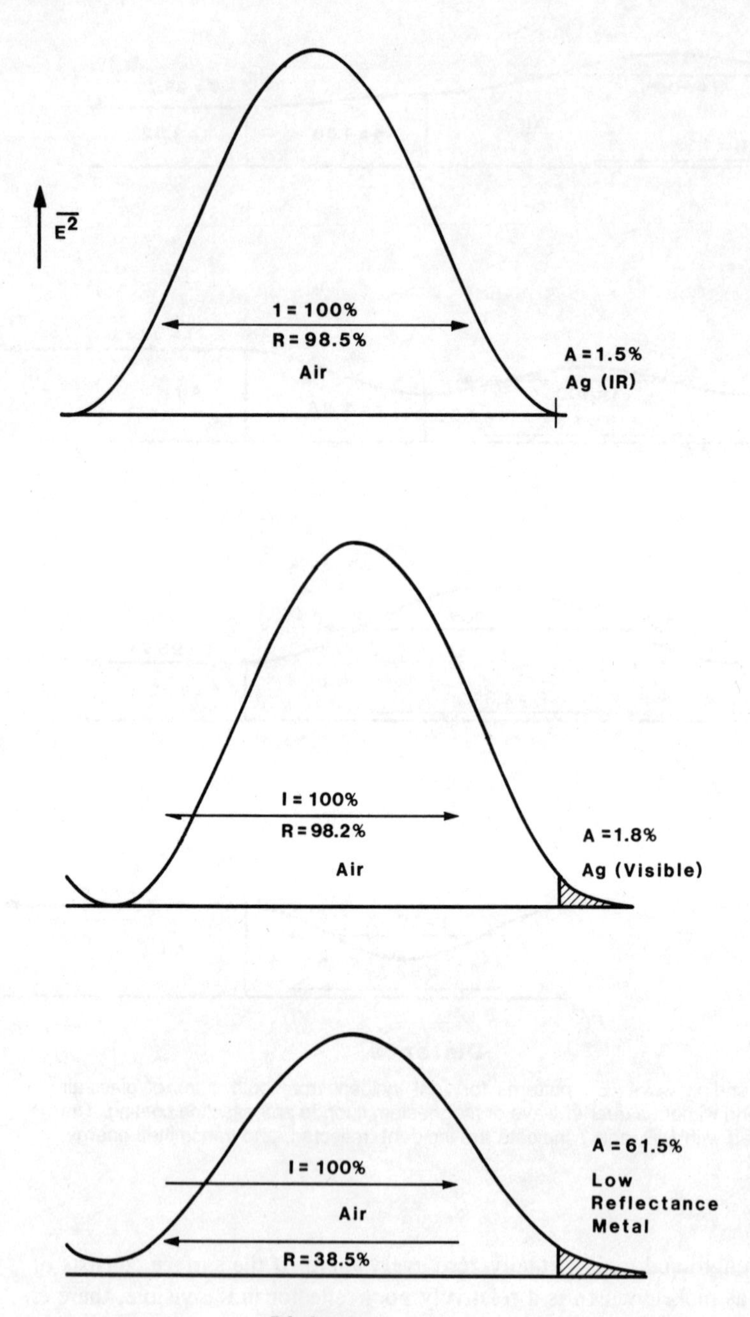

Figure 3.7. Plot of $\langle E^2 \rangle$ showing the *near-node* at silver/air (a,b) and nickel/air (c) interfaces. The wavelengths in the two spectral regions have been scaled to facilitate the comparison. Some of the electric field penetrates into the metals in the visible portion of the spectrum, but in the IR, little of the field penetrates into the silver. The metal indices used were $n_{\text{Ag IR}} = (0.55,15)$, $n_{\text{Ag Vis}} = (0.055,3.3)$, $n_{\text{Ni vis}} = (2.0,2.0)$

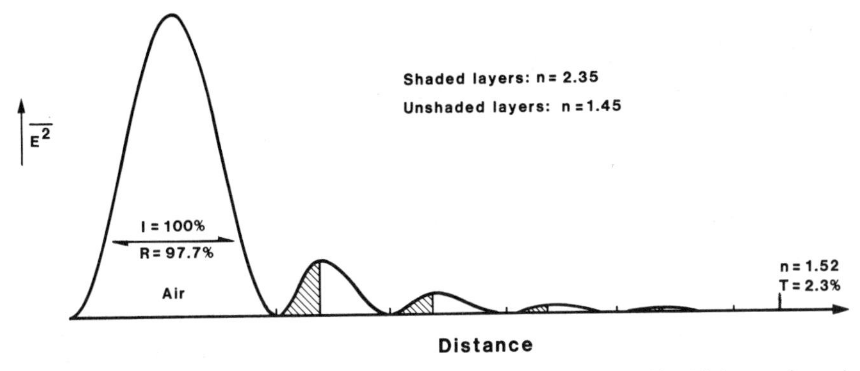

Figure 3.8. Plot of $\langle E^2 \rangle$ for a QW stack. The wavelength of the incident light corresponds to the stack center. The field peaks at the interfaces between layers. The amplitude of the field decays rapidly as it penetrates into the stack.

performance. If the absorption in one of the two materials in a quarter wave stack is higher than the other, it is possible to minimize the field in the higher absorbing material, though at the expense of raising the field in the other material. The result of this is a higher reflecting stack (Carniglia and Apfel, 1980). Another use that has been made of electric field studies is in the design of laser reflectors where laser damage is of serious concern. The modification of the thickness of a few of the outer layers of a quarter wave stack to reduce the peak field in the high index material has been shown to improve the damage threshold (Apfel, 1977).

A classic example of the way the electric field is involved in an absorbing design is the Fabry–Perot cavity filter that has a metal, such as aluminum or silver for the reflector layers. At a reflectance resonance, i.e., the wavelength where peak reflectance occurs, there are *near nodes* at the metal surfaces, and the absorption is relatively low. Away from this resonance, there is a significant amount of absorption, even with silver films, and the electric field is relatively high in the metal layers (Fig. 3.9).

The electric fields in dark mirrors (selective absorbers), which consist of metal/dielectric layer combinations, peak in the absorbing metal layers to increase the absorption of the incident light. This design technique is used in the so-called *trilayer* designs of optical data disks to improve the sensitivity of optical data disks and to control the reflectance (Bell and Spong, 1978). Figure 3.10 shows the standing wave fields for the trilayer configuration, a thick layer of the absorber layer by itself and a thick layer of the absorber material over a high reflector. The absorption values given indicate the effect of the field on the sensitivity. In addition, the amount of material required to obtain a given amount of absorption is reduced, thus reducing the amount of material that needs to be heated to write a bit on the disk.

The fields in antireflection coatings are relatively low by comparison to those found in a high reflector design. Figure 3.6 shows the fields for a single layer AR design, whereas Fig. 3.11 shows the fields present in a multilayer AR design of

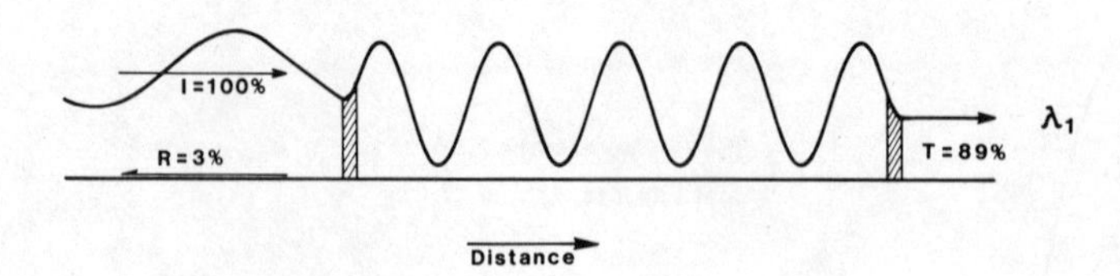

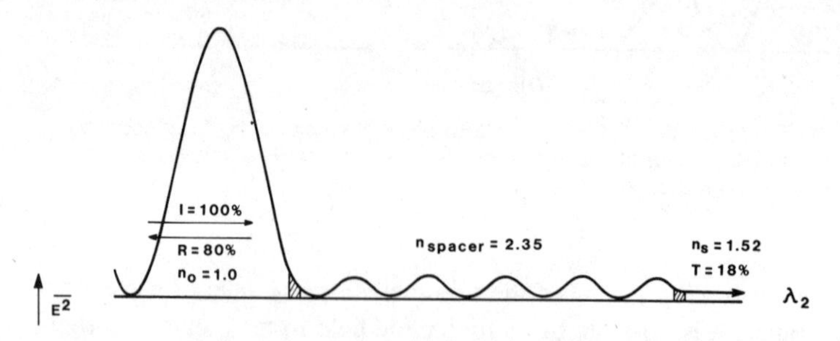

Figure 3.9. Plots of $\langle E^2 \rangle$ for a Fabry–Perot filter with silver reflectors at two wavelengths. The transmission is at a peak at the first wavelength (λ_1), while it is at a minimum at the second wavelength (λ_2). The absorption is at a minimum when the reflection peaks because the field in the metal layers is also at a minimum. The absorption at λ_1 is approximately 9 percent; at λ_2, it is 2 percent.

the quarter-half-quarter (QHQ) type. Here also, the minima and the maxima in the electric field profile can be moved to accommodate specific requirements. In high-power laser systems, it is desirable to reduce the absorbed energy to an absolute minimum to maximize the damage threshold. This approach has been demonstrated experimentally (Apfel, 1977; Demichelis et al., 1984).

3.6. PERFORMANCE SHIFTS WITH ANGLE OF INCIDENCE

In section 1.3.6., we discussed the effect of nonnormal angles of incidence on the effective index of refraction. In this section, we are more specific with respect to the changes in performance of coatings as the angle changes.

The angle-induced changes in filters can be qualitatively divided into two categories: (1) the wavelength shift and (2) the polarization effects.

The wavelength shift is straightforward: performance curves always shift to shorter wavelengths (in the blue direction) as the angle of incidence increases away from the normal. At first glance, it may appear that the performance should shift toward the red because the optical thickness of each of the layers in the design increases as the angle of incidence increases (by a factor inversely proportional to the cosine of the angle of refraction inside the layer). The important parameter, however, is not the total phase shift introduced in the waves by the layers. It is the phase *difference* between the waves that reflect from all of the interfaces that enters

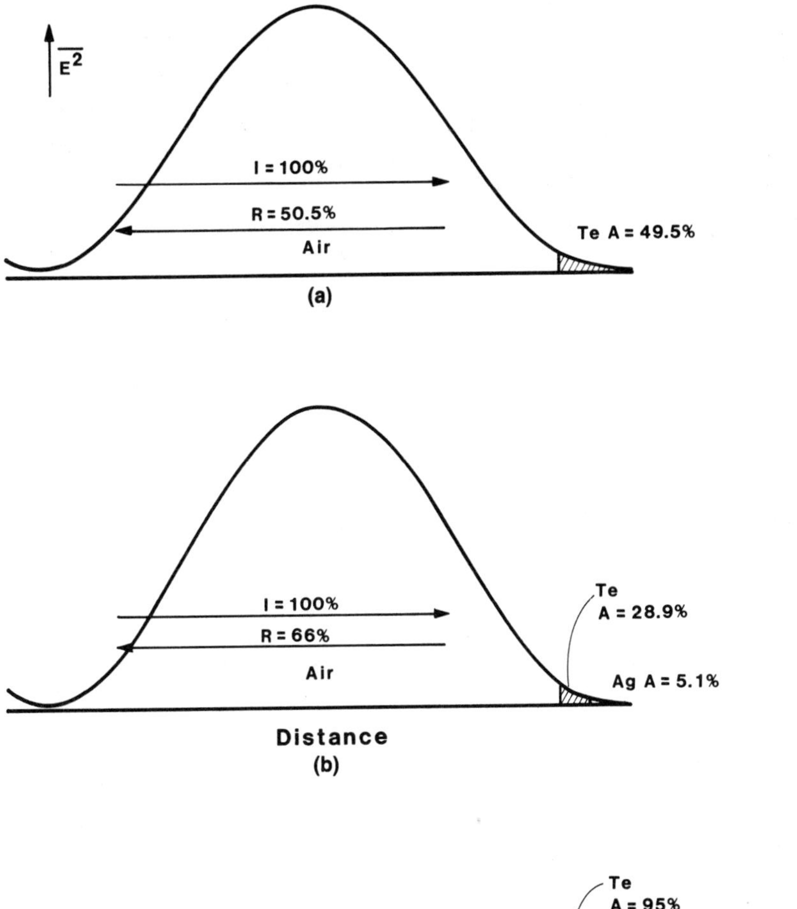

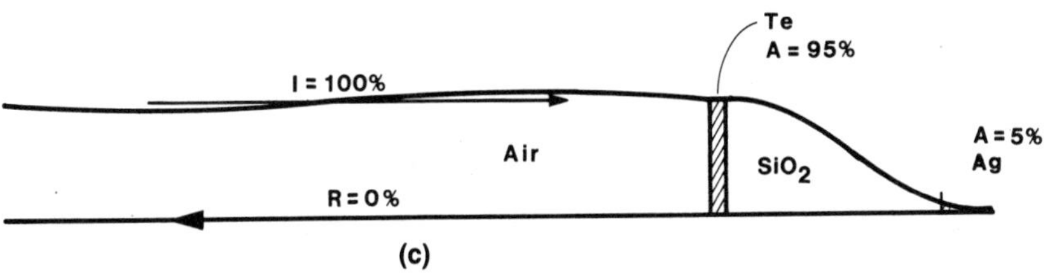

Figure 3.10. The $\langle E^2 \rangle$ in optical data disk recording systems. The reflectance and absorption is shown. In (a), the field is shown at an air/tellurium interface. In (b), the tellurium layer is deposited over a silver mirror (bilayer). In (c), the trilayer design is seen to be the most efficient as the maximum amount of energy is absorbed in the least amount of material.

into the summation of the partial reflected waves. This quantity *decreases* as the angle increases. Thus, the performance shifts toward the blue as the angle of incidence increases because the apparent optical phase difference introduced by each layer is decreasing.

In a bandpass filter, especially narrow ones, this effect can serve as a fine tuning mechanism to position the pass band at a precise spectral value, if the filter

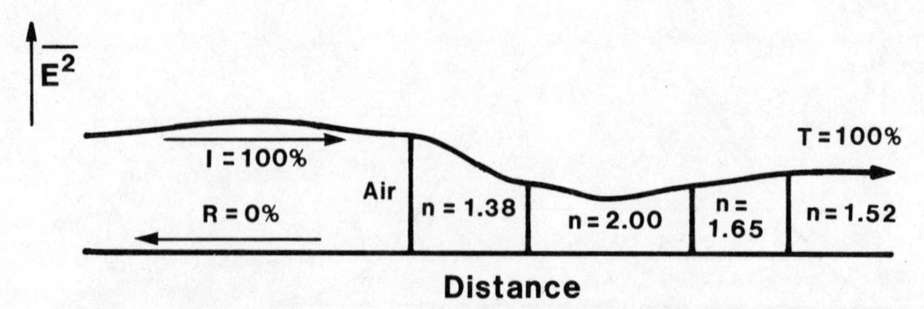

Figure 3.11. Profile of $\langle E^2 \rangle$ for a QHQ antireflection design. In simple AR designs, the field within the coating does not rise above that of the incident or exit beams, whichever is greater (over the wavelength range in which they function as AR coatings).

is originally coated such that the pass band is at a wavelength that is longer than the desired one. The degree of the shift is determined by the average index of the materials used to construct the filter. In a two-material, all dielectric Fabry–Perot bandpass design, the angle shift is less when the higher index material is the half wave thick spacer layer between the high-reflecting quarter wave stacks. Figure 3.12 shows the performance of two designs with approximately equal performance

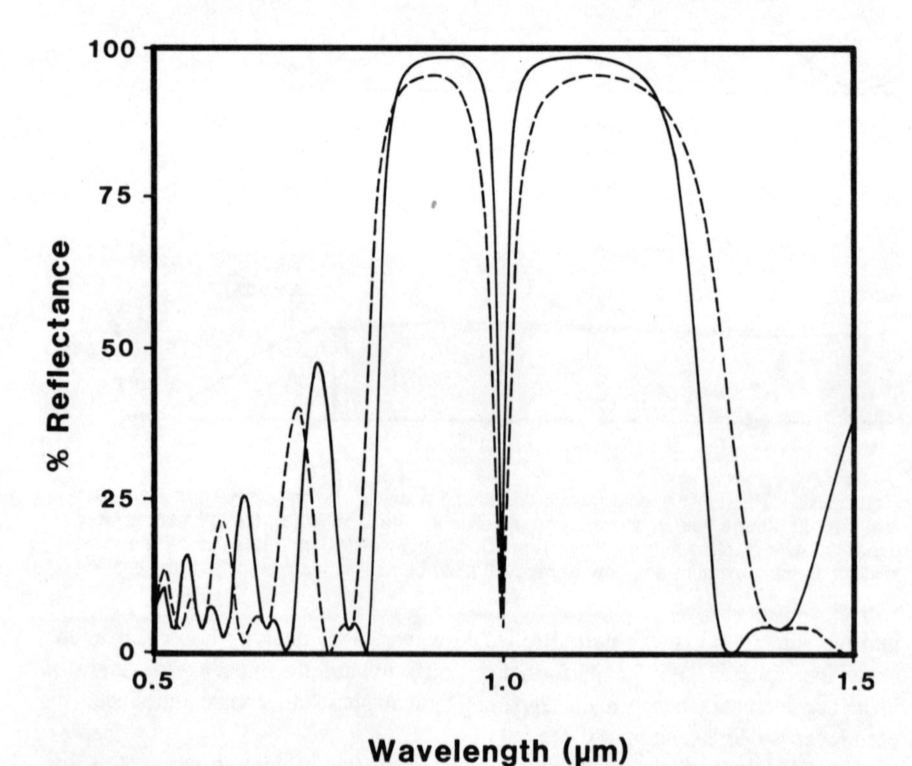

Figure 3.12. Normal incidence reflectance of $(HL)^4(LH)^4$ (*solid line*) and $(LH)^4(HL)^4$ (*dashed line*) as a function of wavelength ($n_H = 2.2$, $n_L = 1.45$). The difference in the performances of the two designs is due to the effect of the index of the outer layer in each design.

at normal incidence (save for a proper AR treatment in the center of the pass band). Figure 3.13 shows the performance of these same designs at a 50-degree angle of incidence. The shift in center wavelength of the coating with the higher index in the spacer layer is significantly less, especially relative to the width of the pass band, than the one for the low index spacer. In addition, the shape of the pass band stays more constant with the higher index spacer material.

The shape of the performance curves of a coating at nonnormal incidence may look quite unlike that at normal incidence. Here, as in other situations in optics, one must examine the performance in the two planes of polarization separately. If each plane of polarization is examined individually, then the angle effects become much clearer. The theoretically calculated performance of a quarter wave stack, as shown in Fig. 3.14, illustrates this well. The stop band for *s* plane light is wider than that at normal incidence, whereas that of the *p* plane is narrower. This is explained by the ratios' of the effective indices at nonnormal angles of incidence being larger for the *s* plane, whereas the opposite is true for the *p* plane.

For some incident angles, it is very difficult to design a coating because the effective indices are all very similar, so that it is not possible to get a reasonably large index ratio. The effective index varies less for the high index materials than it does for the lower index values (see Fig. 1.11). A filter that has to operate over a range of angles with minimum shift in performance should be constructed with the highest indices possible. The infrared region of the spectrum has a wider range of available material indices (both high and low) than does the visible portion. Thus, it is easier to design angle stable filters in the infrared portion of the spectrum than it is for the visible portion, all other parameters being equal. At angles of incidence near 90 degrees, the effective indices of the *s* and *p* components of all materials tend toward zero and infinity, respectively. The reflectance, therefore, approaches 100 percent at grazing incidence. It is difficult to formulate good performance, wavelength sensitive high reflecting designs at high angles of incidence. Antireflection coatings are especially difficult to design at high angles.

The effects of angle shift have been measured and reported in the literature by Baker and Yen (1967). Figure 3.15 is taken from their paper; it shows a range of wavelength shifts in percent as a function of the angle of incidence of collimated radiation. These measured results agree with theory.

3.7. OTHER PARAMETERS THAT AFFECT PERFORMANCE

Water, temperature, and other environmental conditions can affect the performance of thin film filters. The sensitivity of a filter's performance to these conditions is difficult to specify in a general sense. The actual coating processes have a definite influence on the details of the interaction between the coating performance and the environment to which the filter is exposed.

3.7.1. Water and Humidity

It is well known that the performance of optical filters may be sensitive to environmental moisture. The degree to which a filter varies depends on the materials

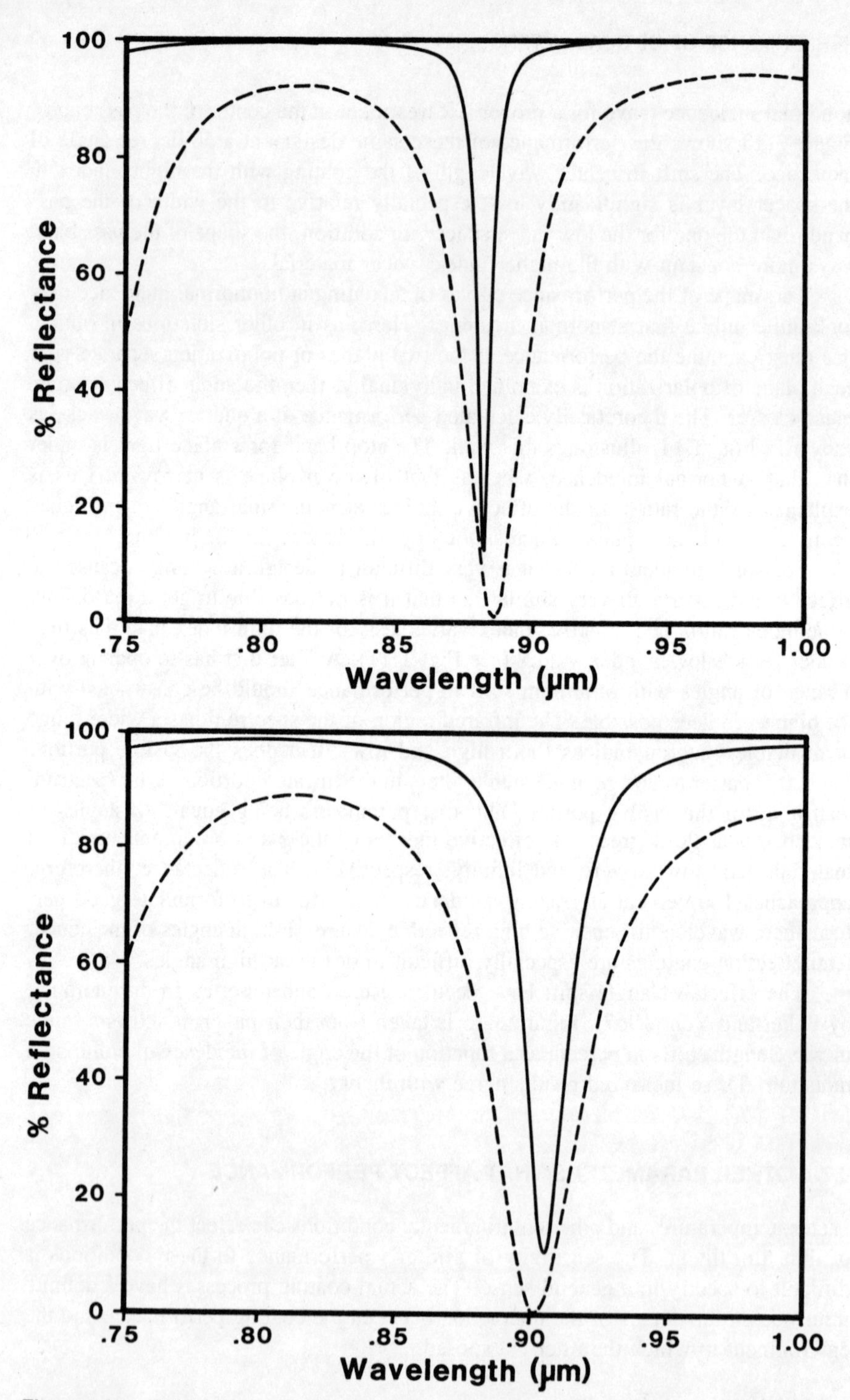

Figure 3.13. Reflectance at 50 degrees angle of incidence of the same designs as in Fig. 3.12: upper. Low index half wave spacer; lower. High index half wave spacer. Dashed curves are for the *p* plane and solid curves are for the *s* plane polarization. Note the expanded scale on the abscissa.

used in the filter as well as on the processing conditions. Gee and co-workers (1985) have measured the change in performance of evaporated titanium dioxide and silicon dioxide filters as they absorbed moisture from the ambient humidity. The general nature of this effect is that the performance of many coatings can undergo a pronounced change upon removal from the vacuum chamber and exposure to the atmosphere. For example, the center wavelength of a design shifts to a longer wavelength as the coating absorbs moisture. If the filter is subsequently dried either by heating in a warm oven or by leaving it in a dry box for a longer period of time, the performance shifts back toward the preexposure values; however, it will never reach the original performance. Thus, there is an irreversible component in the moisture shifting effect. Subsequent cycling of the filter between wet and dry environments shows that there is also a reversible component of the moisture effect. On the basis of this phenomenon, it is apparent that for the most demanding applications, the measurement of a filter's performance should be made after the filter has stabilized in an ambient environment that is similar to that in which it will be expected to operate.

3.7.2. Temperature

Temperature can affect the performance of interference filters because both the index of refraction and the thickness vary as a function of temperature. Baker and Yen (1967) measured the actual temperature shifts in typical infrared filters. Figure 3.16 shows their results.

A temperature-induced shift can be especially significant when one is dealing with a filter with a critical wavelength dependence. The center of a narrow wavelength structure may shift by a large fraction of its width owing to changes in the optical thickness of the layers in the coating. The optical thickness of each layer varies with temperature as a result of (1) the change in index of refraction with temperature (dn/dT) and (2) the linear coefficient of expansion (dl/dT) that affects the physical thickness.

The indices of refraction of most materials become smaller as the temperature increases, i.e., dn/dT is negative. On the other hand, the expansion coefficients dl/dT are typically positive, and these lead to corresponding increases in the path length. Thus, the change in optical thickness change with temperature is a second-order effect. The result is that for most materials, the performance of a filter shifts to longer wavelengths as the temperature is increased.

One solution to the problem of temperature shift is to enclose the filter in a temperature-controlled oven. With this technique, the filter can be tuned over a small spectral range. This technique is useful, for example, to select a specific line from a spectrum. The oven approach, however, does leave something to be desired; it is an active device and it consumes power. In addition, ovens can have a long time constant, i.e., it cannot be varied rapidly.

An alternate solution of the temperature shift problem is to design the filter so that it is temperature invariant. This is possible if a material is available whose index of refraction varies in a way that is opposite to that of the other materials in a design. If this shift is large enough, one can compensate for the temperature-

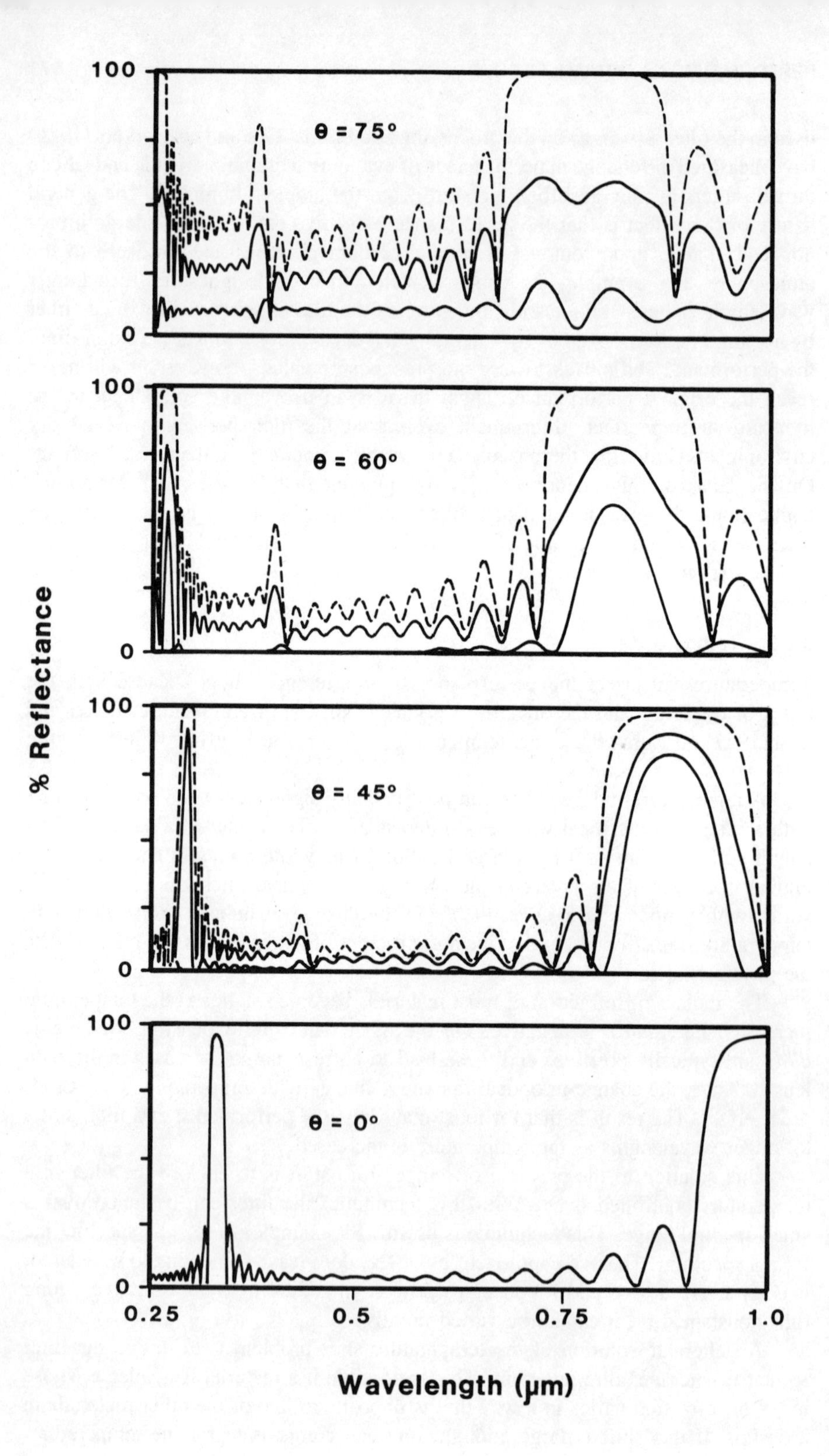
100
0
θ = 75°
θ = 60°
θ = 45°
θ = 0°
% Reflectance
0.25
0.5
0.75
1.0
Wavelength (μm)

induced changes in the refractive index and the thickness so that the overall design is temperature stable.

Some infrared filters can be designed to be temperature insensitive. A material that has appropriate dn/dT and dl/dT in the infrared portion of the spectrum is lead telluride. Seeley and co-workers (1980) have developed this material and formulated designs that put this effect to use. It is now possible to obtain filters commercially that have an extremely small temperature coefficient. These filters have found use in pollution measurement devices. The filters transmit a selected spectral line near 4 μm, which is emitted by carbon monoxide. Unfortunately, no material has yet been found that has the proper characteristics to allow the same technique to be used in the visible portion of the spectrum.

3.7.3. Corrosion

Corrosion can occur in some coatings. In particular, metallic films, such as silver and aluminum, are subject to degradation as they are attacked by environmental agents. The films oxidize and lose their metallic properties. The indices of refraction of oxides are radically different from those of the metallic precursors.

The behavior of metallic silver films is particularly notorious in this regard. Some recent studies at the University of Arizona may be showing the way to tame this very useful metal and make it more durable for thin film applications (Song et al., 1985). Not so well known is the degradation of metallic aluminum film in high humidity environments. Over a period of time, these films can convert completely to their oxide. The presence of ultraviolet light greatly enhances the rate at which this conversion takes place.

Moisture can also lead to the degradation of oxide films. If the oxide is soluble, then the thin film can be dissolved by the ambient water, leading to a loss of the filter. Tests are available that will verify that the filter should not fail in normal use (see Chap. 5).

3.8. WAVEFRONT DISTORTIONS

In most applications, a filter will not degrade the optical performance of the system in which it is placed. In critical applications, however, the effect that the filter may have on the wavefront which is passing through it or being reflected off it could be significant and needs to be considered by the designer. Cases where such conditions may apply are in diffraction-limited optical systems, such as precision camera lenses or in laser systems that are used in interferometry.

When collimated light traverses a filter on a plane substrate, no aberration is introduced into the wavefront. A coating may introduce a uniform phase shift into

---

Figure 3.14. Reflectance of a QW stack as a function of wavelength for various angles of incidence. The design if $(ML)^{10}M$ (n_m = 1.65, n_1 = 1.35) with air as the incident medium and a 1.52 index glass as the substrate. The dashed curve is for the *s* plane of polarization, the lower solid line is that for the *p* plane, and the middle solid line is the average of the light intensity in the two planes.

the wavefront, however, and this may be significant in an optical system in which phases are being compared. For example, a quarter wave thick film deposited on a substrate introduces a 270-degree ($-90°$) phase shift to a transmitted beam at the quarter wave wavelength relative to a bare substrate. When more complicated designs are used, the result is correspondingly more complex. These phase shifts are a function of the angle of incidence and the wavelength. The manufacturer of a filter can supply a calculation of the phase shifts of a given design for use with the user's modeling programs to calculate the overall system performance.

Both the previously mentioned angle shift (see section 3.6.) and the phase shift can affect the performance of diffraction limited or interferometric quality systems in which the lenses have steeply curved surfaces.

On steep surfaces, the thickness of the manufactured coating may vary from the center to the edge because of process limitations. Careful masking can be used to eliminate this run-off, if necessary. On the other hand, it should be noted that even a uniformly thick coating may introduce wavefront distortions because of angle

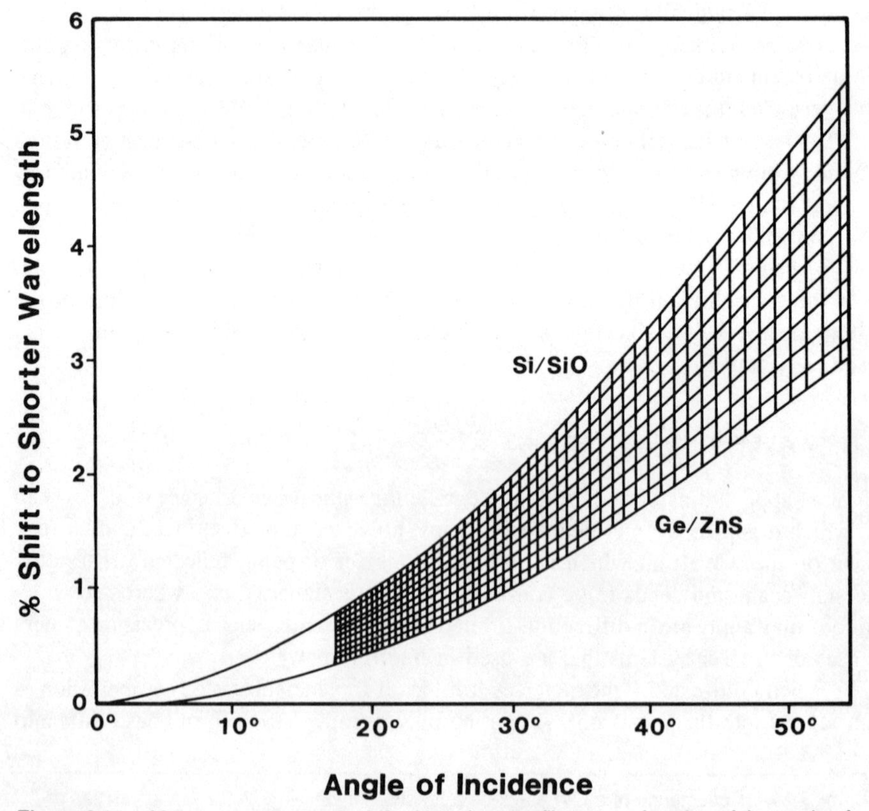

Figure 3.15. Measured wavelength shift of interference filters as a function of the angle of incidence (from Baker and Yen, 1967). The incident radiation is collimated. The filters were infrared types. The upper limit corresponds to an index pair of approximately 3.45 and 1.9, whereas the lower limit pair is approximately 4.0 and 2.2.

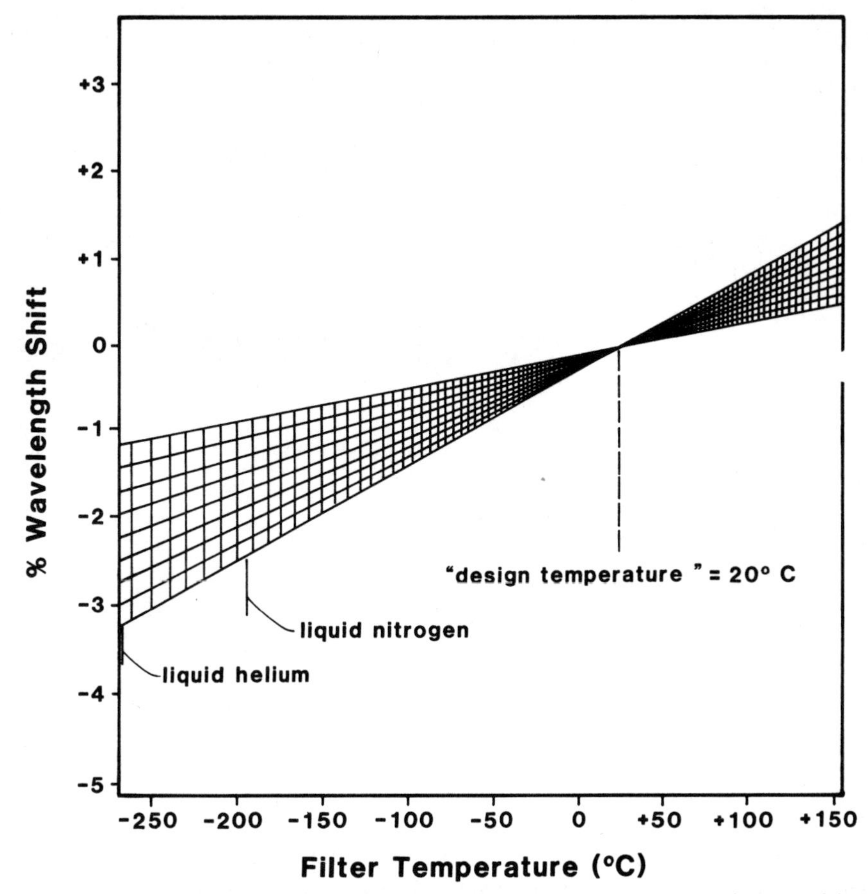

Figure 3.16. Measured temperature shift of infrared interference filters as a function of the temperature (from Baker and Yen, 1967).

effects. When using off the shelf coated optical elements in interferometry, the user should pay attention to these details.

In fast and wide angle optical systems, where steep angles of incidence are commonplace, the operation of a bandpass filter may not be what is anticipated, based on normal angle of incidence measurements. Specifically, in a system with a large spread of angles, the performance is the result of an average of all of the angles incident on the filter. If the filter is especially narrow, this may result in a significant reduction in the light throughput over what the same filter transmits in a collimated beam at a normal incidence.

Chapter 4

Design Types

This chapter catalogs a number of common thin film design types. This list is intended to be generic in nature to give the reader some idea as to the possibilities that are available. It cannot be overemphasized that these are only generic designs, and that they can be, and in most cases *should* be, adjusted and optimized by a competent thin film designer for a particular problem. An almost infinite variety of designs can be made by combining simpler design types. The reader who has an interest in thin film design techniques is referred to a companion volume in this series: *Design of Optical Interference Coatings* by A. Thelen.

The order of the following designs has no particular significance. The more frequently used ones are listed first and the less commonly used ones follow.

4.1. ANTIREFLECTION COATINGS

Antireflection (AR) coatings were the first coatings to be produced by the thin film technologist and are the most commonly used coatings today. Undoubtedly the simplicity and ease of fabrication of a single layer contributed to the early successes of optical thin films. As discussed in Chap. 1, an AR coating for a glass surface can be made with a single low index layer with an optical thickness of a quarter wave. A number of materials can be used for this purpose, and they can be deposited quite easily by heating them with a resistive heater in a vacuum chamber. It is straightforward to judge the thickness of a single low index deposit by eye (Heavens, 1965, p. 243). The precision thus obtained was adequate for the early thin film AR products.

Since the original quarter wave AR coating, many other different types have been developed. The goal of all of these is to make a coating the reflectance of which is as low as possible over as wide a wavelength band as possible. All too frequently, one can get more width at the expense of the reflectance level, and it is difficult to get both simultaneously.

AR coatings reduce the reflection that occurs at interfaces between two optical media. In general, one of these media has an index of unity (air or vacuum), though it could be water or some other medium. Common soda lime glass, with its index near 1.5, reflects approximately 4 percent of the incident light. Dense flint glass, sometimes used in highly corrected optical systems, can have an index as high as 1.8; e.g., Schott Glass type SF6, and a corresponding reflectance of 8 percent. In an optical system with many surfaces, these losses add up to a significant quantity. Worse, the reflected light can contribute to veiling glare in the optical system, which reduces the contrast at the image plane. In infrared systems, with even higher index semiconductor substrates, the situation is even more critical. For example, the reflectance from a silicon surface is more than 30 percent, and that from germanium is more than 35 percent. These losses are intolerable in just about any infrared system and must be suppressed.

Because a multilayer interference coating with zero reflectance at all wavelengths has yet to be designed, a merit function has to be used to evaluate a given design, and to compare its performance to other designs. Typically, a weighted average is used for this calculation. The reflectance is multiplied by a weighting function and the result is summed over the desired wavelength interval. The weighting function is selected to represent the optical system with which the filter will operate; it may be proportional to the human eye response, the response curve of a sensor, or any other function. Thus, the weighting function tailors the optimization to the specific system. Photopically (daylight brightness eye response) weighted systems function well as antireflection treatments for visual applications.

One area that is not generally included in computer refining of AR coatings is the effect of *color* on the visual performance of the coating. In particular, the purity, the dominant wavelength, as well as the integrated brightness are important. Also, the variability of these parameters over a substrate, as well as part to part variability are important. The eye is very sensitive to color shifts, but there is no widely accepted measure of this effect that can be used in thin film design refining and optimizing programs. The difficulty arises from the shift in performance with layer thickness errors. Figure 4.1 shows the variations that can typically be expected if a 2.5 percent standard deviation in the thicknesses is allowed in a three-layer QHQ AR design. The indices used in the calculation are shown in the illustration. Table 4.1 lists the variations in the color space coordinates and the reflected luminance, the dominant color, and the purity of the reflected light. It is clear that the perceived colors are different for each sample, yet the overall reflectance remains

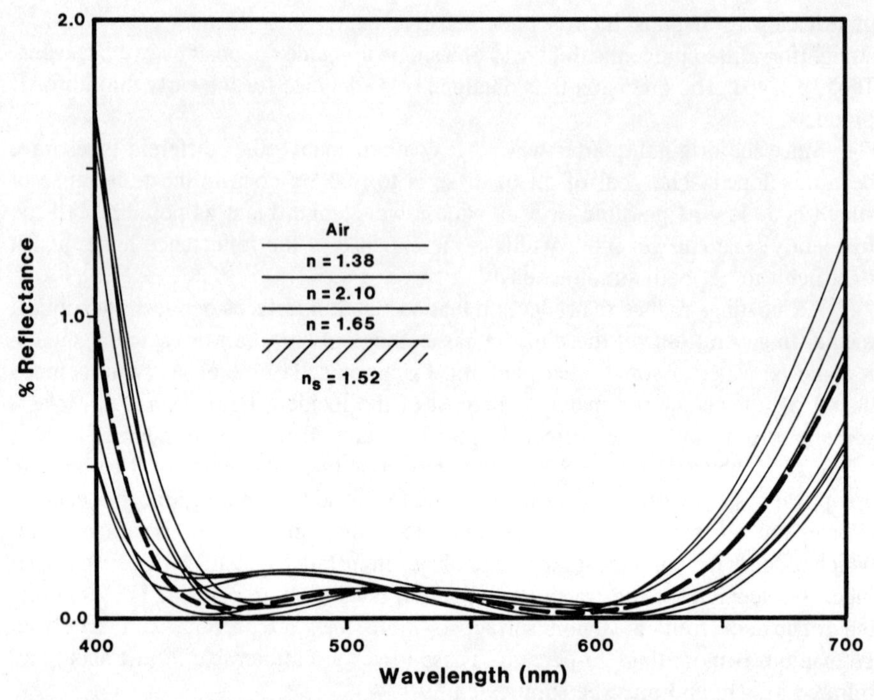

Figure 4.1. Optical thickness tolerance analysis (2.5% standard deviation) or a three-layer AR QHQ antireflection design. The performance of the nominal design is shown by the dashed line. The variations in color purity, dominant wavelength, and photopically weighted integrated reflectance are given in Table 4.1.

Table 4.1 Color Variations as a Result of Layer Thickness Variations

Iteration number	Reflected luminance (%)	Dominant wavelength (nm)*	Purity (%)	C. I. E. (1931)		C. I. E. (1976)	
				x	*y*	*u′*	*v′*
nominal	0.065	565C	16.9	0.318	0.283	0.221	0.442
1	0.066	469	38.1	0.254	0.227	0.195	0.392
2	0.063	472	44.0	0.238	0.218	0.186	0.382
3	0.068	467	44.9	0.242	0.205	0.195	0.371
4	0.070	471	48.6	0.230	0.201	0.186	0.366
5	0.065	556C	43.8	0.328	0.220	0.264	0.397
6	0.068	557C	48.1	0.326	0.207	0.269	0.386
7	0.081	445	22.1	0.295	0.263	0.212	0.425
8	0.061	567C	22.6	0.304	0.263	0.220	0.426
9	0.077	604	18.5	0.392	0.337	0.250	0.484

This data corresponds to the same variations in performance of the design shown in Fig. 4.1.
*A "C" following the wavelength value indicates the complementary color is given.

low compared with the bare substrate value of 4.3 percent. If an isolated substrate has some color in reflection, it may or may not be apparent, depending on how the gradation from one area to the other occurs. When adjacent panels with different coating errors (colors) are compared, the differences may be quite noticeable and unacceptable on the basis of cosmetics.

Generally, AR coatings are fabricated with nonabsorbing materials; as a result, any reduction in the reflected light results in an increase in the transmitted energy. It is also possible to reduce the reflectance with the use of absorbing films in the design. The absorption reduces the transmitted light, but in some applications, the reduction in the reflected light is more important than keeping the transmitted light at a high level.

One application of an absorbing AR is the contrast enhancement filter used on computer terminals with light emitting ("active") displays such as cathode ray tube, plasma, and electroluminescent displays. (Absorbing AR designs are not useful with liquid crystal or other "passive" displays because to function, these types of displays depend on ambient light reaching them.)

Alternative techniques exist that reduce the reflected glare from surfaces, but there is usually a reduction in optical performance associated with each of these. For example, the surface of the display may be etched or frosted lightly; this treatment destroys the specular reflection from the surface by scattering it in non-specular directions. Although etching and frosting reduce the direct image problem, it also reduces the sharpness of the screen image, and the dark areas are lighter than they would otherwise be. In high-contrast, high-visual fidelity systems these approaches can detract from the other system improvements (see Haisma et al., 1985).

Given no constraints on indices of refraction, it is possible to design an AR coating that has zero reflectance at all wavelengths. It is a matter of grading the refractive index at the interface between the substrate and the other medium. In other words, the index should vary continuously from that of the substrate to that of the incident medium so that there is no discontinuity. The actual shape of the index gradation function is not especially critical, though the thickness of the graded region should be at least a quarter wave thick, and preferably of the order of a half wave thick. Such an AR layer has indeed been realized in practice (Minot, 1976; Yoldas and O'Keefe, 1979). The process begins with a glass, which consists of at least two separate chemical phases. An etching process dissolves one phase, while leaving the other untouched. As the etchant attacks a surface, the phase that is attacked is completely removed near the surface, but as the removal process works from the surface and proceeds inward, the percentage removed decreases as the depth increases. The microstructure left behind at the surface is mostly air. Thus the surface of the layer has an index close to unity, and the index increases with depth. Because there is almost no discontinuity in the index function at the interface so treated, the reflection is very close to zero and not a function of wavelength. Unfortunately, the porous surface consists of many fine capillaries that can retain liquid contaminants, such as oils from fingerprints, thus making the surface very difficult to clean.

A major challenge exists for the designer of thin film AR coatings. It is to arrive at a design, the performance of which meets the original specifications, and that only incorporates indices available in Nature. A number of papers have been published reporting studies on the index ranges that are most satisfactory for constructing AR filters in a number of special cases. The designer has to make compromises and use special tools to synthesize the indices needed. The synthesis of unavailable indices can be done by using combinations of layer thicknesses and available indices. Epstein (1952) and Herpin (1947) have published papers explaining mathematical formulations that enable a designer to generate a specific equivalent index. These approaches function adequately over relatively narrow wavelength ranges, and the challenge is how to make the synthesized index function over ever wider ranges.

Multilayer AR coatings are highly tuned structures. The design of an AR coating for a low-index glass will not likely be satisfactory if applied to high-index glass.

A general rule in noncritical uses of thin film filters is that the performance does not change markedly with small changes in the angle of incidence; e.g., in changing from normal incidence to 30 degrees. Reflectance measurements are not frequently made at exactly normal incidence, but at some small angle away from normal, often 10 degrees. This small angle will not generally cause a significant error in the measurement.

4.1.1. Single-Layer Designs

As previously mentioned, the single layer AR coating was the earliest application of the art of thin film design and fabrication. The pioneer thin film coaters used cryolite (sodium fluoaluminate, Na_3AlF_6), chiolite ($Na_xAl_yF_z$), or magnesium fluoride (MgF_2) as materials for coating AR layers on glass because they are easily evaporated from resistance sources and have indices that are suitable to antireflect glass. The optical thickness required to obtain a minimum reflectance is approximately a quarter wave. The spectral performance of single layer films is given in Fig. 4.2. It should be noted that (in the absence of dispersion) the reflectance never rises above the bare substrate reflectance at any wavelength when a single quarter wave film with an index lower than the substrate is used. Conversely, when the film index is greater than that of the substrate, the reflectance never drops below that of the bare substrate. The half wave points serve as indicators of whether the film has absorption or inhomogeneities in it. Unless the reflectance returns to that of the bare substrate, one, or the other, or both of the assumptions of no coating absorption or inhomogeneity in the layer is violated.

The reflectance of a single quarter wave thick film is given by:

$$R = \left[\frac{n_f^2 - n_o n_s}{n_f^2 + n_o n_s}\right]^2, \tag{4.1}$$

where n_f is the index of the film, n_o is the index of the incident medium, and n_s is the index of the substrate.

In a film of magnesium fluoride, which has an index of approximately 1.38,

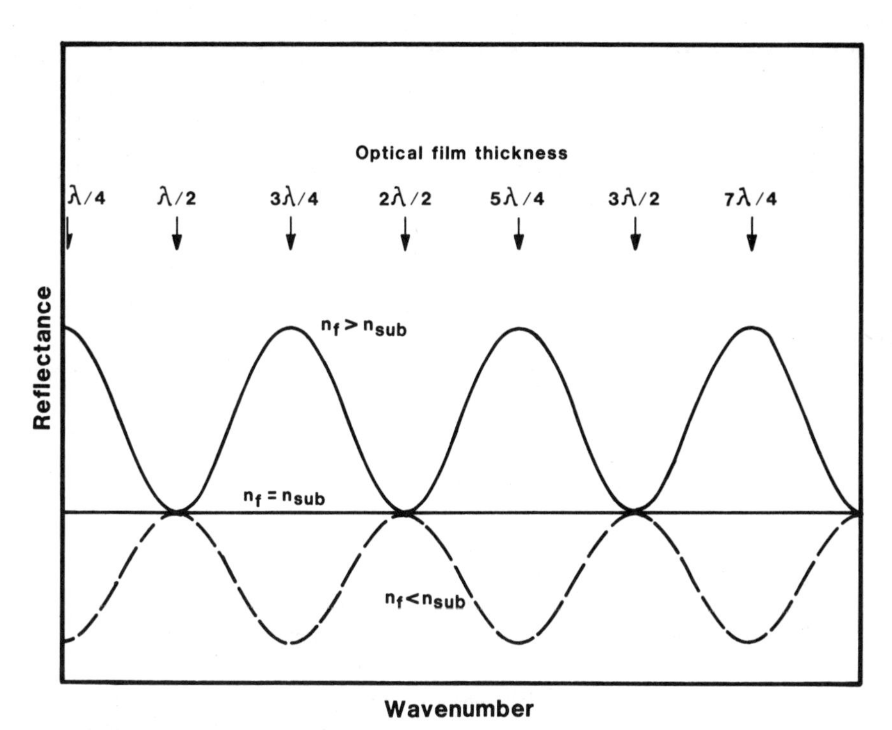

Figure 4.2. Reflectance of single layer dielectric coatings with an index n_f on a substrate with an index n_{sub}. The reflectance extrema occur at the points where the films are a multiple of a quarter wave thick (marked with the respective fraction of a wavelength λ). Note that the reflectance returns to exactly that of the substrate when the film is an even multiple of a quarter wave, i.e., a multiple half wave.

deposited on glass with an index of 1.52, the reflectance is reduced from that of bare glass, 4.26 percent, to 1.26 percent. Equation 3.1 is also applicable when the index of the film is higher than the index of the substrate.

The minimum reflectance is exactly zero when the index of the film is the geometric mean of the incident and substrate indices:

$$n_f = \sqrt{n_o\, n_s} \tag{4.2}$$

In the case of a glass with $n_s = 1.52$, the optimum film index is approximately 1.23. No material exists in nature that satisfies this index requirement and is also stable in a typical environment. The nearest index material that is durable is magnesium fluoride. Other materials, such as cryolite and chiolite, have lower indices, but they are to some extent water soluble, and films of these materials will not survive in humid environments.

Equation 3.1 applies to all odd multiples of quarter wave thick films.

4.1.2. Narrow-Band AR Designs

A zero reflectance design is the two-layer "V-coat." In fact, for a given combination of materials that leads to a valid design, there are two such designs. The two designs

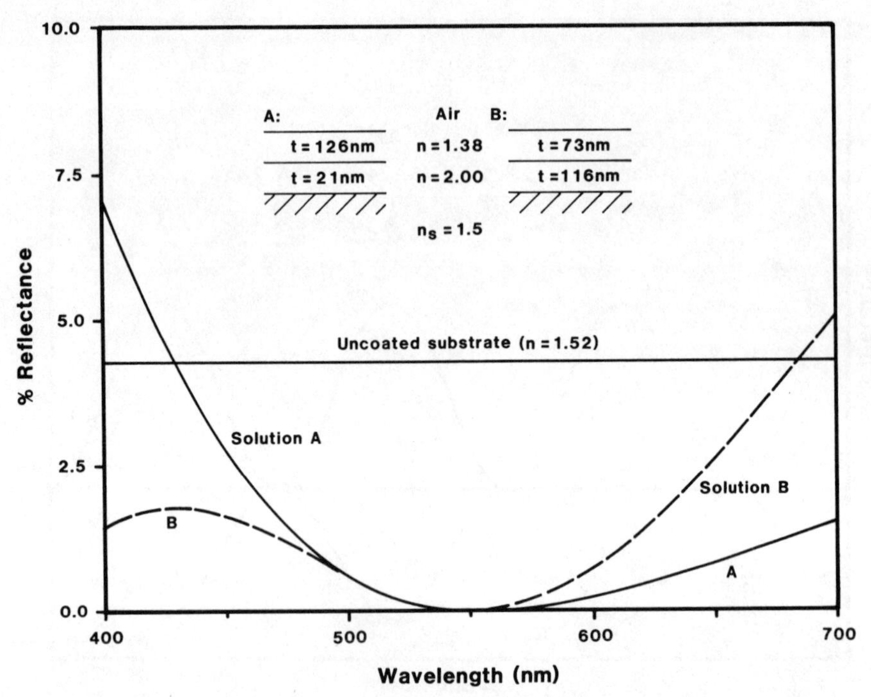

Figure 4.3. Reflectance of both V-coat solutions. The reflectance is zero at the minimum for both designs. The shape of the relatively narrow low reflectance band gives this design its name. The physical thicknesses of the layers in each design are shown in the figure.

for selected indices and a glass substrate are shown in Fig. 4.3, along with the performances. Occasionally, there will be a single degenerate solution; the thicknesses of the layers are both equal to a quarter wave optical thickness. It is not always possible to come up with a solution for a two-layer design given two films of arbitrary indices, nor is it always possible to find materials that have the optimum indices.

The criterion used to select which of the two solutions is to be used is based on stress considerations, all other things being equal. The difference in the thicknesses of the two solutions sometimes causes the stress level in one solution to be much better balanced than the other solution.

These designs are useful in high-power laser applications. Because they have few layers, they can have low stress. The materials that are used in the design can be selected for their low absorption characteristics and high damage thresholds, rather than their optical index of refraction as index in this type of design is not a critical parameter.

For many applications, the width of the low reflection region is too narrow and other types of designs need to be used.

4.1.3. Wide-Band AR Coatings

The limited performance of quarter wave AR coatings led to the development of more complex designs. The quarter-quarter (QQ) solution consists of two quarter wave thick films of appropriate indices. The performance of a typical design is shown in Fig. 4.4. The addition of a half wave thick layer between the two quarter wave layers results in even better performance over a wider wavelength range. Such a design is commonly referred to as a QHQ design. Modern AR coatings can easily achieve an average reflectance of less than 0.5 percent over the visible portion of the spectrum. Figure 4.4 also shows the theoretical performances of a QHQ design. A multitude of papers on the selection of indices and the design of such coatings can be found in the literature.

Source and detector weighting functions become important when evaluating broad band performances such as these. In particular, these designs have a tendency to act like the proverbial "toothpaste tube," i.e., when one point is pushed down, another point rises. If the width of a design is increased, the reflectance tends to rise. Thus, the appropriate type of weighting function must be used in designing

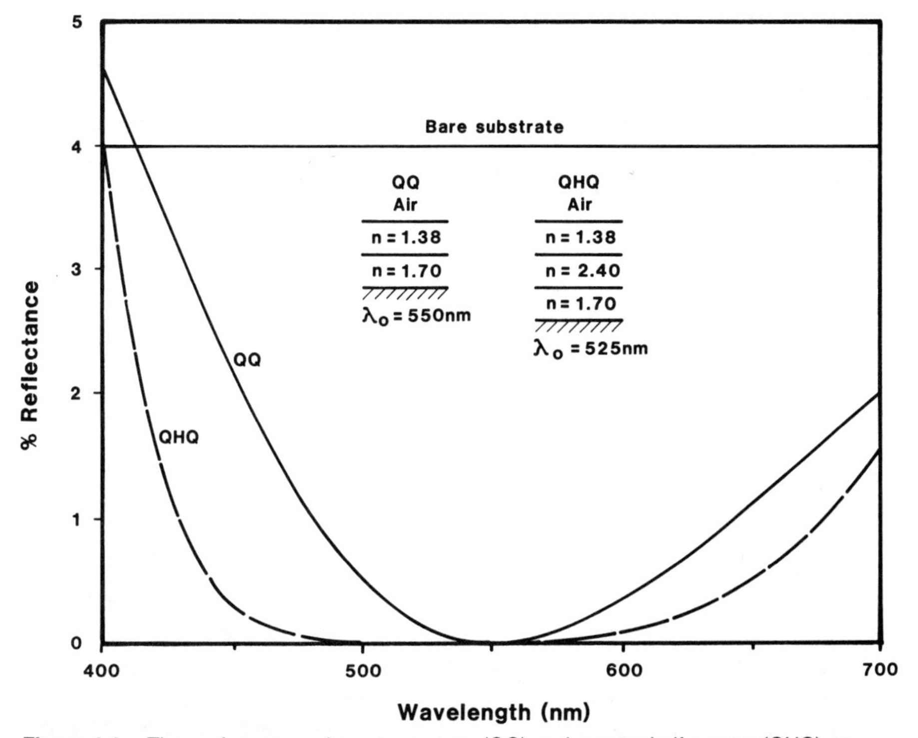

Figure 4.4. The performance of quarter-quarter (QQ) and quarter-half-quarter (QHQ) antireflection coatings. The reflectance of an uncoated substrate is shown for reference. Note the additional width that these designs have over the V-coat shown in Fig. 4.3.

and evaluating these coatings. When one specifies a wide-band antireflection coating for visual use, the designer should give consideration to the purity and the dominant wavelength of the reflected light, as well as the reflected brightness.

4.1.4. Superwide Infrared AR Coatings

With the larger range of material indices available in the infrared as compared with the visible portion of the spectrum, it is possible to obtain much wider infrared AR designs. Coatings that span several octaves of the spectrum are not uncommon. The design of a 4 to 1 wide AR ($\Delta \lambda/\lambda_{min}$) can consist of a stack of approximately quarter wave thick layers, the indices of which progressively decrease from the substrate interface to that of the incident medium. Figure 4.5 shows the performance of one such design at normal incidence. The original reflectance of the bare substrate is shown for comparison.

4.2. HIGH REFLECTION DESIGNS

High reflectance designs fall into three general classes: (1) metallic, (2) all dielectric, and (3) enhanced metallic. Each of these is covered in turn.

4.2.1. Metallic Reflectors

Metallic thin film reflectors antedate optical interference thin film coatings by many centuries. By far the earliest manufactured optical thin films, metal films find use today in telescope mirrors and in other optical instruments. They are often fragile and cannot be contact cleaned. The common metals used are silver, gold, and aluminum. Silver is sometimes used in astronomical telescopes because it can be deposited by an electroless process on mountain tops without the need for expensive coating equipment. Silver tarnishes quite readily, therefore, this process has to be repeated periodically to keep the reflectivity high. Silver has the highest reflectance of any metal in the visible and infrared portions of the spectrum, and it makes an excellent material for this application when it is fresh and untarnished. Recent work be Song et al. (1985) indicates that a solution to the tarnishing of silver may be near at hand. Such a breakthrough would be extraordinary and a boom to the art of high reflector manufacture. Silver is not a good reflector in the ultraviolet portion of the spectrum because it has an absorption band that starts at 350 nm and extends to shorter wavelengths. This absorption band is caused by electron plasma resonances in the metal lattice, and it is not likely that it can be eliminated as it is a basic material property.

When light has to be reflected over a large range of angles of incidence and for a wide band of wavelengths, it is difficult to find a better reflector than a metallic surface such as silver, gold, or aluminum. Most dielectric stacks have high reflectances over a relatively narrow band of wavelengths. Although several of these can be used one atop the other to cover a spectral region wider than that obtainable with a single stack, a point of diminishing returns is reached. It is not possible to improve the reflectance beyond a point because of stress buildup in the coating, scatter, or absorption in the coating materials.

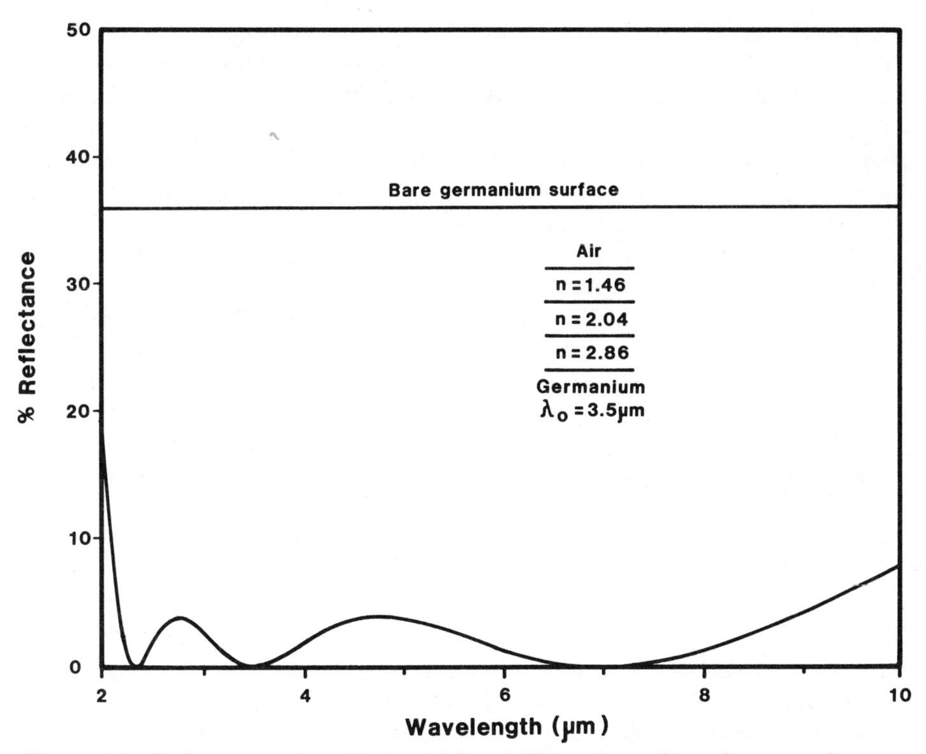

Figure 4.5. Performance of a wide band infrared AR on the surface of a germanium substrate. All layers are nominally a quarter wave optical thickness in the center of the band. Each surface of uncoated germanium has a reflectance of 36 percent, and a substantial improvement in energy throughput is achieved by adding an AR coating to the surface.

Gold is used frequently in optical instruments designed to operate in the near and far infrared portions of the spectrum. In these regimes, its reflectance is almost as high as that of silver, and, it does not tarnish. Gold evaporated in a standard vacuum is not very adherent to glass. Over the years, techniques have been developed to solve this problem. The most common solution in an evaporation process involves the use of a thin chrome layer under the gold. The chrome is adherent to the glass, and the gold adheres well to the chrome layer. The use of an ultrahigh vacuum coating system or a sputtering process can deposit gold without the intermediate layer. Although these techniques solve the problem of adhesion, they do nothing for the problem of the fragility of the gold itself. In a sealed optical instrument where access is limited and the chance of the optics' getting dirty is minimal, the cleaning and damage considerations are not a drawback.

Both gold and silver mirrors can be overcoated to give them more durability than they otherwise may have. There are several limitations with this approach. Overcoats are generally not transparent over the entire range that the metal is reflective, thus the overcoat limits the spectral range of the mirror. For wideband instruments and optics, this can be a serious problem. On the other hand, all is not as bleak as it may seem. Figure 4.6 shows the normal incidence standing wave electric field at the surface of a metallic reflector with a thin overcoat. The node at the surface limits the amount of energy absorbed in the overcoat. This is illustrated in Fig. 4.7, which shows the infrared absorption of a free-standing fused silica layer in air. Figure 4.8 shows the reflection of silver, both bare and overcoated with a silicon dioxide overcoat, the thickness of which is the same as that in Fig. 4.7. The absorption added by the overcoat is the difference between the two curves (note the expanded scale). The reflection over most of the spectrum is only slightly affected by the presence of the coating.

The situation is much different at nonnormal incidence. Figure 4.9 shows the reflectance at a 60-degree angle of incidence for both planes of polarization. It is obvious that there is much reduced reflection especially in the 9- to 10-μm region. This corresponds to the reststrahlen reflection region of the overcoat material. In this wavelength region is a resonance frequency: the molecules that make up the crystal lattice undergo a collective oscillation that is driven by the electromagnetic field. Thus, the lattice that makes up the film can absorb a significant amount of the energy in the incident light beam if it is properly coupled into the film. The electric field plots in Fig. 4.10 show the effect of the angle of incidence on the standing waves. Although the pattern in the *s* plane resembles that at normal incidence, the *p* plane pattern is rather different. The field in the overcoat is extremely high, and this leads to a very high absorption level. The vast majority of the energy is deposited in this layer. That these theoretical plots represent the real world has been confirmed with measurements made on actual mirrors.

For the overcoat on a metal mirror to be durable, it has to be one or more

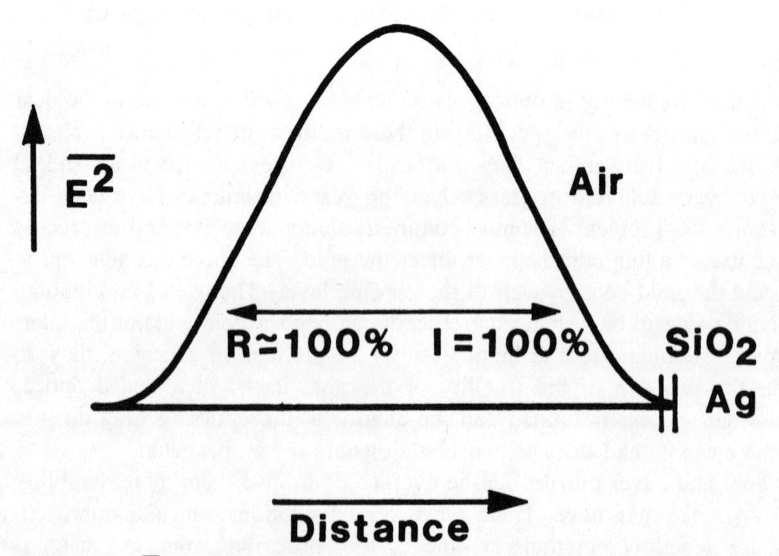

Figure 4.6. $\overline{E^2}$ plot for overcoated silver in the center of the reststrahlen band (9.3 μm). Because there is a node at the surface of the metal, little energy is absorbed in the overcoat.

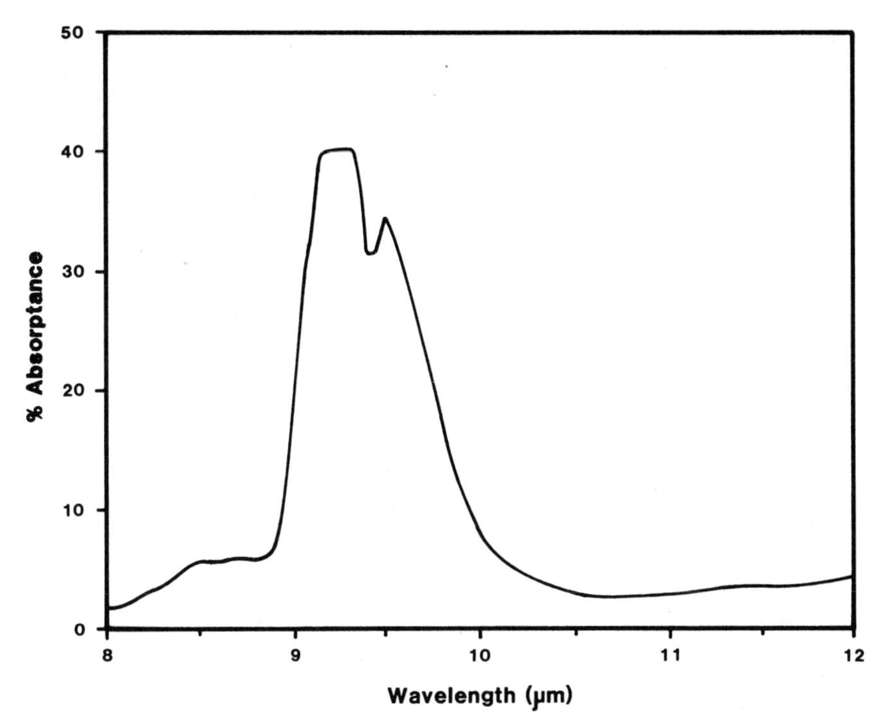

Figure 4.7. Absorption of a free-standing silicon dioxide film. The absorption occurs in the reststrahlen band and is significant when compared to a similar film used as an overcoat on a high reflecting surface.

visible quarter waves in optical thickness. Thinner overcoats will provide proportionately less protection. Thicker films may have stresses high enough to cause the entire film system to fail by crazing. It should be noted that metal films do not have the shear strength of dielectric films, and therefore metal films will fail at stress levels much less than that which cause the failure of dielectric films. In addition, the previously mentioned standing wave electric fields have antinodes inside the protective layer. This enhances the absorption in those regions where there is some absorption present in the protective overcoat, as discussed previously.

The protective overcoat, which may consist of only one layer, will reduce the reflectance of the metallic surface if it is nearly a quarter wave thick, as shown in Fig. 1.7. This problem may be minimized if the thickness is selected to be a quarter wave at a wavelength remote from the region of interest. An example of this is the design of Fig. 4.8, where the overcoat has a physical thickness that is only about 1 percent of the nominal wavelength of 9 μm. Another solution is to select a film that is a half wave thick at the wavelength of interest, but once again both stress and absorption have to be considered. A trade-off needs to be made between several desiderata to arrive at the best solution.

Aluminum metal mirrors are common because they are relatively inexpensive,

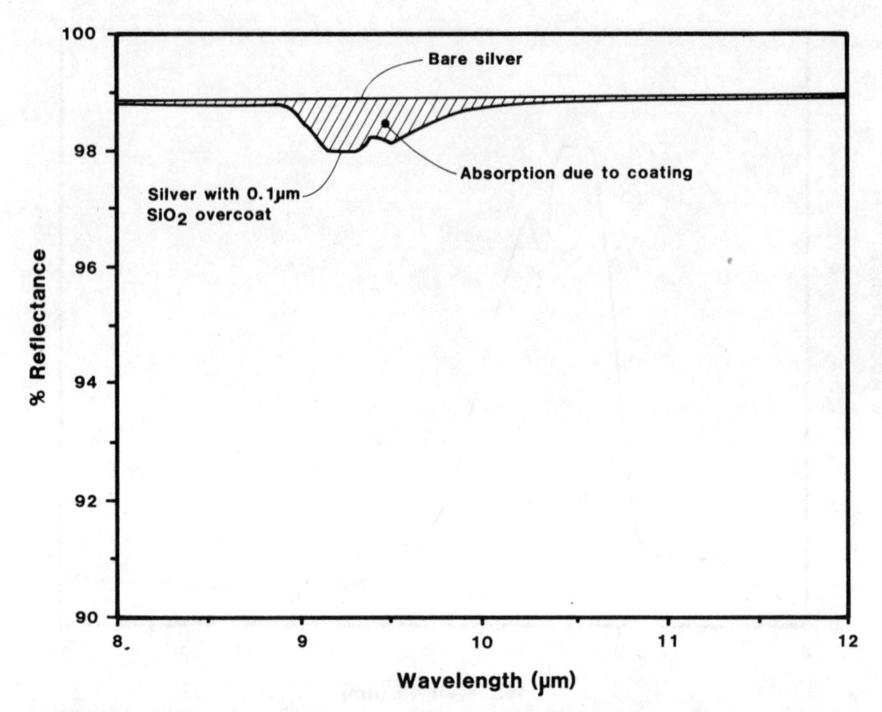

Figure 4.8. Reflectance of a silver substrate with and without a silicon dioxide overcoat at normal incidence as a function of wavelength. The reflectance scale is expanded. Note the small amount of absorption as compared to the configuration used in Fig. 4.7.

and can be made with a minimum investment in coating equipment. This may include a vacuum chamber, a heavy stranded tungsten filament wrapped with aluminum wire, and a low voltage AC or DC power source to supply a current to heat the filament. Aluminum adheres well to glass, therefore, the film does not require a preparatory adhesion layer as does gold, for instance.

The control of the thickness of aluminum films is important to get the optimum specular reflectance from them. Overly thick aluminum films, on the order of 100 nm or more, begin to scatter light because of surface roughness. This roughness increases rapidly with thickness beyond this point. On the other hand, films that are too thin will transmit light, and the reflectance will not be the highest possible. The optimum reflecting aluminum mirror will just barely transmit light from a strong incandescent light bulb as viewed with the naked eye.

An aluminum film oxidizes with time, starting immediately upon contact with air. The oxidation continues, albeit at a reduced rate, for the entire life of the film. In some situations, this oxidation can proceed at an accelerated rate, as when both humidity and UV light are present simultaneously. The oxide film does have its good points. It does slow down the oxidation rate as it gets thicker, and does serve to protect the remaining metal film. Because aluminum oxide is transparent over a very wide spectral range, it does not interfere with the metallic reflectance, at least not in thin thicknesses.

Although aluminum reflectors have less reflectance than silver, they are very useful because they are good ultraviolet reflectors, whereas silver is quite poor. On the other hand, there is a dip to 86 percent in the reflectance of aluminum in the near infrared at about 825 nm. Furthermore, the average reflectance of aluminum in the visible portion is around 92 percent, as compared with silver's 98 percent. Thus, when there are many reflecting surfaces in a system, it is obvious that there are going to be significant light losses if 92 percent reflectance is all that is available: 5 surfaces will reduce the available light by almost 34 to 66 percent. If these 5 surfaces had a fresh silver coating with a 98 percent reflectance each, the combined reflectance would be about 90 percent.

4.2.2. All Dielectric Designs

4.2.2.1. Nonsymmetric Stacks. The alternating high and low index stack of layers, the optical thicknesses of which are all a quarter wave, is one of the oldest multilayer designs. The performance of such an arrangement of layers is shown in

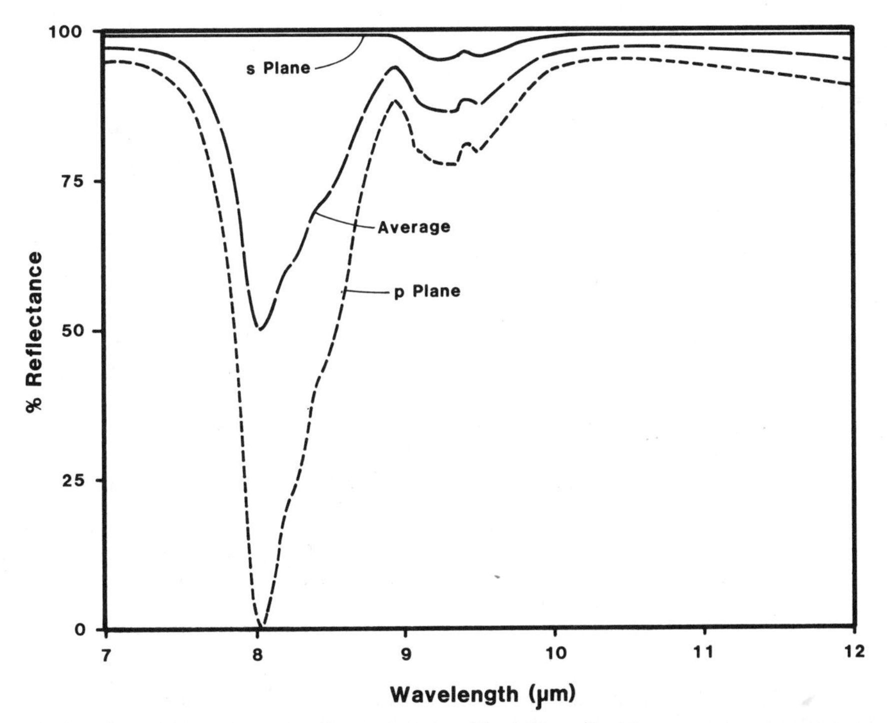

Figure 4.9. Reflectances of a silver substrate with a silicon dioxide overcoat on an opaque silver substrate at a 60-degree angle of incidence. The reflectances for both planes of polarization are shown along with their average.

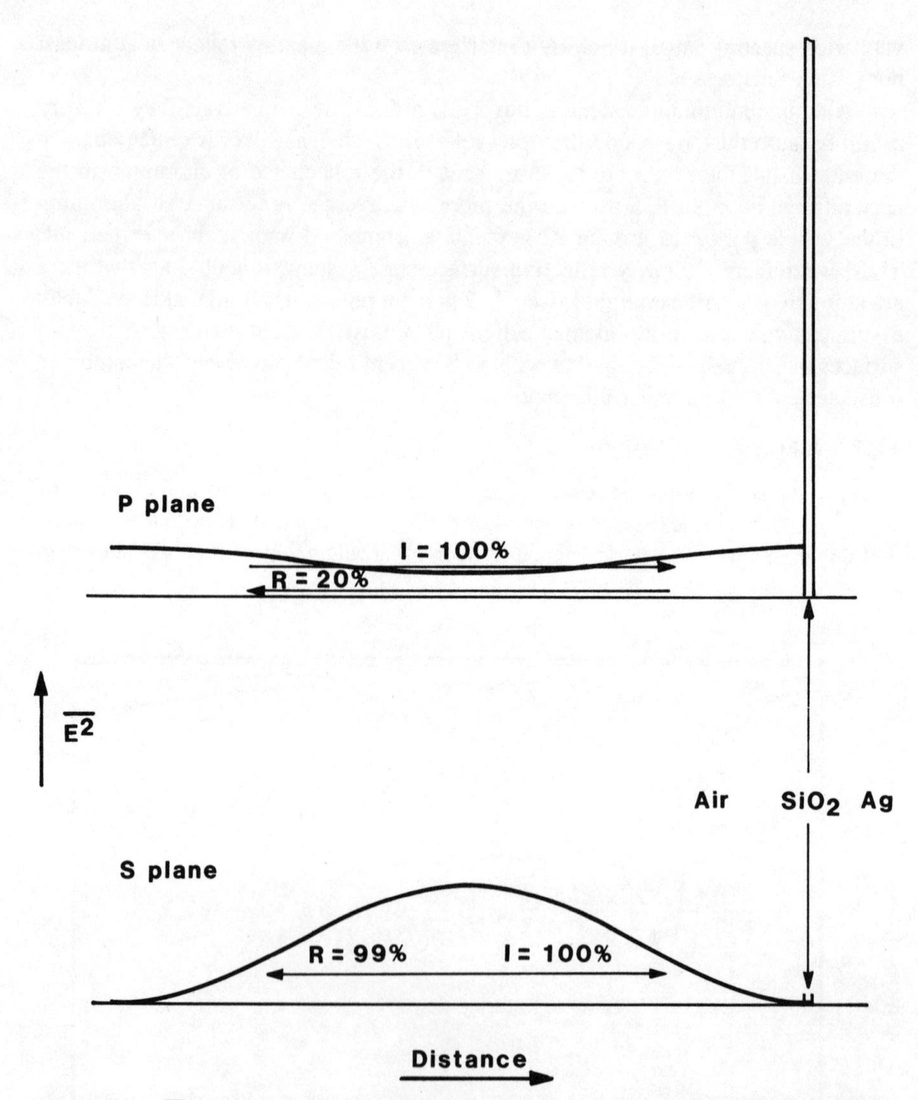

Figure 4.10. $\overline{E^2}$ at 60 degrees for both planes of polarization for a silicon dioxide overcoated silver mirror. The wavelength and the thickness of the overcoat are the same as that of Fig. 4.6; the two figures can be used to compare the standing wave fields. The penetration of the *s* plane field into the overcoat and the substrate is minimal, whereas in the *p* plane, there is a great deal of interaction.

Fig. 3.3. There are several parameters that characterize such a design. The index ratio of the high and low index materials is probably the most important one as its value determines the width of the high reflection band, and, together with the number of pair of layers in the design, determines the peak reflectance of the coating.

The width of the high reflectance band is given by Eq. 3.1. This high reflectance band is sometimes referred to as the *stop band*. This terminology derives from the fact that the transmission is blocked (*stopped*) by the high reflection in this wavelength band.

The definition of the edges of the stop band is generally given as those wavelengths for which the reflectance only increases (theoretically) as the number of layers in the stack increases. This definition is very practical, in that it specifies just how wide the design will be in the limit of an infinite number of layers. The peaks and valleys between stop bands also have an envelope that does not depend on the number of periods. Macleod (1969, Eq. 5.9) shows that the peak transmittance at the center wavelength of a quarter wave stack $(n_s|(LH)^p|n_0)$ in air is given by

$$R = \frac{\left(1 - g^{2p}\dfrac{n_H^2}{n_s}\right)^2}{\left(1 + g^{2p}\dfrac{n_H^2}{n_s}\right)^2}$$

where p is the number of layer pairs, g is the ratio of the high index value to the low one, and
n_H is the index of the higher index material.

A conservative estimate of the number of layers required to attain at a minimum a given level of transmission can be obtained by dropping the (n_H^2/n_s) factor from this equation. The effect of this term is to increase the reflectance over that calculated on the basis of the exact equation as the value of this term is greater than unity in most practical situations. The result of this simplification is

$$R = \left|\frac{1 - g^{2p}}{1 + g^{2p}}\right|$$

The plot of this function is shown in Fig. 4.11. The importance of the index ratio is readily apparent by the rapid increase in the reflectance with a relatively small increase in index ratio.

The reflectance of a quarter wave stack is frequently shown plotted against a wavenumber or a normalized inverse wavelength abscissa. A typical plot of this type is shown in Fig. 3.5. The advantage of this type of presentation is readily apparent: the periodicity of the stop band patterns are very evident and predictable, therefore, only one order needs to be plotted, and all of the other orders can be calculated from that one. It should be noted that dispersion in the index and the absorption coefficient has been ignored in these theoretical calculations. These parameters can result in significant deviations between the calculated performance and that obtained in a real coating.

Theoretically, the number of stop bands in a quarter wave design is infinite. An order number is associated with each stop band. The order of the stop band that occurs at the longest wavelength, i.e., where all of the layers are a single quarter wave thick, is 1. Note that there are no stop bands at wavelengths longer than the wavelength at which the stack layers are a quarter wave.

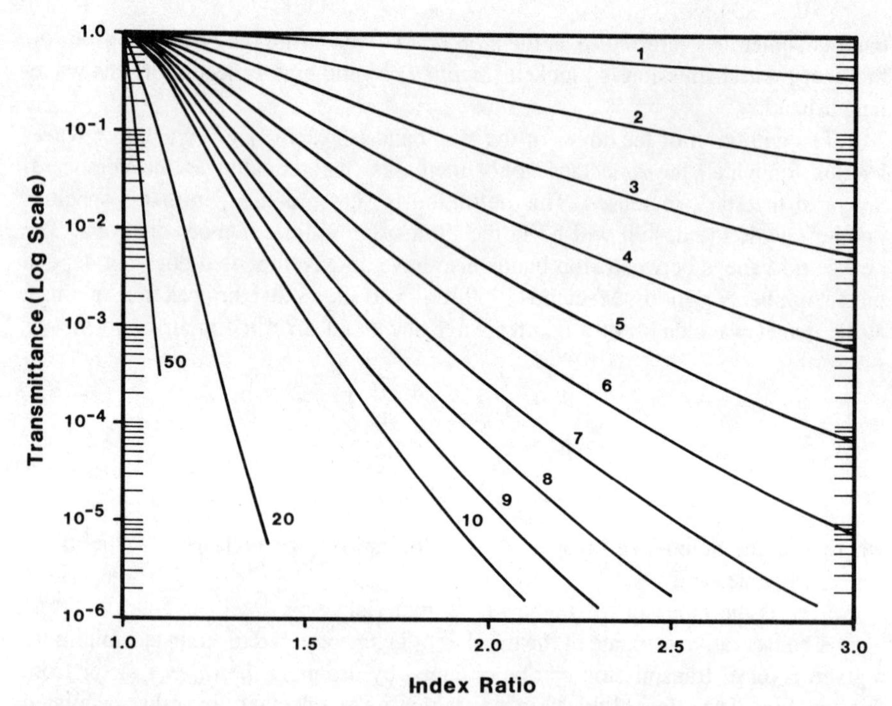

Figure 4.11. Approximate maximum transmittance as a function of index ratio at the stack center wavelength for a stack with quarter wave thick layers. The substrate and incident medium indices are both 1.0. The numbers next to the curves indicate the number of pairs of layers in the stack.

A stop band can exist wherever the total optical thickness of a period is equal to a half wave. Whether it actually exists or not depends on the individual layer thicknesses in the period. At those wavelengths where the layer thicknesses are all an integral multiple of a half wave, the entire stack is absentee. At the wavelength that corresponds to the second order of a quarter wave stack, all of the layers are a half wave thick. Consequently, no stop band exists at the second order location for this type of stack. The same is true for all even orders for this type of design; only the odd orders actually have stop bands. Note that in the plot of the theoretical performance (Fig. 3.5), only the stop bands with odd order numbers exist.

For a given basic stack period, the width of the stop band is constant on a *wavenumber* scale. The width does not depend on the order number. The percent width decreases as the order increases. On the other hand, on a *wavelength* scale, both the width and the percent width decrease as the order increases.

It is straightforward to design stacks that skip any given order, such as skipping every third one. To do this, one replaces the basic HL period with one that has a different proportion of one material to another; e.g., a stack with a basic period of HLH will have the first and second order high reflectance bands present, but every third stop band will be missing. In practice, because of dispersion that occurs in most materials, only one order can be made exactly absent by ascertaining that the optical thicknesses are exactly correct at that one wavelength. At other wavelengths, the match will not be exactly right, and the order will be more or less evident.

When a basic period other than the simple HL type is used, the peak reflectance will not be as high as that of the basic HL design, given the same number of periods and the same kind and amount of coating material.

Thelen (1963, 1973) has published papers indicating how one goes about designing a filter that has a number of orders suppressed. The reader who is interested in this type of design is referred to these papers and to the companion volume by Thelen (1987).

4.2.2.2. Symmetric Designs. In the previous section, we discussed designs in which the optical thicknesses of all of the layers were equal. The resulting periods that made up the stack, however, were not necessarily symmetric. For example, a (HL) stack does not have symmetry, though it has periodicity. In this section, we treat the case where the basic period is symmetric about its center. The center of a symmetric period can be either in the middle of a layer, such as in (HLH), or it can be at an interface, such as in (HLLH). Herpin (1947) and Epstein (1952) have shown that if the basic stack period is symmetric, then the mathematical representation of the period is especially simple. The entire symmetric structure can be reduced to an equivalent *single* layer with a real equivalent index in wavelength regions outside the stop band. Such a period is referred to as a Herpin layer, with a Herpin equivalent index, n_H. An equivalent thickness exists for this basic period that is analogous to the thickness of a single layer. This is a useful concept for designing thin film filters. Figure 4.12 shows the performance of a symmetric design. Note the similarity of this performance to that of the quarter wave stack. As previously noted, the quarter wave stack (all layers are a quarter wave thick) gives the highest reflectance for a given number of layers. The symmetric stack always has somewhat less reflectance than a comparable quarter wave one. The major difference between them, however, is in the transmission zones between the stop bands. In this wavelength range, the Herpin equivalent index is real, and the transmission can be high or low, depending on the relative indices of the medium and the substrate and the equivalent index of the Herpin period. It should be pointed out that the Herpin equivalent index is dependent only on the thicknesses of the layers in the basic period and the indices of these layers. It is *not* dependent on the number of periods in the stack. On the other hand, the Herpin equivalent thickness does depend on the number of periods in the coating. Figure 4.13 shows on a wavenumber abscissa the reflectance of a symmetric stack over a broad wavelength range. The repeat pattern of the performance is evident. The equivalent index is different on either side of the stop band.

The regularity seen in Fig. 4.13 for the performance of a periodic stack may not be observed if the number of periods is small and the index of the substrate and the incident medium are not equal. Figure 4.14 shows the performance of a single period of a quarter wave stack and that of a larger number of periods. In this figure, the index of the incident medium is unity, whereas that of the substrate is 1.5. It is obvious that the performances of the two designs are quite different.

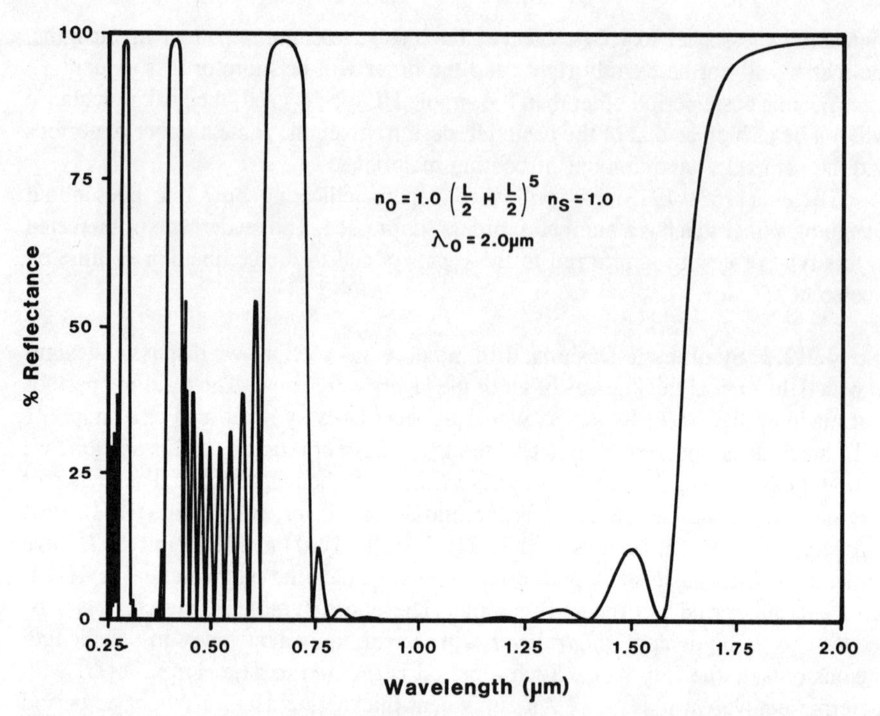

Figure 4.12. Theoretical reflectance of a $(L/2\ H\ L/2)^5$ stack on a wavelength scale. There are no other high reflection bands at wavelengths longer than that of the first order. The pattern of high reflectance bands repeats as the wavelength decreases.

The dispersion in the indices of refraction of the materials used in a stack can disrupt the regularity of the stop bands and the high transmission bands. In addition, a change in the index of refraction can lead to a *mismatch* in the optical thickness of some layers. Normally, the thicknesses of the layers of a quarter wave stack are adjusted so that they are all matched in optical thickness at some wavelength. Dispersion in the index of refraction can make orders appear where otherwise they should be suppressed as the layers are not all a quarter optical thickness at wavelengths other than the match wavelength.

Orders can be suppressed in symmetric designs in the same way they are in the previously discussed nonsymmetric ones. Figure 4.15 shows a design where the third harmonic has been suppressed, whereas the second harmonic has a high reflection stop band.

4.2.3. Enhanced Metallic Reflectors

To improve the performance of metallic mirrors, coatings have been developed that enhance their reflectance. In the simplest case, the design consists of a stack of layers with alternating high and low indices. With the exception of the first layer next to the metal, each layer has an optical thickness of a quarter wave at the wavelength at which the highest reflectance is desired. Figure 4.16 shows the performance of a design that enhances the reflectance of aluminum in the visible. The first layer's optical thickness is slightly less than a quarter wave to account for the phase shift that occurs at the metallic interface.

A trade-off is necessary in as much as the higher reflectance in this type of design is obtained at the price of bandwidth. At wavelengths outside the stack's high reflectance band, the reflectance of the combination of metal and coating is less than that of the bare metal. The electric field profile in such a design is interesting. Figure 4.17 shows such a calculation, and it is seen that the electric field does not reach down to the metal with any significant amplitude, thus indicating that the resulting mirror will have little absorption. At wavelengths away from the dielectric stack center, the situation is rather different. Figure 4.18 shows the electric field at the minimum of the reflectance curve in Fig. 4.16. It is obvious that at this wavelength much of the field penetrates down to the metal, and that absorption is enhanced over that which would take place for the bare metal. Some filters are

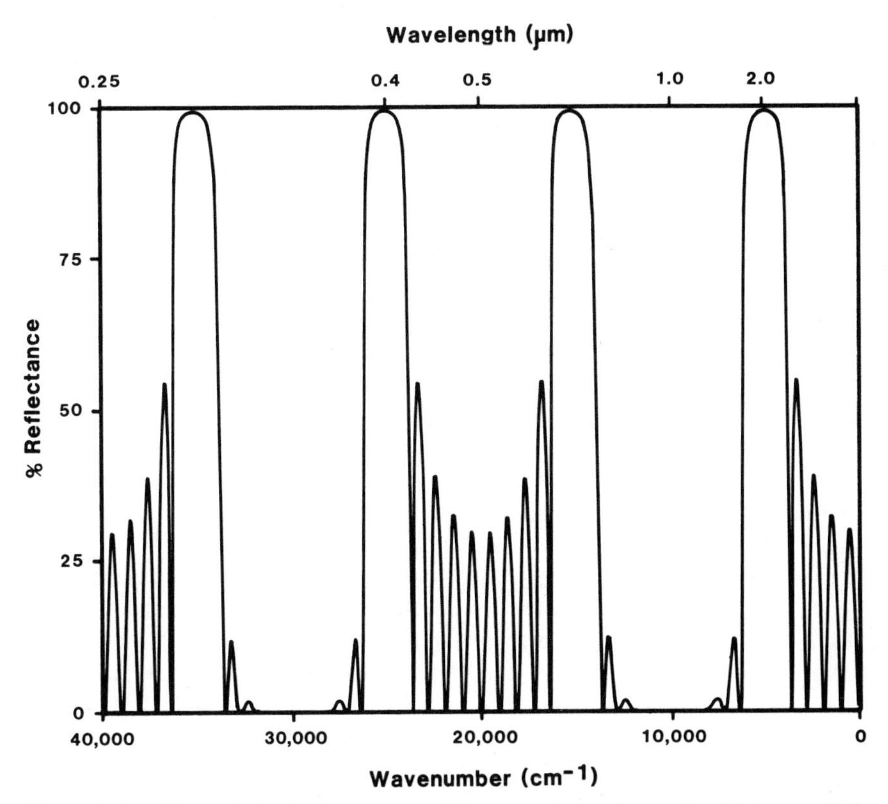

Figure 4.13. Theoretical reflectance presented on a wavenumber scale of the same $(L/2\ H\ L/2)^5$ stack shown in Fig. 4.12. Notice that with this representation the stops bands produce a periodic pattern.

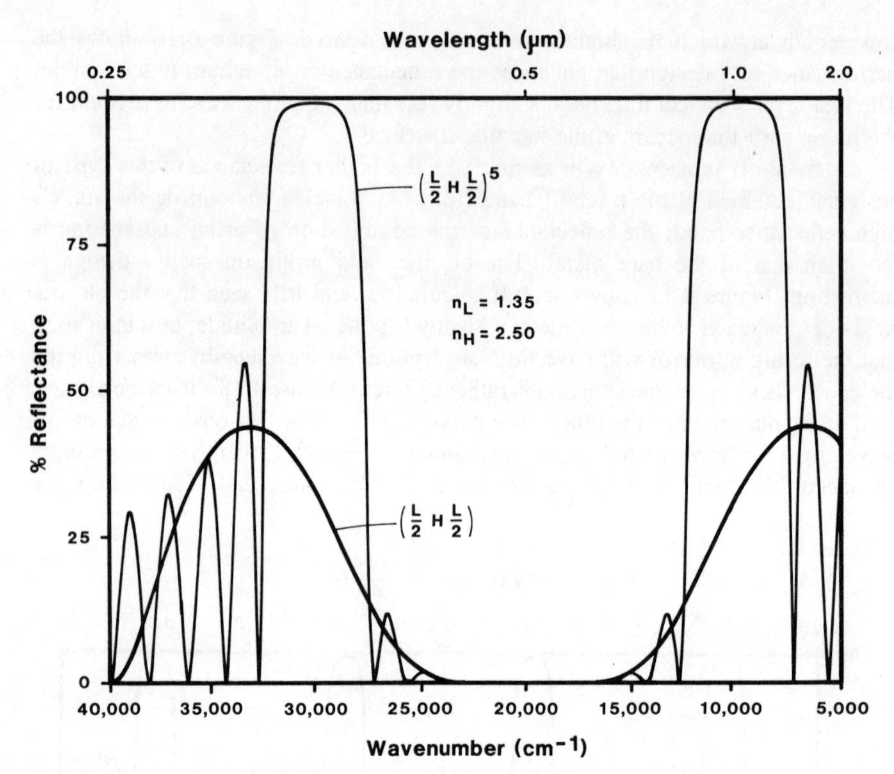

Figure 4.14. Reflectance of 1- and 5-period symmetric stacks on a wavenumber scale. Both the incident medium and the substrate have an index of 1.0. When the number of periods in a symmetric design is low, the peak reflectance does not occur in the center of the stop band.

optimized for this feature and are called *induced absorbers;* they are discussed in further detail in section 4.6.1.

The advantage of this type of high reflector design is in its economy of layers. Because a full quarter wave stack does not have to be built up to obtain high reflectance, the resulting mirror will generally be less costly. As the enhanced metal design requires fewer layers and total coating material than an equivalently reflecting design built up only with a stack of quarter-wave thick dielectric layers, the amount of scatter in the coating can be less. In practice we can, therefore, obtain a higher absolute level of reflectance from one of these designs. Because most computer programs that calculate thin film performance do not include scattering in their mathematical model, this aspect does not show up in theoretical plots.

In addition to the reduced width of the high reflection band, the other consideration that needs to be evaluated is the environmental durability of the coating. In general, a metallic coating will not be as durable as one with equivalent reflectance built up of all oxide materials. Thus, if the coating is going to be exposed to a harsh environment, such as salt fog or high humidity, then perhaps the all dielectric design would be the preferable choice.

The reduction in the high reflectance zone can be overcome by adding stacks to the design. The center wavelength of these additional stacks can be located to optimize the reflection over a broad range of wavelengths.

If absorption and scatter are left out of the computer model, any arbitrarily high value of reflectance can be obtained. Carniglia and Apfel (1980) have shown how to maximize the reflectance of a dielectric quarter wave stack when one of the materials has more absorption than the other.

4.3. EDGE FILTERS

Edge filters are similar to the symmetric period high reflector designs introduced in the previous section. The region of interest in this case is the edge of the pass band, rather than the peak reflection and the width of the stop band. In this type

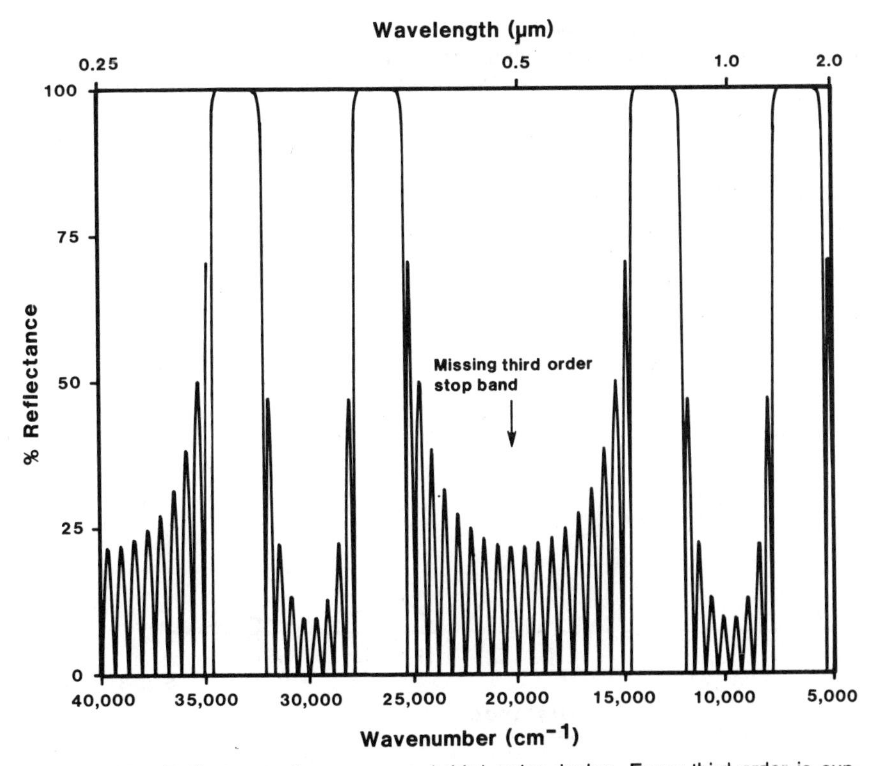

Figure 4.15. Reflectance of a suppressed third order design. Every third order is suppressed in this design, the basic period of which is (*L H L*). The multiple reflection peaks between the highly reflecting stop bands can be effectively reduced to allow good transmission between a pair of stop bands.

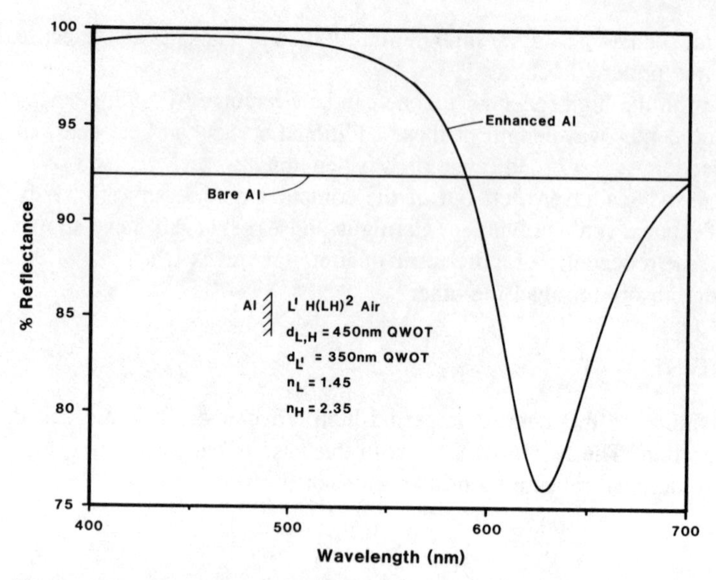

Figure 4.16. Theoretical reflectance of bare and enhanced Al reflectors. Note that the ordinate is an expanded scale. Although a higher reflectance can be obtained in one part of the spectrum with enhancing layers, other portions of the spectrum have a reduced reflectance.

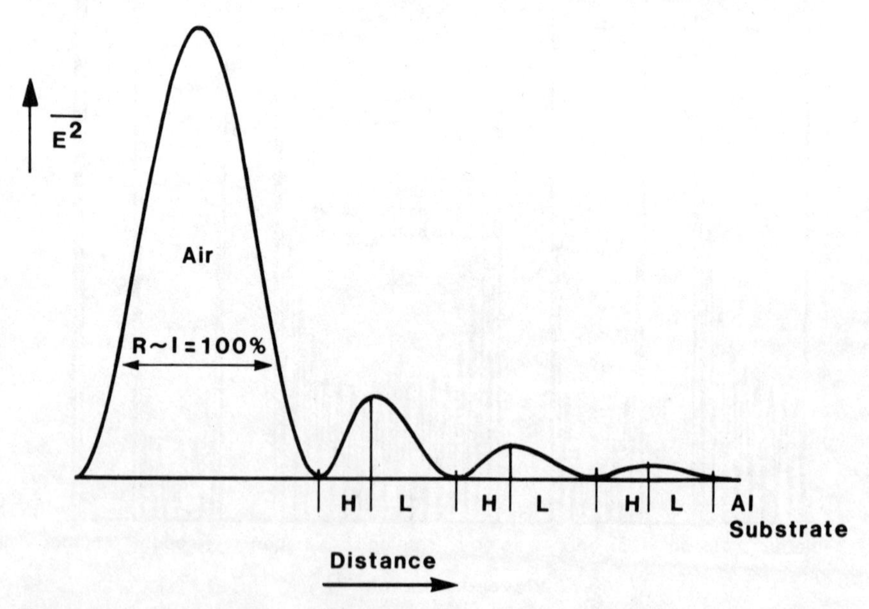

Figure 4.17. Plot of the time average of the electric field in an enhanced mirror. The field is very close to zero at the metal interface. The wavelength corresponds to that where the reflectance reaches a peak in Fig. 4.16. The indices of refraction are $n_L = 1.45$ and $n_H = 2.35$.

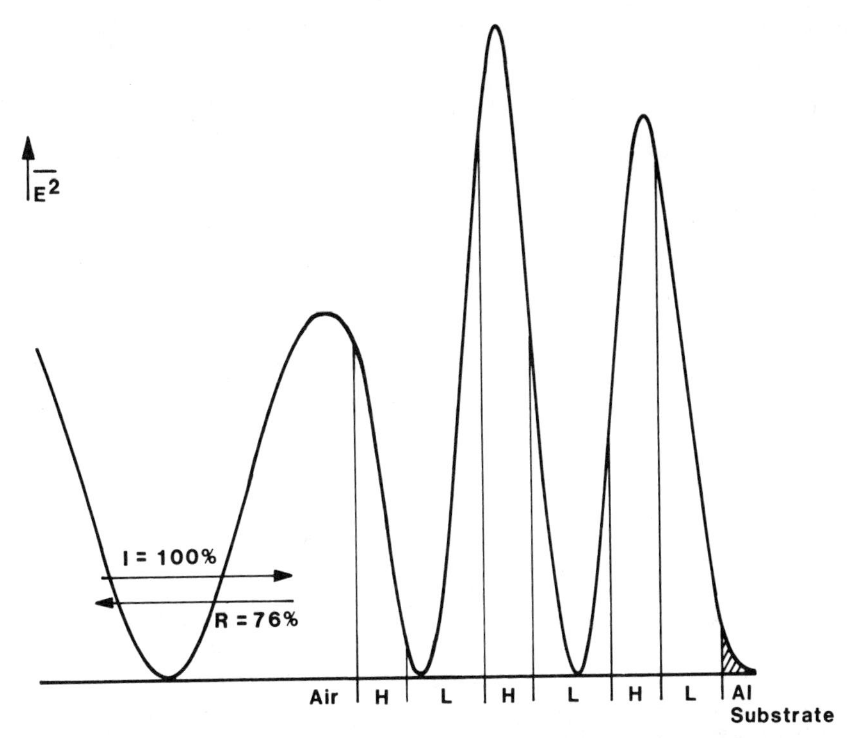

Figure 4.18. $\overline{E2}$ for an enhanced Al reflector at the wavelength corresponding to the first reflectance minimum point. The scales of this plot are the same as that of Fig. 4.17. Note the relatively large amplitude of the field in the absorbing substrate, which leads to the enhanced absorption of almost 25 percent. At this wavelength, the coating tends to act as an absorption inducing or antireflection design.

of filter, it is generally desired that (1) the transition from reflecting (stop band) to transmitting (pass band) be as sharp as possible and (2) the transmission zone be as close to 100 percent as possible. Figure 4.19 shows an expanded view of an edge filter. The basic design in this filter is the same as in the previously mentioned symmetric high reflector, except that some additional layers may be added to reduce the reflection losses in the transmitting region. These additional layers can be designed using antireflection concepts as the high reflector stack can be represented by real Herpin equivalent indices and thicknesses in this region.

A nomenclature problem sometimes exists with edge filters. The use of terms, such as *short pass* and *long pass,* are to be avoided as they can mean different things to different people, depending on their background. On the other hand, the use of terms, such as *short wave pass* or *long wave pass,* leaves no room for ambiguity.

Another terminology exists with these filters. A *hot* or *heat* mirror is one that

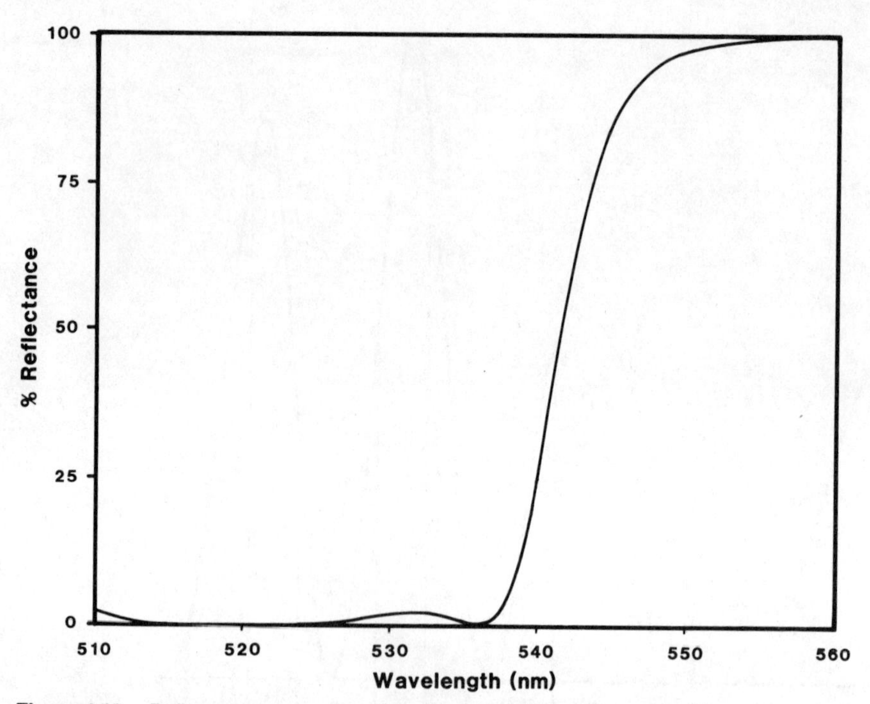

Figure 4.19. Reflectance of an edge filter. Note that the wavelength scale is expanded to show the shape of the edge. The steepness can be increased by adding periods to the stack, but then humps can begin to appear in the transmitting portion.

reflects the infrared and transmits visible light well. Conversely, a *cold* mirror reflects visible light and transmits infrared heat. In this latter case, the reflected light is *cold,* i.e., it contains relatively little infrared energy.

There is one major distinction between short wave and long wave pass edge filters. The long wave pass filter that has been designed at the edge of the first order stop band has no stop bands appearing in the pass band (at wavelengths longer than the edge). There is no limit to how far the pass band extends in this case. Dispersion in the real coating materials generally limits the range over which the coating will be useful. On the other hand, the short wave pass filter has a pass band constrained by a higher order stop band. Some of these can be suppressed to widen the pass band (see Thelen, 1963, 1973).

Edge filters are generally made from all dielectric materials, and therefore, exhibit the typical angle shift described in section 3.6. The shift is dependent on the average index of the materials that make up the design. As a result of the higher index materials available in the infrared portion of the spectrum, infrared filters tend to shift less than equivalent filters for the visible portion of the spectrum, where available indices are much lower.

In addition to the shift in the performance of a filter as a function of the angle of incidence, another effect on the performance is introduced by polarization. Because the width of the stop bands are different for each of the two planes of polarization at nonnormal angles of incidence, the filter has polarizing properties near the edge. This can be useful if a polarizing filter is desired. If a nonpolarizing edge is the requirement, however, it can be built according to a recipe given by Thelen (1981).

4.4. BANDPASS FILTERS

Bandpass filters allow only a relatively narrow portion of the spectrum to pass through them, while blocking other portions, either by reflection or by absorption. Edge filters can be considered bandpass filters of sorts, but generally this terminology is reserved for filters with transmission bands that are narrower than 5 to 10 percent of the central pass band wavelength.

The following discussion of these filters is organized along the lines of their construction, and a given width of pass band may be obtained with more than one design. Other considerations, therefore, have to be used to determine which filter is best for a given application. Bandpass filters come in nonabsorbing and absorbing versions. In either case, it is important for the user to know just how much transmission can be expected from a typical production filter. Even though a filter is fabricated entirely with dielectric materials, this does not mean that it will have 100 percent transmission. In narrow bandpass filters on the order of 0.25 to 0.5 of a percent in width, the peak transmission can be well under 50 percent.

4.4.1. Two Edge Filters in Series

When a relatively wide pass band is desired, it can be obtained by placing two appropriately positioned edge filters in tandem. One of these will be a short wave pass, whereas the other will be a long wave pass filter. A suitable set of edge filters enable an experimenter to assemble a bandpass filter to suit the application. Pass bands in the range of 5 or 10 percent to as little as 2 percent may be obtained with this method. The primary limitation on the minimum bandwidth that can be obtained this way is the steepness of the slope of the edge filters.

If custom filters are to be made by placing two bandpass filters in series, there are several options possible: (1) put both filters on the same side of one substrate, (2) put the filters on the opposite sides of one substrate, and (3) put the two filters on separate substrates. Cost considerations enter into this decision: two substrates cost more than a single one as four sides have to be prepared and polished. If two substrates are used, then it may be necessary to put an antireflection coating on the other two surfaces. Thus, the choice of two substrates is not an especially inexpensive one. There are exceptions to all rules, and this one is no different. If the filters are especially difficult to make, i.e., the edge filters have to be precisely positioned, then it may be less expensive to coat separate substrates to obtain adequate yields. Additionally, the use of two substrates allows the use of the angle shift to fine tune each edge separately.

Putting both filters on the same side of the substrate is theoretically the least costly, as only one surface needs to be cleaned, and only one vacuum cycle is required to complete the filter (if an AR coating is not required on the second surface). If the substrate is thin and stresses are a problem, this approach may result in a warped substrate, and the prudent approach is to coat the two-edge filters on opposite sides of the same substrate.

A bandpass filter made of two edge filters has the highest transmission characteristics because the two halves of the filter can be designed independently to have good transmission. Transmission values of close to 100 percent can be expected. On the other hand, the width of the filter cannot be made very narrow.

4.4.2. Fabry–Perot Filter

The classical Fabry–Perot filter, frequently called an etalon, consists of two plane parallel surfaces with silver reflecting coatings; both of these surfaces require a slight amount of transmittance. These mirrors are separated by a small air space. Only those wavelengths that are approximately integral multiples of the air gap spacing are transmitted by this element. To function properly, the surfaces must be parallel to very tight tolerances, and therefore, are difficult to adjust. A *solid* version of this etalon can be made by replacing the air gap with a transparent thin film. Although such a device is no longer adjustable as the *gap* between the two reflectors is not variable, it does retain the filter performance of its predecessor and is much more durable.

Table 4.2 gives the prescription for a thin film (solid) Fabry–Perot filter and Fig. 4.20 shows its performance. The mirror layers consist of silver films that absorb some light; thus, the transmission does not reach 100 percent in the pass band. Away from the pass band, the absorption is reduced as can be seen in Fig. 4.21.

The width of the pass band in a Fabry–Perot filter is given by

$$\gamma = \frac{2}{\sqrt{\psi}}$$

where

$$\psi = 4\frac{R_f}{(1 - R_f)^2}$$

The reflectance of the two cavity mirrors, R_f, is assumed to be equal; the transmission through these mirrors is represented by T_f. The peak transmission through the etalon is

$$T_{\max} = \left(\frac{T_f}{1 - R_f}\right)^2.$$

A final parameter that is useful with this type of filter is the finesse, which is the ratio of the separation between the peaks to the width of the pass band. This is given by

$$F = \frac{\pi\sqrt{\psi}}{2}.$$

A more complete discussion of Fabry–Perot filters is given in Chap. 16 of Stone (1963) and Chap. 7 of Born and Wolf (1970).

Table 4.2 Design of a Solid Fabry–Perot Filter*

Incident Medium:	$n = 1.0$	
Silver ($n = 0.05$, $k = 3.0$)		15 nm
Spacer ($n = 2.35$)		511 nm
Silver ($n = 0.05$, $k = 3.0$)		15 nm
Substrate:	$n = 1.52$	

*The performance is shown in Fig. 4.20.

4.4.2.1. Variable Spacer Designs. An air spaced Fabry–Perot filter can be used as a tunable filter by varying the spacing or index between the mirrors. Piezoelectric transducers can be used to vary the mirror spacing smoothly, thus eliminating vibration-induced manual adjustments. The index can be varied by changing the gas mixture or the gas pressure between the mirrors.

The spacing between orders can be increased by using two cavities in tandem. These cavities must not be coupled; i.e., there cannot be multiple reflections between

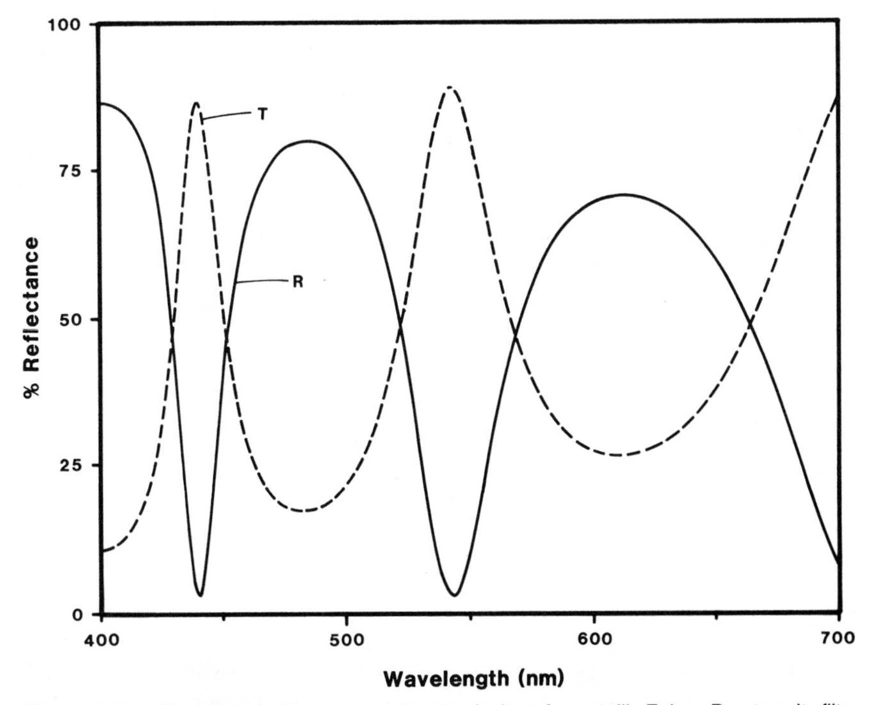

Figure 4.20. Theoretical reflectance and transmission of a metallic Fabry–Perot cavity filter as a function of wavelength. The peaks do not correspond exactly with the optical thickness of the spacer layer because of the effect of phase shifts that occur at the spacer/metal interface.

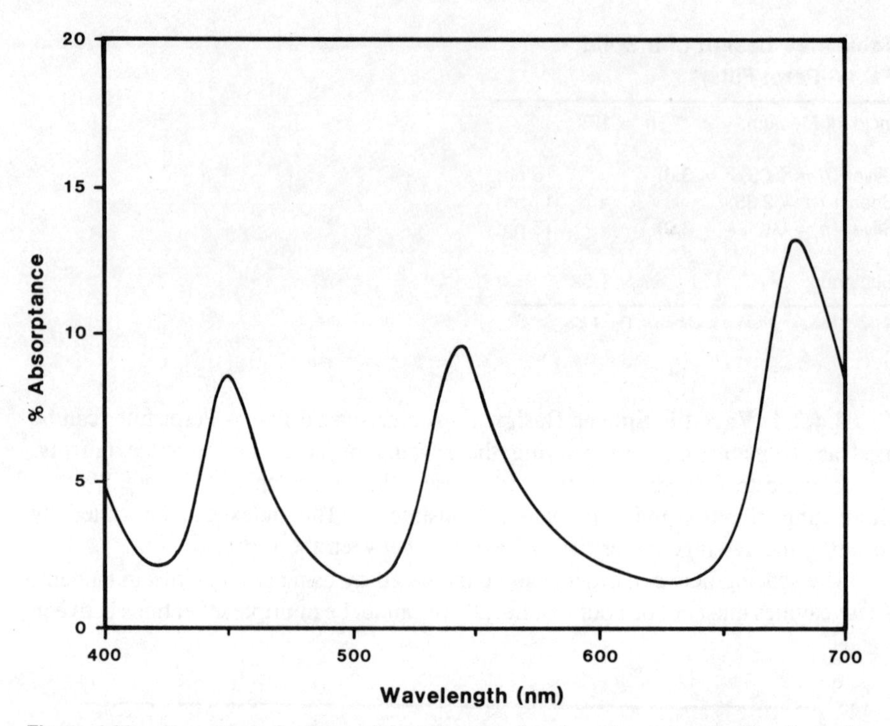

Figure 4.21. Absorption of a Fabry–Perot filter as a function of wavelength. The design is the same as that used in Fig. 4.20. The absorption takes place at those wavelengths where the transmission is high as it is at these wavelengths that the electric fields penetrate into the coating to the absorbing (metallic) layers.

a mirror in one cavity and a mirror in the other cavity. This can be easily obtained by introducing a slight tilt between the two etalons. If a pass band of one is chosen to be at the same wavelength as the other, and they have different cavity spacings, then many of the other harmonics do not coincide and are blocked by one filter or the other.

The shape of a Fabry–Perot pass band has a roughly triangular shape; the exact shape for normal incidence is given by the Airy formula:

$$T = \frac{T_{\max}}{1 + \psi \sin^2 \varepsilon},$$

where

$$\varepsilon = \phi + \frac{nd}{\lambda_0},$$

ϕ is the phase shift upon reflection from the metal mirror surface, d is the mirror spacing, n is the index of the spacer medium, and λ_o is the wavelength. The use of thicker gaps only makes the width narrower without changing the overall shape very much. As the width is narrowed, the peak transmittance decreases also.

Silver is the only practical metal to use to make Fabry–Perot etalons for use in the visible portion of the spectrum. All other commonly used metallic thin films have too much absorption. In the infrared, gold and a few other metals can be used. All of the metals that make good, low loss reflectors are delicate and easily damaged. This mirror problem, coupled with the alignment sensitivity problem, make air spaced Fabry–Perot filters with metal mirrors useful in only a few special applications.

To circumvent the problems that one encounters with metal mirrors, all dielectric mirror coatings have been developed to replace the thin film metal coatings. Several benefits arise from this: (1) higher reflectivities can be achieved; (2) etalons can be made that can operate in spectral regions where there are no good metal film reflectors, such as in the near UV; and (3) the dielectric films are much more durable than the metal ones and can be cleaned without damage. Of course, the same limitations that were described in the section on the bandwidth of dielectric mirrors apply here also. If a wide band of wavelengths is to be covered, then several stacks are required to get the reflectance to a high value over the entire band.

A problem common to all real etalons, whether they use metal or dielectric mirrors, is dispersion in the optical constants. Over a small portion of the spectrum the wavelength is proportional to the space between the reflectors. As the dispersion in the materials and designs vary, different phase shifts are introduced at different wavelengths, so that the wavelength that fits exactly in the cavity (resonance) is a function of the phase shift as well as the mirror spacing. Typically, the only variable that is considered when one uses a Fabry–Perot filter is the cavity spacing. In many instances, this is adequate as they are used only over a small wavelength range. If a large range is to be covered, then calibration of the wavelength versus spacing function should be undertaken at known wavelengths over the entire spectral range of interest.

4.4.2.2. Solid Fabry–Perot Filters. As mentioned in the previous section, the spacer medium between the mirrors in a Fabry–Perot filter does not have to be air. If the gap is a thin solid film, the performance will be the same with the exception that the index of the gap is changed. In particular, this has an effect on the angle sensitivity of the filter. The higher index the spacer layer is, the less the angle shift, as shown in Fig. 4.22. Of course, these filters do not have the normal incidence tunability that an air gap Fabry–Perot filter has, but the filters can be tuned over a narrow range by tilting them. If such a fine tuning technique is to be employed in a filter, it is necessary that the normal incidence pass band be at a wavelength longer than that at which one desires to operate the filter, as the angle shift is only toward the blue.

If more than one Fabry–Perot cavity filter is to be used in tandem, they must be placed on separate, nonparallel surfaces or different optical elements so they do not add coherently. If they are placed on the same surface, they will interact by forming an additional uncontrolled Fabry–Perot filter in which the substrate thickness serves as the cavity spacer.

When narrow band filters are used, the coherence of the light increases. This can be a source of difficulty if the light is sufficiently coherent that the thickness of the substrate is less than the coherence length of the light. In such a case, even coatings on opposite sides of a substrate can interact in a coherent manner and give

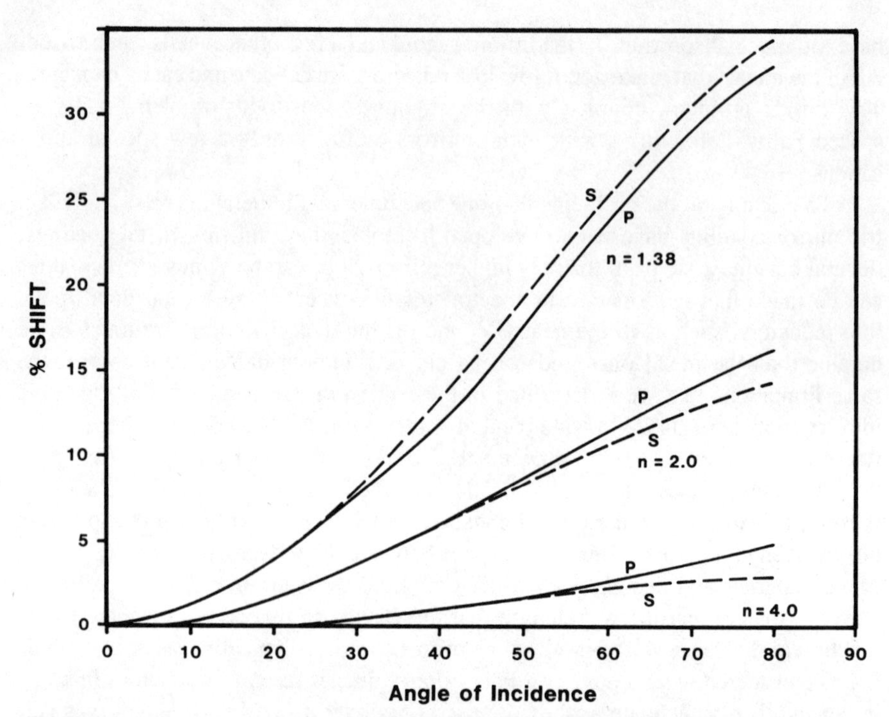

Figure 4.22. Peak transmission wavelength shift (given as a percentage of the peak wavelength at normal incidence) of a Fabry–Perot filter as a function of angle of incidence. The shift is illustrated for a number of spacer layer indices. The index of the metal reflector layers is (0.05. 3.0). The direction of the split of the peak positions for the two planes of polarization depends on the index of the spacer layer and that of the reflector metal.

rise to unexpected effects. Thus, it is prudent, when one is using very narrow bandpass filters, to treat the system as if it were illuminated by a laser, and to use precautions similar to what would be used to eliminate interference effects in such a system.

4.4.3. Narrow-Band Dielectric Filters

Narrow-band dielectric filters resemble solid Fabry–Perot filters, however, the shape of the pass band is much different. These filters have squarer tops to their transmission band, thus allowing more energy in the pass band through the filter. Design techniques allow the shape to be squarer still if three materials are used in constructing the filter. Figure 4.23 shows the performance of an infrared square bandpass design; a comparable simple thin film Fabry–Perot filter is also included in the illustration for comparison. This type of design is ideal for an application where maximum energy in a band is to be transmitted and maximum rejection of energy in an adjacent band is desired. This contrasts with the Fabry–Perot design that may be a good choice for isolating an atomic emission line, for example, and where the emission line shape is the dominant factor in determining the filter's performance.

The design of these square-topped filters generally consist of coupled Fabry–Perot cavities, with 3, 5, or more cavities in a single filter. The cavities are at least one, and frequently more than one, half wave thick. The remaining layers are usually all a quarter wave thick at the center wavelength. By incorporating a third material with an appropriate index into the design, the *super square* types of filters can be achieved. The details of the methods used to arrive at these designs is beyond the scope of this book (see Jacobs (1981) for more information).

Square bandpass filters are sensitive to the angle of incidence similar to all other dielectric interference filters. The performance shifts to shorter wavelengths as the angle of incidence increases from normal incidence. To minimize this effect, the designer usually selects a high index of refraction material for the spacer layer, as this is the layer that affects the shift the most.

Several parameters are required to describe the performance of narrow bandpass filters. In addition to the transmission at some wavelength, it is common to specify the width of the filter at specific transmission levels, such as the width at the 10 percent points, or the full width at half maximum (FWHM). Figure 4.24 shows some common specifications.

The standing wave electric fields set up in these filters by the incident light can become quite high. Figure 4.25 shows the standing wave fields for a three-cavity filter. If there is some absorption in the layers with high electric fields, the

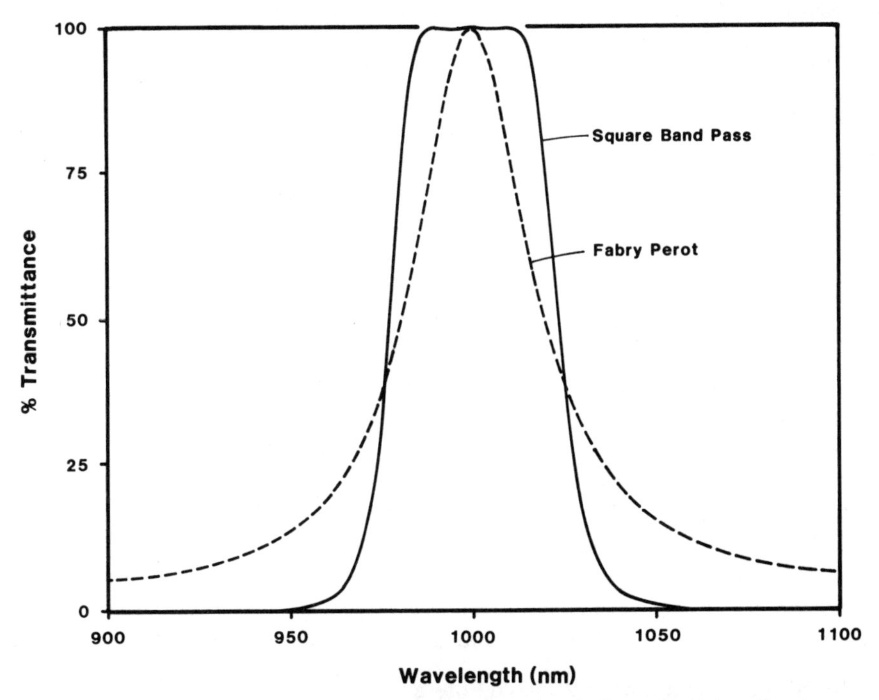

Figure 4.23. Transmission of a square bandpass and a single cavity Fabry–Perot as a function of wavelength. The design of the square bandpass filter is $(LH)^2\ L^2\ (HL)^4\ (LH)^4\ L^2\ (HL)^2$ and that of the Fabry–Perot is $(HL)^2\ L^2\ (LH)^2$. The indices of the high and low layers are 3.5 and 2.0, respectively.

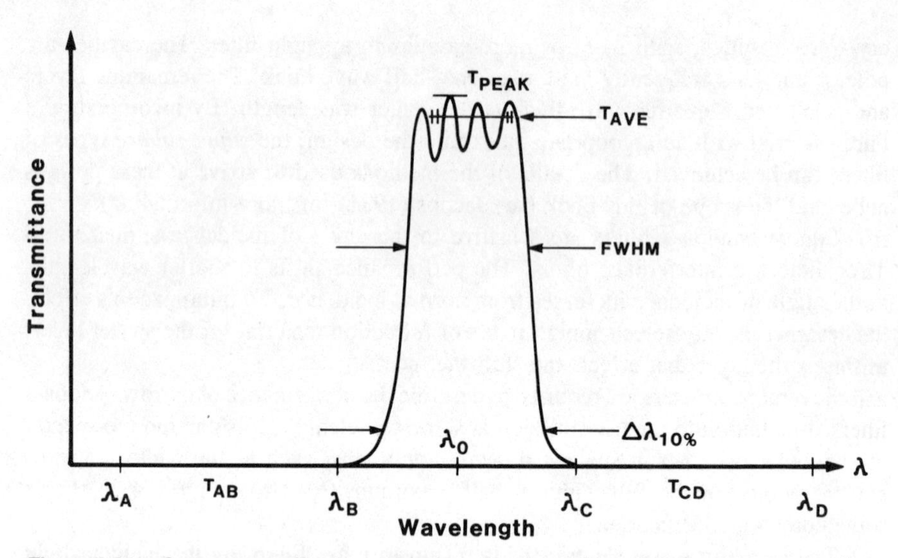

Figure 4.24. Common specification points for a bandpass filter. Not all points are usually specified. T_{AB} is the blocking range on the short wavelength side of the pass band, whereas T_{CD} is that for the long wavelength side. The center wavelength, λ_0, can be specified on the basis of an arithmetic average of points on the slopes of the filter. The result will be different if the abscissa is a wavelength or a wavenumber scale. The width is typically given as the full width at half of the maximum value (FWHM). The steepness of the slope can be specified by the width at the 10 percent point ($\Delta\ \lambda_{10\%}$).

total filter absorption becomes quite large owing to the amplification of the absorption by the field (Eq. 3.2). This absorption reduces the peak transmission significantly, as shown in Fig. 4.26, where the peak transmission is plotted against the value of the absorption coefficient for the previously mentioned three-cavity filter.

The high fields can be used to determine the absorption coefficient in low index materials. A Fabry–Perot filter is coated and the material to be evaluated serves as the spacer layer. By varying the spacer thickness and measuring the height of the transmission peak, the absorption in the spacer layer can be determined (see Southwell, 1982).

Because the performance of the filters is a result of thickness-sensitive spacer layers, small changes to these layers can have a significant effect on the overall performance. For example, if we assume that there will be a 2 percent error in thickness in the coated spacer layer, the pass band of the resulting filter can vary greatly as shown in Fig. 4.27. If the spacer layer is prone to absorb water, which in turn changes its optical thickness, then, at wavelengths near the pass band, the transmission will be a function of the relative humidity (Gee et al., 1985).

4.4.4. Circular and Linear Variable Filters

Although the solid Fabry–Perot filter is not adjustable, except over a fairly narrow range by angle or by temperature tuning, it is possible to use a wedged film configuration to obtain a pass band that varies as a function of the incident beam's position on the surface of the filter. Such filters have been fabricated in both linear and circular geometries. The pass band generally has the shape of the classical Fabry–Perot filter.

A linear version of this type of filter was outlined in a paper by Wirtenson and Flint (1976). The structure of such a film is shown in Fig. 4.28. As the distance from a reference edge increases, the thicknesses of the layers increase as well. Thus, the center wavelength of the pass band varies with position without changing its shape.

Another variation on the same theme is the circular variable filter (CVF). In this geometry, the wavelength transmitted by the filter varies as a function of the azimuthal angle through which the filter is rotated. Complex rotating fixtures allow linear thickness variations as a function of filter azimuth angle (Apfel, 1965; Illsley et al., 1969; Thelen, 1965). By proper gearing, the mask and the substrate rotate at precise speed ratios with respect to each other. This enables the formation of a well-defined wedge for the spacer layer. The transmission of such a CVF is shown in Fig. 4.29. These can be made in visible or infrared versions and find uses as

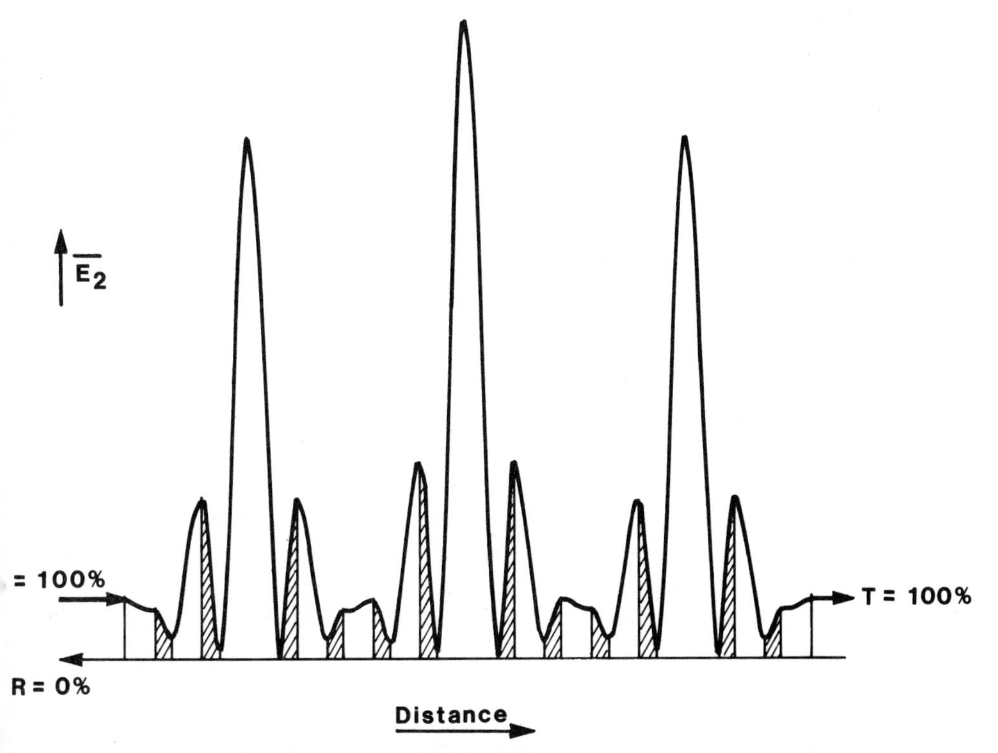

Figure 4.25. The time average of the square of the electric field in a 25-layer, 3-cavity square bandpass filter at the center wavelength of the pass band. The spacer layers consist of the low index material. The high fields in the spacer layers make the design very sensitive to any absorption in these layers. The high index layers are shown hatched.

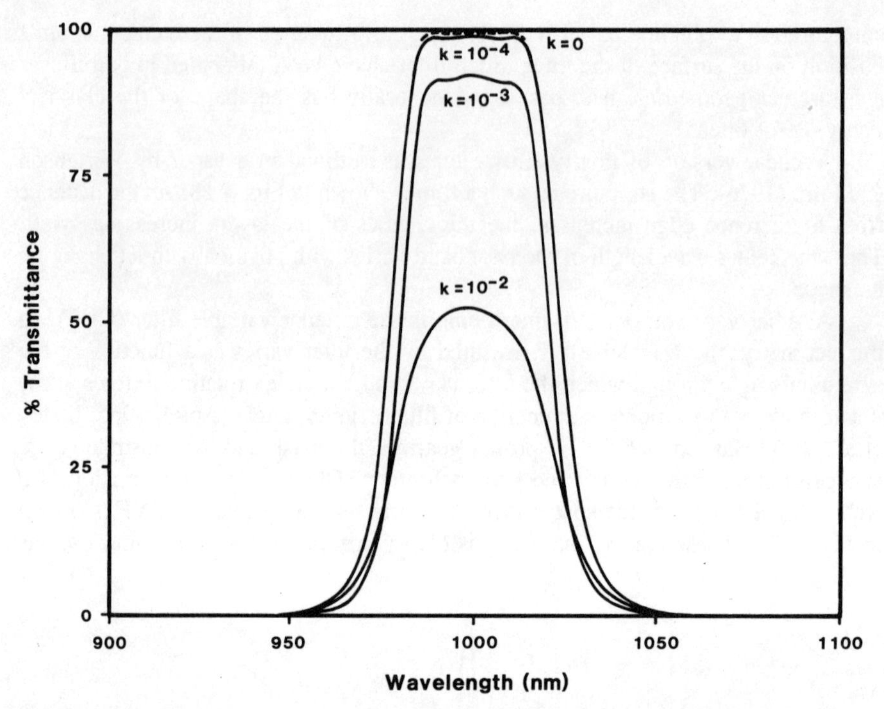

Figure 4.26. Transmission versus wavelength for the 3-cavity filter shown in Fig. 4.25. The k values were introduced into the low index layers only.

wavelength selecting elements in field instruments, such as pollution monitors, where portability and ruggedness are prime requirements. When silver is used for the mirror layers, the CVF is frequently laminated with a cover glass and sealed to improve the environmental durability.

4.4.5. Reflex or Phase Conjugate Filter

The manufacture of a narrow bandpass filter can be quite difficult. For example, if a filter with a pass band tolerance of 0.25 percent is desired, then the thickness of the spacer layer must be controlled to this accuracy. This is quite difficult to achieve in practice, and very narrow bandpass filters are frequently obtained on the basis of an individual selection process, or are made singly on a custom basis with the performance of the filter monitored in real time during the coating process to ensure accurate placement of the transmission band.

In contrast, the phase conjugate filter design allows the pass band deviation to be determined on the basis of the *average* error in the entire coating, rather than the absolute error in a single layer. Thus, it is much more likely that the phase conjugate filter will be more successful than the standard approach.

The design of a phase conjugate filter was published by Austin (1970). Its prescription is given in Table 4.3, and its performance is shown in Fig. 4.30. Figure 4.31 shows the variation in performance expected for a standard Fabry–Perot type

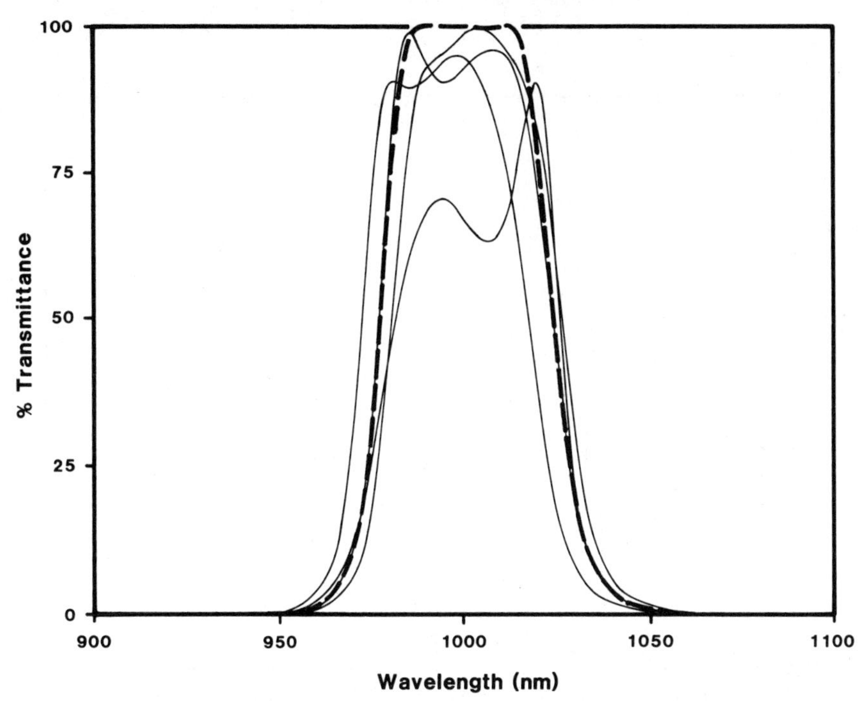

Figure 4.27. Effect on bandpass shape caused by thickness variations in the three half wave layers. The design is the same as in Fig. 4.25. The standard deviation of the thickness variations is 2 percent. The dashed curve shows the nominal performance.

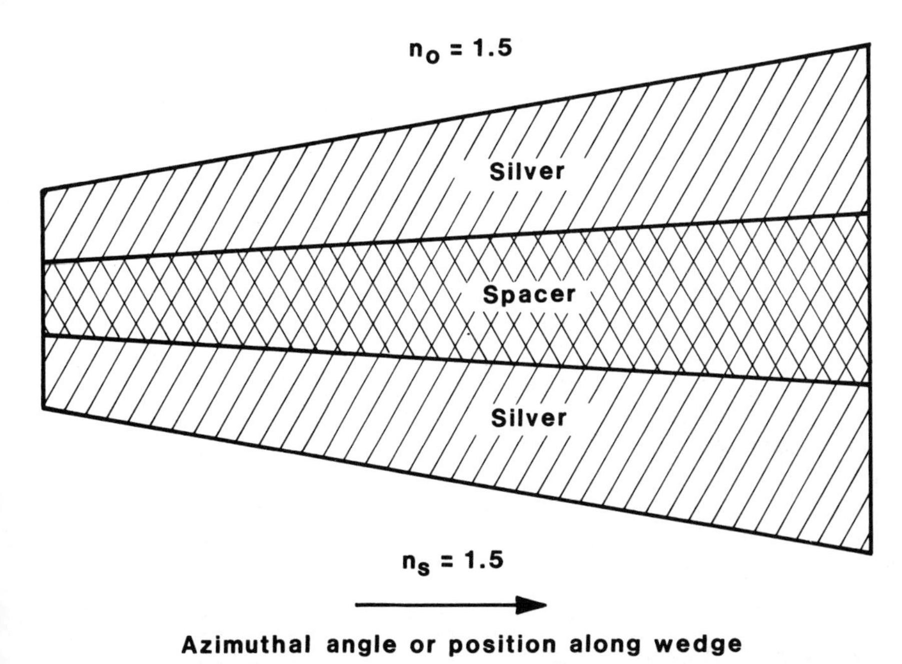

Figure 4.28. Cross-section of a circular variable filter (CVF) or a linear variable filter. The vertical scale is exaggerated for clarity. All layers have some degree of thickness gradation. These filters are often cemented between two glass plates for durability.

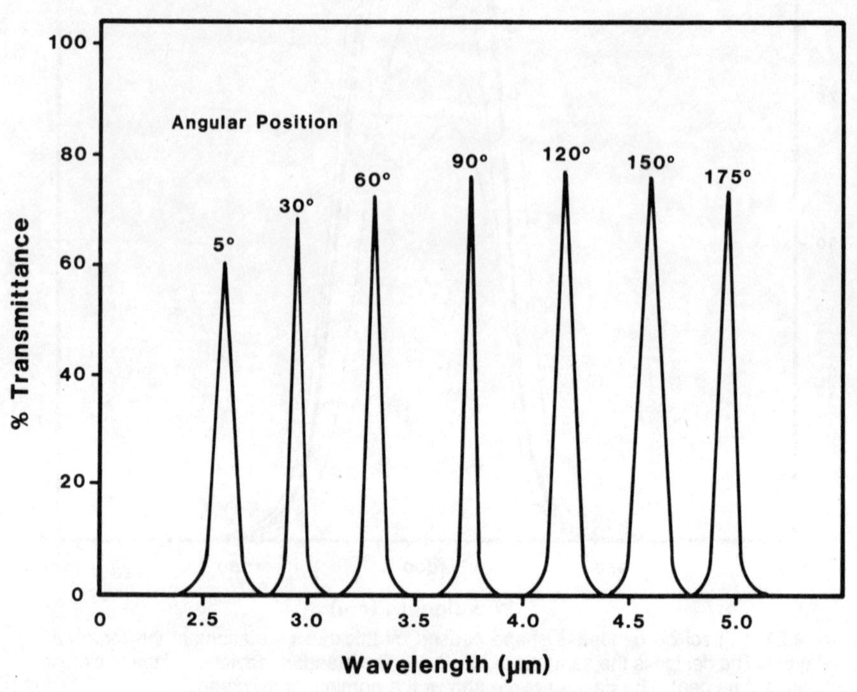

Figure 4.29. Transmission versus wavelength plots for a CVF. The wavelength position of the pass band depends on the azimuthal rotation or position along the wedge.

of narrow bandpass filter when a 2 percent standard deviation Gaussian distribution of thickness deviations is applied to the thickness of each layer (including the spacer layer). Figure 4.31 also shows the same thickness variation factors applied to the layers of a comparable phase conjugate type of design. It is apparent that the phase conjugate design has superior stability as compared with the traditional design approach. A natural characteristic of these designs is to have an extremely narrow (but stable) pass band. It appears difficult to adjust the bandwidths of these filters to an arbitrarily wide value for a given index ratio of coating materials.

Table 4.3 Design of a Phase Conjugate Narrow Bandpass Filter*

$n_0 = 1.00$
$n_L = 1.45$
$n_H = 2.00$
$n_s = 1.00$

$$n_0(HL)^3\left(\frac{H}{2}L\frac{H}{2}\right)\left(\frac{L}{2}H\frac{L}{2}\right)(LH)^3 n_s$$

*The performance is shown in Fig. 4.30.

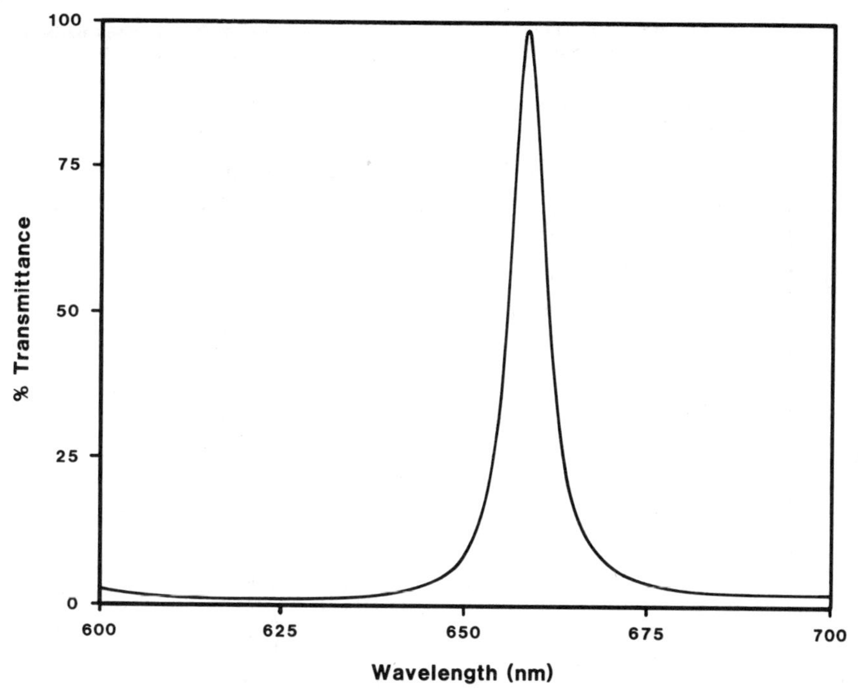

Figure 4.30. Transmission of a phase conjugate filter as a function of wavelength. The prescription for this design is given in Table 4.3.

4.5. BEAM SPLITTERS

4.5.1. Wide-Band or Neutral Density Beam Splitters

Beam splitters divide a ray of light into two beams. The performance of wide band or neutral density designs is relatively independent of wavelength when compared with that of dichroic types that have a strong wavelength dependence (see section 4.5.2.). Beam splitters are available in two configurations. The simpler one has the coating exposed to air, whereas in the other, the coating is immersed in a medium with an index close to that of the substrate. The latter type is frequently in a cube geometry with the coating deposited on a face diagonal surface. Figure 4.32 shows typical configurations of both types.

Each type of beam splitter has its advantages. The angle of incidence can be easily varied and only one surface needs an antireflection coating when a plane parallel substrate with a beam splitting coating on each side is used. This type of design is relatively inexpensive when compared with a cube type. In the cube configuration, the system is symmetric, and this can be an advantage in optical systems where (1) the path lengths need to be matched or (2) the light returns through the beam splitter and symmetric splitting is desired. In addition, the coating is protected against the environment by the cement. Up to four surfaces may require an AR coating in a cube design.

Figures 4.33a and b show beam splitters designed for 45 degrees, which

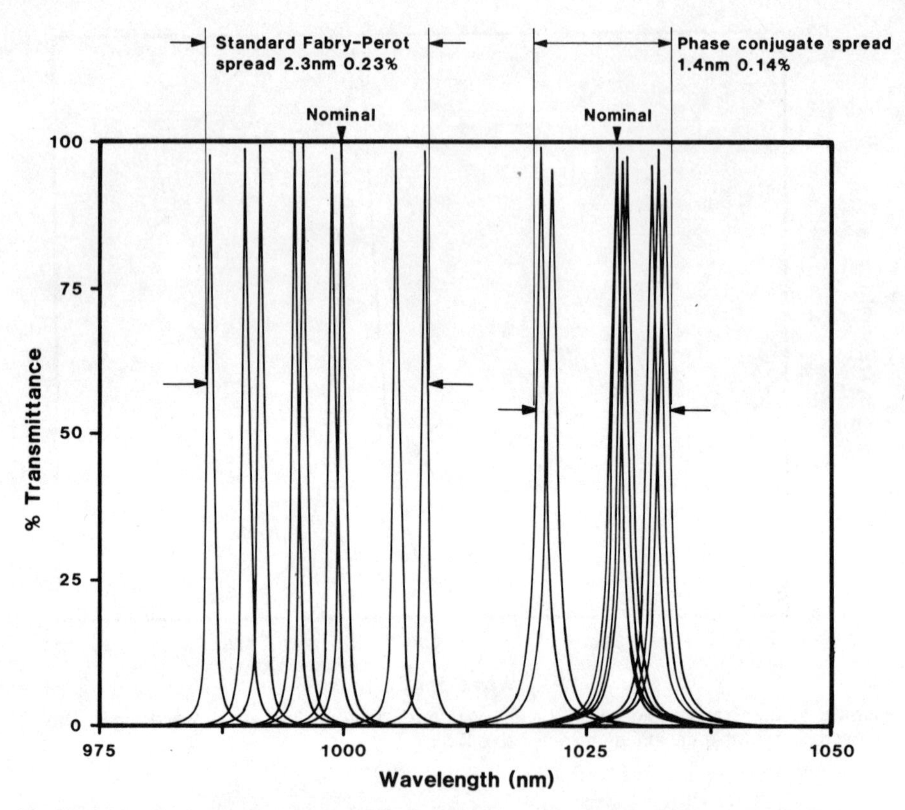

Figure 4.31. Performance variation of two types of bandpass filters when random errors are present in the construction of the filter. The standard deviation of the thickness errors for both designs is 2 percent. The center wavelengths of the two designs were slightly offset so that the performance of the two could be displayed together without overlap. The width of the spread of bandpass centers is about 0.23 percent for the Fabry–Perot design, and 0.14 percent for the phase conjugate one.

provides for orthogonal reflected and transmitted beams. Other angles can be used as long as the coating is designed properly. Each design type is discussed separately later.

4.5.1.1. Slab Beam Splitter. Because the surfaces and coatings of these beam splitters are usually oriented such that the light is incident at nonnormal angles, polarization parameters of the optical system need to be considered. Figure 4.33 shows the theoretical performance of a 45-degree beam splitter with equal reflectance and transmittance at about 550 nm for the *average* of the two planes of polarization. The coating could consist of a single layer of metal such as silver. At this wavelength, however, the reflectance does not equal the transmittance for either plane of polarization. The reflected and the transmitted beams in the p plane are equal at about 455 nm, whereas the intensity of the reflected s plane beam is almost 2.5 times larger than the transmitted one at that wavelength. At 690 nm, the two s plane beams are equal, whereas in the p plane, the transmitted component has about 2.5 times the intensity of the reflected one. These unequal reflectances can cause difficulties when a laser with a randomly polarized output is directed at such a device. The beam splitter will convert the fluctuations in the plane of polarization into intensity noise.

Beam splitters are generally polarizers, and therefore, need to be used with care in an optical system. Multiple polarizing beam splitters in a system with no other polarizers can lead to difficulties as the polarizing components are not explicitly identified.

The coating used in beam splitters can be either a dielectric stack or a metal layer. Most metal films can be used as beam splitter coatings, though some have specific properties that are better than others. Figures 4.34 and 4.35 show the performance of typical metal and dielectric designs, respectively. They both have approximately the same (relatively low) reflectance at normal incidence at 500 nm. The metal coating has the flatter curve, indicating that it will function well over a

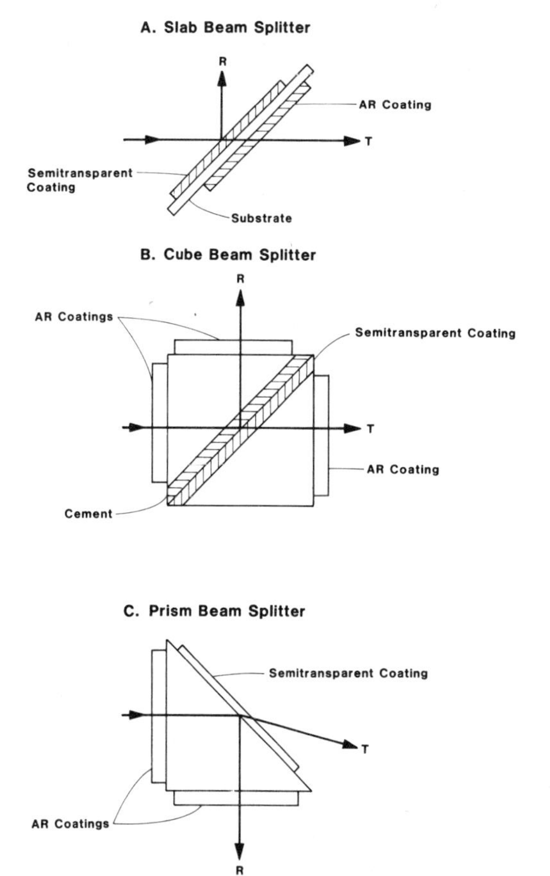

Figure 4.32. Beam splitter configurations. The types shown in illustrations A and B can produce orthogonal incident, transmitted, and reflected beams. Type C cannot produce rectilinear beams because of refractive effects at the exit surface of the transmitted beam.

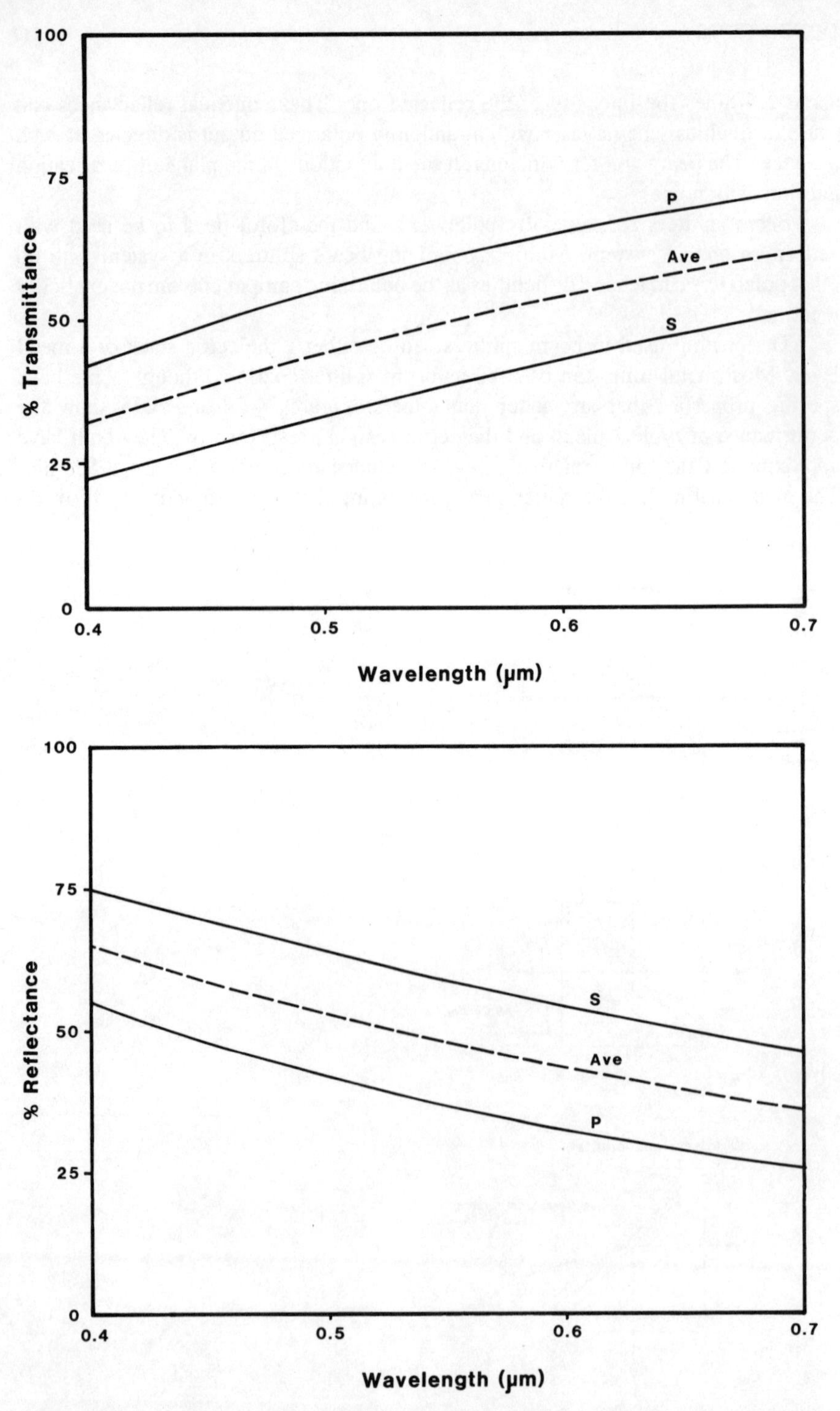

Figure 4.33. Performance of a silver film slab beam splitter as a function of wavelength. The transmittance (a) and reflectance (b) cannot be equal for the two planes of polarization at the same wavelength. The angle of incidence is 45 degrees.

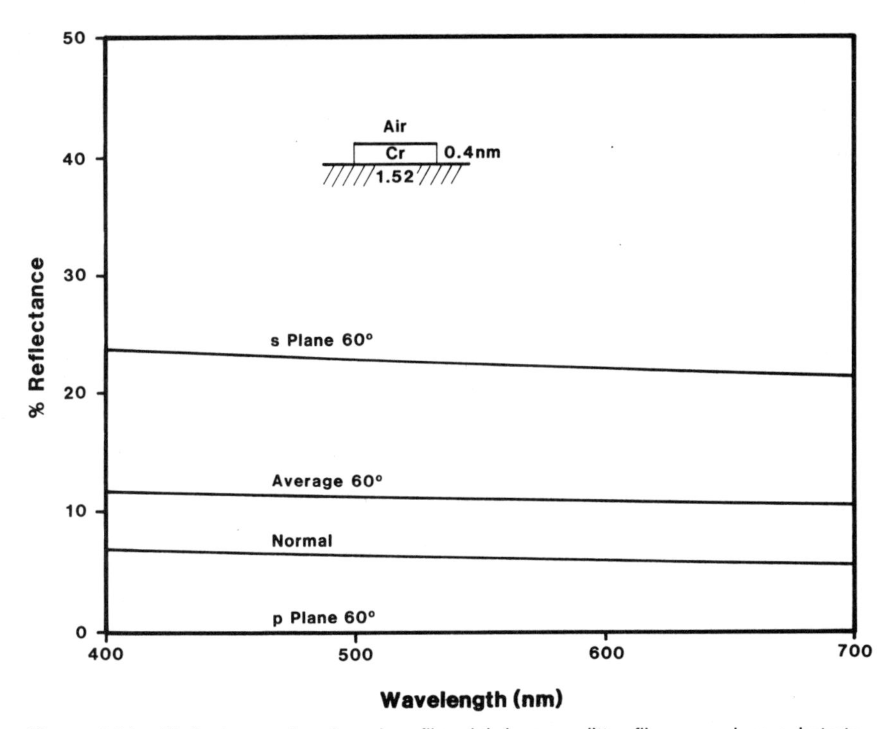

Figure 4.34. Reflectance of a chromium film slab beam splitter film on a glass substrate as a function of wavelength for 0 and 60 degrees angle of incidence. No dispersion is included in the calculation. Note that this design makes a very good polarizer.

broad range of wavelengths. At an angle of incidence of 60 degrees, the performances of the two designs change dramatically. In the metal coating, we find that the *p* plane light reflects poorly or not at all, whereas *s* plane light reflects well. Note that in this example the light is incident on the coating at approximately the Brewster angle.

In the dielectric design, the *p* plane is relatively flat, whereas there is significant chromatic variation in the *s* component. In some applications, such as in high-power laser systems, the fact that the dielectric layers absorb no energy is a most desirable characteristic.

A study of the angular performance of slab beams splitters is informative because they may frequently be used at variable angles of incidence. Figures 4.36a and b show the transmittance and the reflectance of a low loss metal beam splitter as a function of the angle of incidence in air. It splits the incident light equally into the reflected and transmitted beams at normal incidence. At angles above 45 degrees, the intensity imbalance between the two beams becomes quite large. At angles greater than 20 degrees, the intensity differences between the two planes of polarization become significant. The losses in this system are small, as shown in Fig. 4.36c.

Figures 4.37a and b show the transmittance and reflectance of another metal beam splitter coating as a function of angle. The coating material used in this design

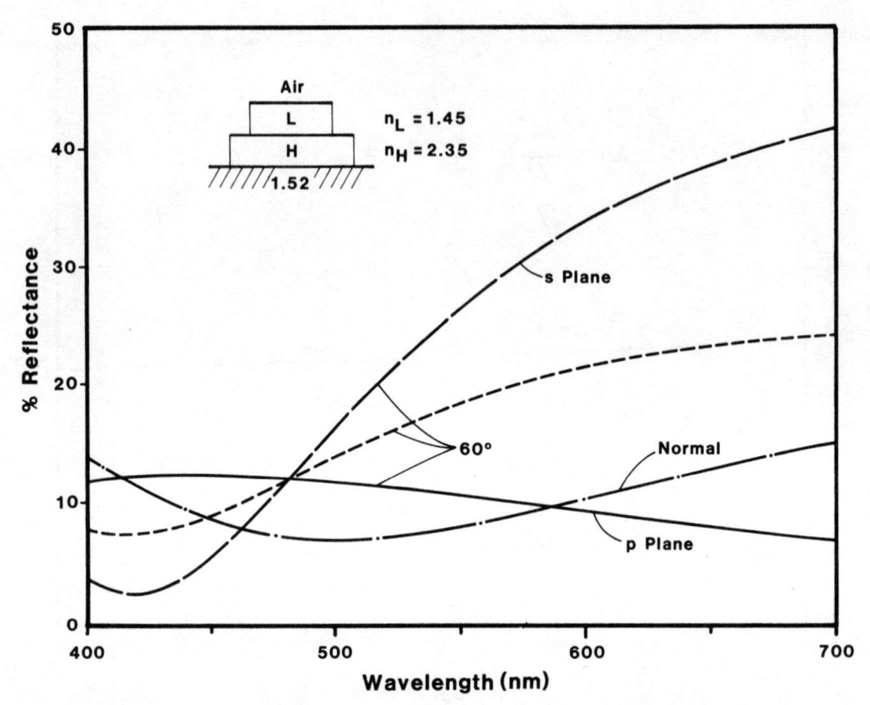

Figure 4.35. Theoretical performance of a dielectric slab beam splitter as a function of wavelength. Both layers in the design are a quarter wave thick at 500 nm. Note the substantial variations in reflectance of the two planes of polarization as the wavelength varies.

is much more absorbing and consequently the value at which the reflected and the transmitted intensities are equal is much lower than that in the previous design. The average reflectance of the two planes of polarization is equal to the average transmittance from normal incidence to about 70 degrees. Figure 4.37c shows the absorptance of this beam splitter. A substantial amount of the incident energy is lost in this beam splitter. If the beam splitter is turned around so that the incident light passes through the substrate before encountering the metal layer, the absorptance increases by 10 percent or more. In addition, the two beams will not have equal intensities when used reversed. Thus, it is important to use these beam splitters so that the light is incident on the metallic coating from the air side.

On occasion, the beam splitter coating is deposited on the hypotenuse of a prism (see Fig. 4.32c). This geometry allows the light to strike the coating at any arbitrary angle. When the coating is a metal, the reflectance and the transmittance as a function of angle of incidence can have unexpected characteristics. Figure 4.38 shows the transmittance as a function of the angle of incidence. At the critical angle, the transmitted intensity is reduced to zero, and no light can leave the prism at angles greater than this angle. The intensity of the reflected beam is quite different, as shown in Fig. 4.39. There is a range of angles between 50 and 60 degrees where there is very high absorption of the *p* plane light. If the prism had no coating on the hypotenuse, the reflectance for both planes of polarization would increase to 100 percent at the critical angle, and it would remain at that value for all larger angles. Such components need to be used carefully.

4.5.1.2. Cube Beam Splitters. By their very nature, these beam splitters are designed to function at a nominal angle of incidence. Many times, optical systems require equal path lengths in both beams to minimize aberrations or other differences between the two beams. This is especially true when the beams are recombined in a second pass through the beam splitter. In those instances, a cube beam splitter is the best solution.

The cube beam splitter is inherently rugged as the coating is not exposed to the environment. The symmetry of the cube ensures that there can be no problem with orientation of the incident light with respect to the beam splitter. The performance of this type of beam splitter is very similar to that of a similar slab design.

Most of the comments that were made about the slab beam splitter also are pertinent to the cube version. The same effects can be observed at high angles of incidence as are seen with the slab, and the polarization effects are similar.

4.5.2. Dichroic and Trichroic Filters

Dichroic and trichroic filters separate wavelength bands, transmitting certain wavelengths while reflecting others. This can be accomplished with edge filters, but since in color splitters they are frequently used at nonnormal angles of incidence, they are considered separately here. Because they are used at angles where polarization can be severe, a primary consideration should be the state of polarization of the incident light beam. If the light is mostly polarized in one plane, then the designing process can at times be simplified over what may be necessary if the incident light's polarization state has to be assumed to be random or unpolarized.

If the incident light is polarized in the p plane and the filter has to operate near Brewster's angle for the interface between the two coating materials, then a severe design problem exists because, as mentioned in section 3.6, the effective index ratio is close to unity in this regime. With no index ratio with which to work, the thin film designer has a very difficult time in conjuring up a filter. During the layout of an optical system, it is useful to keep in mind the difficulties polarization may cause so as to avoid a situation in which the desired filter cannot be built.

The most extensive use of dichroic filters is in the separation of infrared (heat) radiation from visible light so that the reflected beam is *cold;* this is the origin of the term *cold mirror*. Many projector lamps have an integral reflector that has a coating that is transparent to heat and reflects the visible light onto the film or slide.

The converse of the cold mirror coating is one that reflects the infrared wavelengths and transmits the visible light. Such a filter can be included between the lamp and the film plane to reduce further the heat load on the film. Because these mirrors reflect heat, they are sometimes called *hot mirrors* or *heat mirrors*.

Another application of these filters is in the reflector of a medical or dental lamp where (1) the heat could be injurious to tissues and (2) patient comfort is

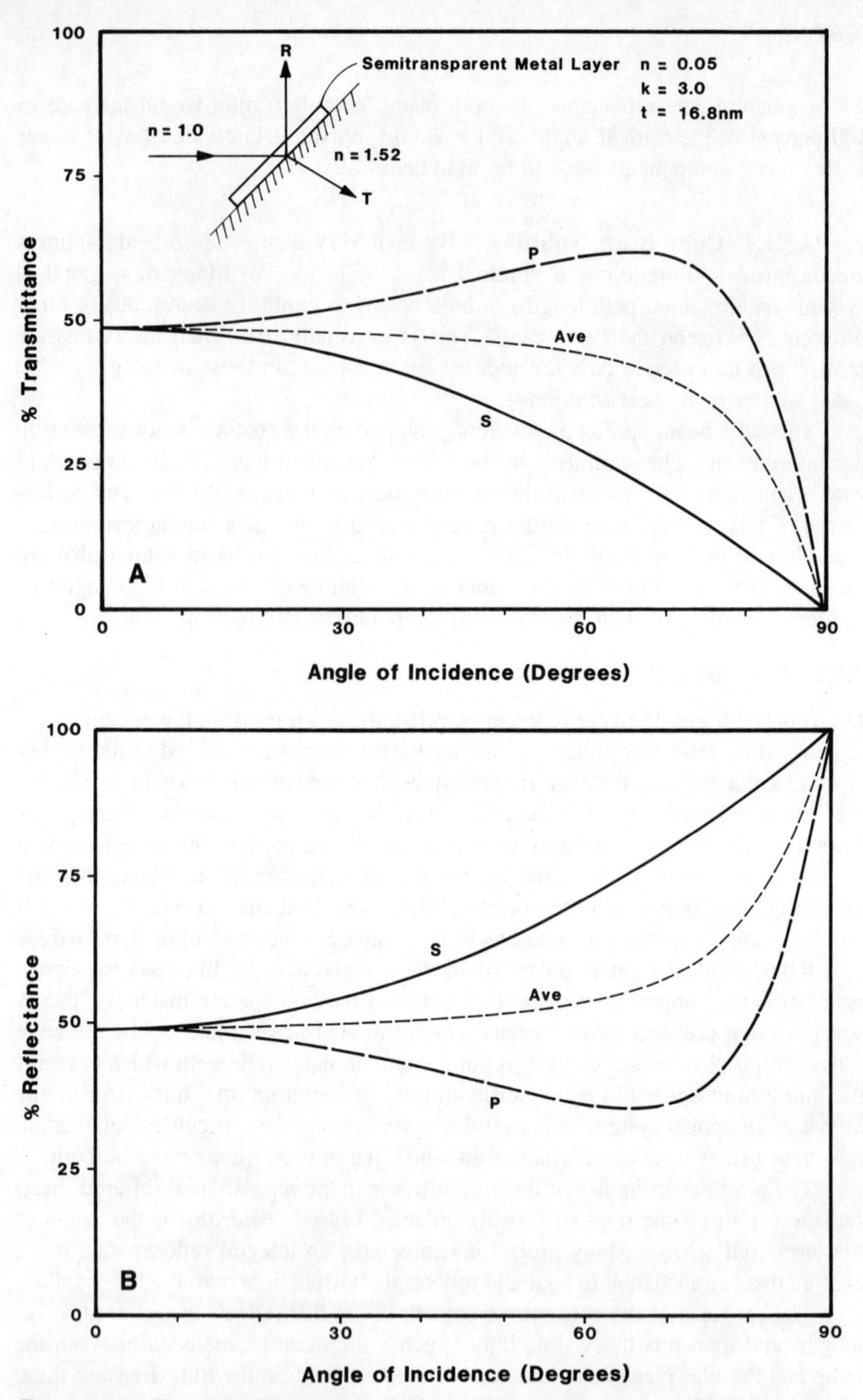

Figure 4.36. The transmittance (A), reflectance (B), and absorptance (C) of a low loss silver slab beam splitter as a function of the angle of incidence in air. The reflectance and transmittance are approximately equal at normal incidence. Note the expanded scale on the absorption plot.

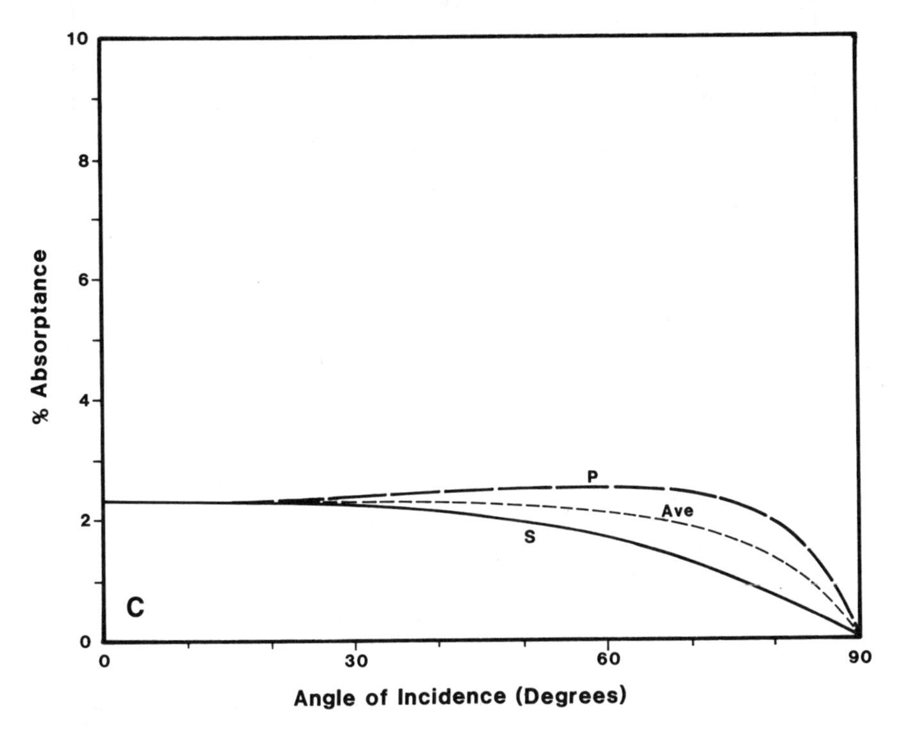

Figure 4.36 (*Continued*)

important. An additional benefit of this type of arrangement is that the apparent color temperature of the light source can be altered so that the reflected light matches sunlight more closely, thus greatly reducing the problems associated with metamerism (see section 4.6.).

4.5.3. Filters for Use in Harmonic Generation Experiments

A crystal lattice can convert photons from a fundamental laser excitation frequency into photons that are multiples of this frequency. This phenomenon occurs when a sufficiently intense excitation energy density drives the electrical field strength into a nonlinear portion of the dielectric function. Certain crystalline materials and proper experimental conditions can efficiently generate second, third, and higher harmonics of the fundamental wavelength. Interference filters are required in these harmonic generators to separate the various harmonics in the output beam. Specifically, the detectors used for some low energy content harmonics can be very sensitive to other, high power, harmonics. Thus, the wavelength separation and blocking become very important. To obtain signals for a wavelength without interfering signals from other wavelengths, the rejection level of a filter is critical. Currently, the major application of these filters is in diagnostic subsystems that follow an experimental chamber.

In addition to the directly transmitted light, scattered light may be strong enough

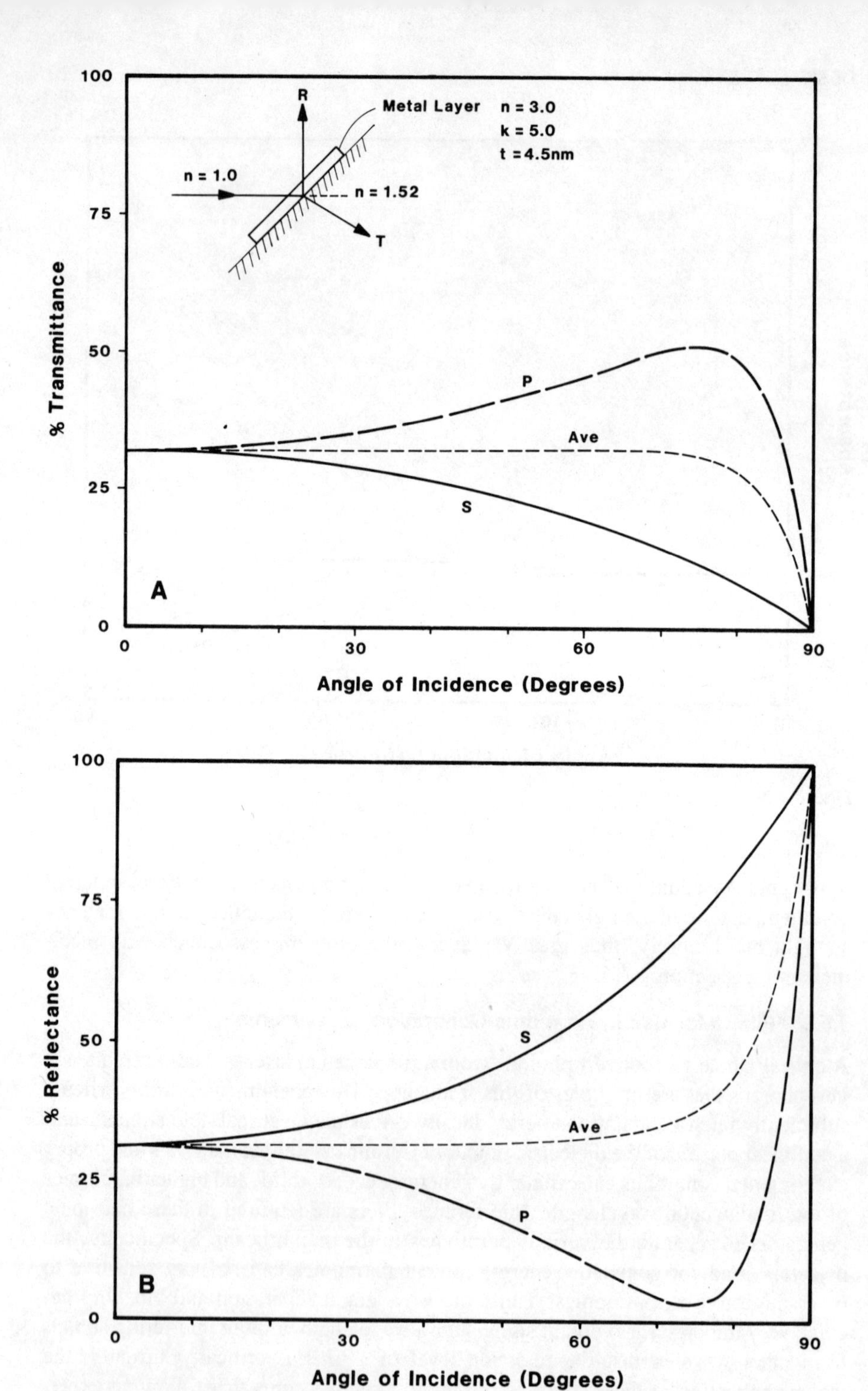

Figure 4.37. The transmittance (A), reflectance (B), and absorptance (C) of a metal (Cr) beam splitter with equal reflectance and transmittance at normal incidence. The absorption losses are much greater than that obtained with a silver coating.

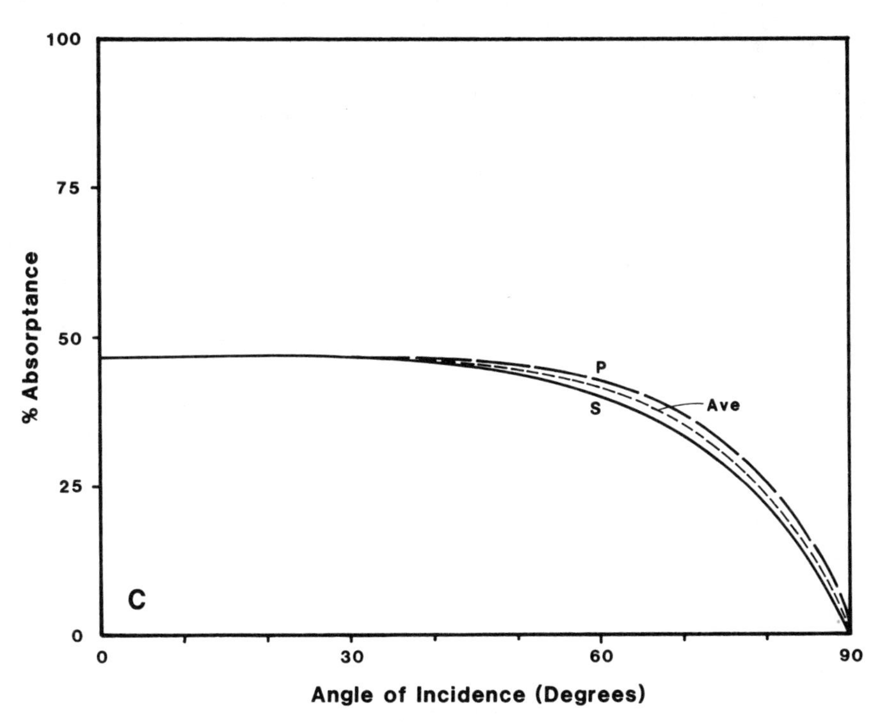

Figure 4.37 (*Continued*)

to swamp a sensitive detector. Therefore, the coatings need to have very low scatter in addition to all of the other requirements.

Typical diagnostic separation filters operate with an angle of incidence in the range of 10 to 40 degrees. Consequently, polarization is a parameter that needs careful attention. The incident light is often polarized as the conversion efficiency of crystals is strongly dependent on the orientation of the polarization vector.

In third harmonic generation, diagnostic filters are typically used in pairs. This makes the experimental setup more flexible and at the same time simplifies the control of the phase retardance introduced by the mirrors. The phase retardance must be controlled when ellipsometers and interferometers are a part of the diagnostic subsystem. The beam-directing mirrors and wavelength separating filters should not modify or add a component to the polarization of the beam.

If a laser with a wavelength different than any of the harmonic wavelengths is used to align the optical system, then there may be some requirements at that wavelength also. The greatest challenge facing the fabricator of these filters is the wide range of wavelengths of interest. Because thin film interference filters depend on the matching of the optical thicknesses of the layers, dispersion in the refractive index makes it difficult to match the thicknesses at several widely separated wavelengths. The lack of a good match reduces the rejection level available at wavelengths away from the one at which the optical thicknesses are matched.

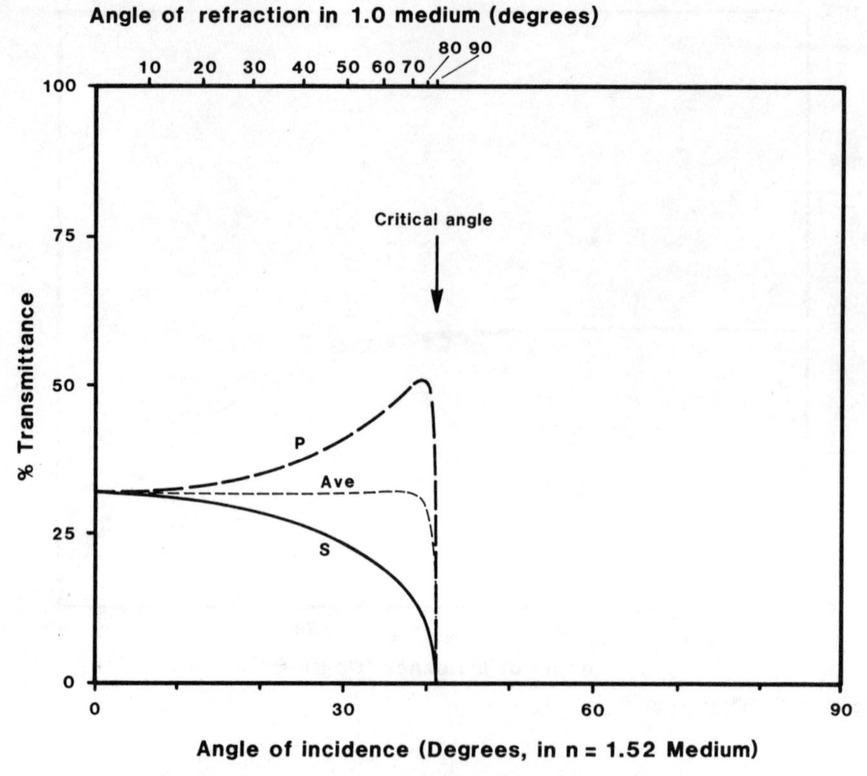

Figure 4.38. The transmittance of a prism beam splitter as a function of the angle of incidence on the hypotenuse from within (*lower scale*) and from outside (*upper scale*) the prism. The index of the prism is 1.52. There is no transmission beyond the critical angle.

Because very high-power densities are used in harmonic generation systems, the damage threshold of these coatings must be high.

The thickness uniformity of the coating must be such that variations in the optical thicknesses of the deposited layers do not introduce an unacceptable performance degradation. This can be caused by an optical mismatch or a shift in pass bands or rejections bands away from the center wavelengths as a function of position over the surface of the filter.

A typical filter set used for third harmonic work may have a transmission of 3 percent at the fundamental frequency, 3 percent of the second harmonic, and 50 percent of the third harmonic.

4.6. OTHER FILTERS

4.6.1. Dark Mirrors, Selective Absorbers, and Induced Absorbers

Dark mirrors and induced absorbers belong to a class of designs in which a primary performance parameter is the absorption of light. The difference between the types is principally a matter of semantics. In dark mirrors, the reflection is low in the visible portion of the spectrum, but is high elsewhere. Because the substrate is generally opaque in the visible, all of the incident energy in this spectral range is absorbed, either in the opaque substrate or within the coating itself. Induced absorbers are a related type of design, but in these there is no high reflecting region. Selective absorbers absorb only in selected spectral regions.

In some applications, there is a need for a coating that completely absorbs the light at one or more wavelengths. These coatings all contain at least one absorbing layer. Additional dielectric layers enhance the amount of light absorbed for a given thickness of absorber. Generally, these coatings have metallic layers that may be susceptible to environmental degradation. It is necessary to use care in designing these coatings into a system because of the inherent potential for their degradation. For example, these coatings should be used on surfaces that are not susceptible to scratching, as such damage leads to light scatter. The dark background of the coating makes the scratched area highly visible because of the very high contrast ratio.

The reflection from any substrate can be reduced to zero at a single wavelength

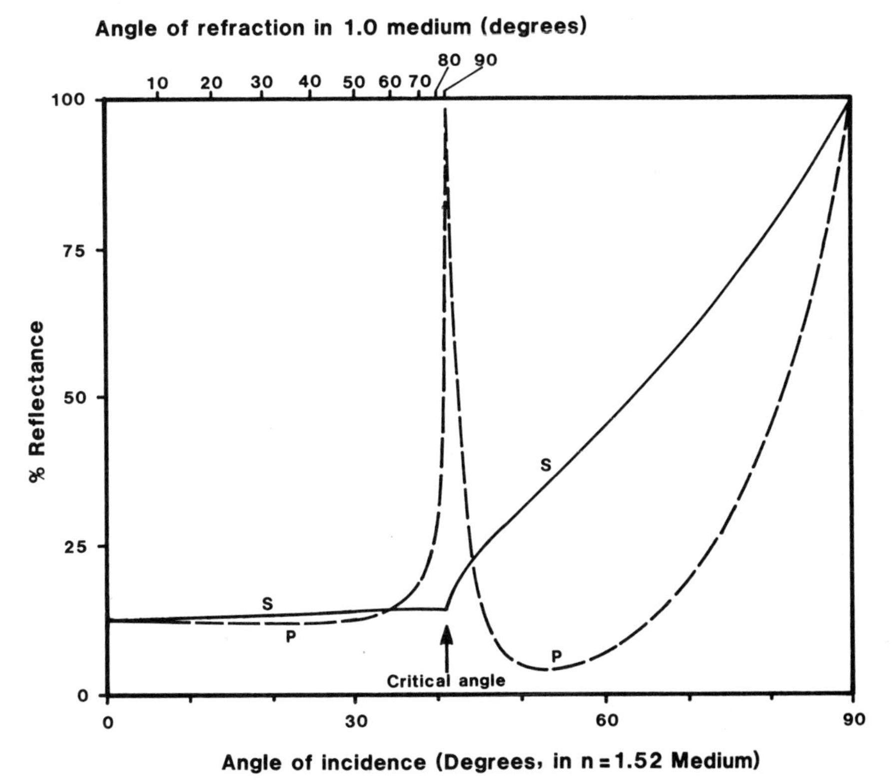

Figure 4.39. The reflectance of the same prism beam splitter shown in Fig. 4.38. The behavior of the reflectance is more complex than that of the transmitted beam. There is a range of angles over which the *p* plane light is more strongly reflected than *s* light (the opposite is more typically the rule). The difference between the reflectance of the two planes of polarization is relatively small from normal incidence up to nearly the critical angle.

with two thin film layers: one layer is an absorber, and the other, a dielectric. The complex index of refraction of the absorbing layer can be any value. For a dielectric substrate, the order of the two layers is straightforward: if the index of the dielectric film is less than the geometric mean of the index of refraction of the incident medium and that of the substrate, then the absorbing layer is located next to the incident medium. Otherwise, the absorbing layer is placed next to the substrate. The latter situation is obviously the more desirable as the metal layer is then protected to a certain extent from environmental degradation. If the substrate is metallic, the selection of the dielectric, and the derivation of the design, is more complicated. These designs are akin to the exact solution one can obtain with the two-layer AR V-coat. Figure 4.40 shows the reflections from two-layer dark mirrors, one using a dielectric with an index greater than the square root of the substrate (the incident medium has an index of unity), and one that is less; the same metal indices are used for the metal layers in both cases.

Frequently, the need exists that a range of wavelengths be absorbed; this requires the addition of more layers, usually in the form of added periods of metal/dielectric pairs.

It is possible to combine designs to obtain certain special effects. For example,

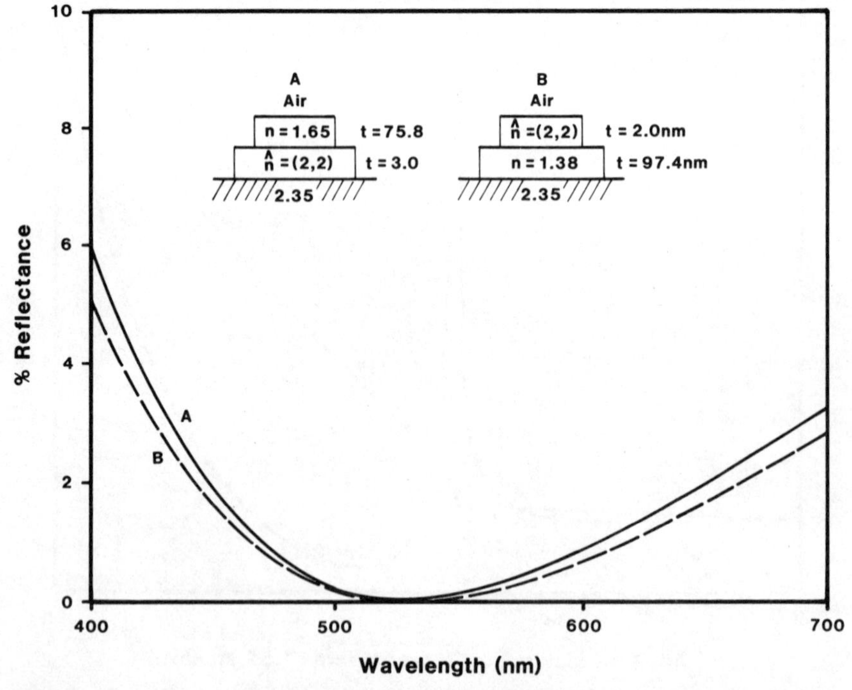

Figure 4.40. Reflectance as a function of wavelength for two-layer dark mirrors. When the index of the dielectric layer is greater than the square root of the substrate, the metal is placed next to the substrate, whereas the reverse is true when the index of the dielectric layer is less than that of the substrate. The designs are given in the figure.

it is possible to combine a dark mirror with an edge filter; the reflectance is then high where the edge filter reflects and very low where it transmits. The transition between the reflecting and the absorbing regions can be very sharp, and can be controlled quite precisely with the edge filter transition.

A good example of a use for an induced absorber is in optical data disk recording media. In write once, read many times (nonerasable) disks, a pulse of heat generated by a focused laser beam forms a hole or other perturbation in the surface of a rotating disk. Because the light needs to be absorbed to generate heat, the recording medium requires a metal or other absorber. Metals have the highest absorption coefficient per unit thickness. This is an advantage in that only a small amount of material needs to be heated, thus minimizing the input energy required to write a data bit. Unfortunately, metal films tend to have a fairly high reflectance, and much of the incident energy is reflected back toward the source, rather than being absorbed to heat the recording material.

Bell and Spong (1978) described a recording medium that had a coating that reflected only enough light to allow the focusing system to function. It consisted of a metallic reflector, a spacer layer approximately one quarter wave thick, and a thin absorber layer. The effect of this structure is to position the absorber near the peak of the standing wave electric field so that the recording layer absorbs the maximum amount of light possible. This is shown in Fig. 3.10. About 95 percent of the absorbed energy is deposited in the recording layer, whereas the rest goes into the reflector. This is a very efficient scheme for absorbing the energy and depositing the heat in precisely the right location. In terms of the energy absorbed per unit mass of absorber film, it is far superior to an earlier system in which a single, relatively thick layer of the absorbing material was used as the recording medium (Rancourt, 1981).

4.6.2. Solar Absorbers

Solar thermal systems collect the sun's heat and make it available for use in thermomechanical systems. The higher the temperature of the working fluid, the more efficient and economical is the overall system. To absorb and keep as much of the solar energy as possible, the surface that absorbs the energy must have some fairly well defined characteristics. Specifically, the absorption spectrum of the surface should be high for sunlight, whereas the emittance at the operating temperature should be as low as possible. Because the solar spectrum and the blackbody emissivity spectrum at the system operating temperature do not overlap significantly, as shown in Fig. 4.41, a coating can theoretically have the desired optimal characteristics as shown by the dotted line in the illustration.

The objective for solar absorbers is to absorb the solar radiation, while losing as little energy as possible through unwanted conduction, convection, and radiation. Heat removal is through the hot fluid outlet port. Usually, the absorbing surface is insulated from convection and conduction losses by a vacuum jacket, therefore radiation is the primary heat loss mechanism.

A first approximation to this coating is perhaps an oxide of indium and tin (ITO) coating on black glass. A better, though more complicated solution, may be

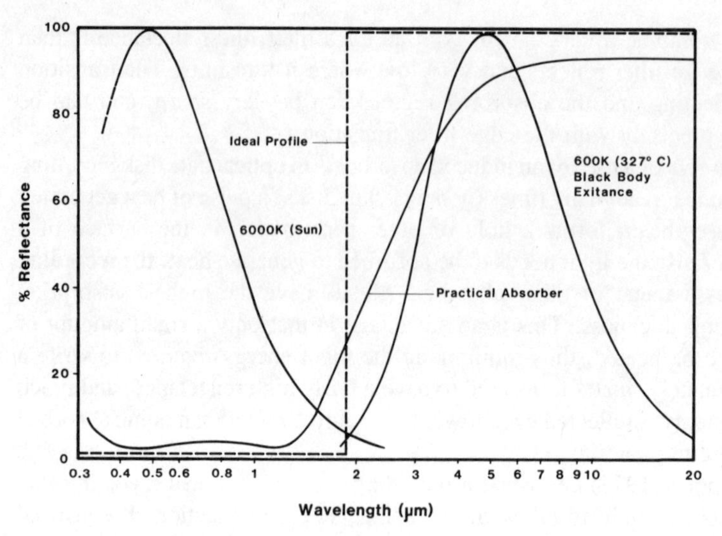

Figure 4.41. Curve for solar absorptance and reflectance of ideal solar collector (from Hahn and Seraphin, 1978).

a thick layer of silicon over a silver or other infrared reflective material. An AR coating may be necessary to reduce the first surface losses at the silicon/vacuum interface. A number of papers have been published on this subject in the solar energy literature. So far, the major stumbling block has been the economical manufacturing of durable, long-lasting coatings so that the overall system costs can compete with the other sources.

4.6.3. Notch Filters

Notch filters have a narrow rejection band within a much wider high transmission band. Such filters are not easily designed since the width of the rejection region is defined by the index ratio of the materials that make up the coating. Narrow rejection zones imply a low index ratio, and this in turn implies a large number of layers to obtain a significant rejection level. In addition, a simple quarter wave stack has side lobes in the transmission region near the main rejection region. Several papers (Baumeister, 1981; Thelen, 1971; Young, 1967) show how the side lobes can be suppressed by varying the thickness of the periods in the stack.

4.6.4. Transparent Conductive Coatings

These coatings are used for a variety of purposes. The more familiar ones are (1) heating glass surfaces for defogging and defrosting, (2) conducting electrical signals for transparent displays such liquid crystal display (LCD) units, (3) providing electrical shielding for electromagnetic or radio frequency interference (EMI/RFI) suppression, and (4) serving as an infrared reflector and a visible transmitter in oven windows.

Because these applications usually call for the coating to be on a transparent substrate that is exposed to all manner of environments, they have to be very durable. In LCD devices, the coating must have certain etching characteristics so that the user can fabricate patterns in the coating to operate the device.

A number of materials are used to provide both transparency and conductivity. Gold and silver are good examples that have been used for a number of years. From a durability standpoint, gold is the preferred choice between these two metals as it is less chemically reactive. Both metals, however, are very soft and are easily scratched unless protected. Other metal and doped semiconductor films can be used, but their performances do not approach that of the best silver films.

In the past few years, a mixed oxide of indium and tin has been used. The exact composition depends on the manufacturing process, and it is usually tailored to the coating's ultimate use. The conduction mechanism is of a semiconductor nature. The material's properties appear to be consistent with what one would expect of an indium oxide matrix doped with tin. Many conducting films have patterns etched in them, and the etch time and other etching characteristics of the film are of vital interest to the user of conducting films.

The requirements placed on conductive coatings include the usual ones placed on optical coatings, plus some that are unique to the conductive aspect. Conductivity is rarely specified explicitly; rather, the sheet resistivity is specified, usually in terms of ohms per square, since with this form, it is not necessary to specify the thickness of the layer. The relation between resistivity and sheet resistance is

$$r_s = \frac{\rho}{t}$$

where r_s is the sheet resistance in units of ohms per square,
ρ is the resistivity of the coating in ohm cm,
and t is the thickness of the coating.

The resistances of transparent conducting films may have a significant temperature coefficients if they consist of semiconductors, such as ITO.

Hall effect measurements of a conducting film can provide useful diagnostic information on its physical properties by providing data on the number density, sign, and mobility of the charge carriers in the film.

The optical transmission of the film constrains the maximum thickness of the film. Thus, the film specifications are given in terms of the functional parameters of sheet resistance and transmission. This approach to specifying conductive films take into account the fact that the physical properties of metal films can vary. They depend on the coating process, especially in the transparent conductive material most commonly used, indium–tin oxide. The coated film has a composition that varies, and its conductivity (or resistivity) is determined by the volume density of the charge carriers in the film, as well as by their mobility. These parameters can vary significantly, and depend on the composition, the level of oxidation, and the deposition process parameters.

The transmission and reflection of a thick conductive coating of indium–tin oxide is shown in Fig. 4.42. The doping level has an especially strong effect on

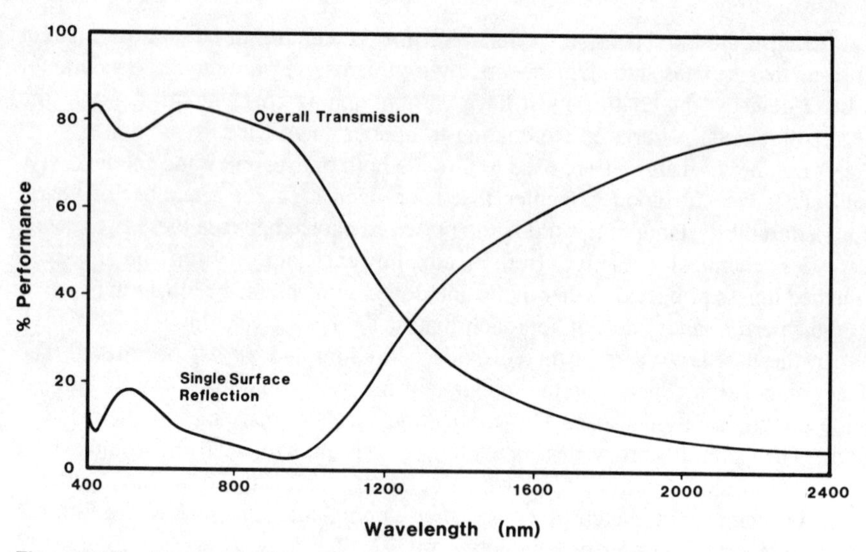

Figure 4.42. Measured reflectance and transmission of an ITO film on a glass substrate in the visible and the near IR. The physical thickness of the film is approximately 325 nm.

the slope in the wavelength region where the material switches from transmitting to reflecting, i.e., the 1- to 2-μm range.

The shape of the curve in Fig. 4.42 approaches the ideal one for solar thermal uses. The conductive coating is used as a reflector to return the infrared radiation back to a receiver, whereas the sunlight passes through the coating to heat the receiver. Unfortunately, the position of the edge of this coating material is not quite correct for the kinds of efficiency needed to make this design viable in solar thermal energy applications. In addition, the infrared reflectance is not as high as might be desired.

ITO has a high index of refraction when compared to glass. As a consequence, the reflectance of a coated area appears quite reflective to an observer, as compared with the bare glass where the ITO was removed. This is frequently objectionable, and additional thin film layers are used to reduce the contrast in reflected light between the two areas. These layers afford additional environmental protection to the conducting films. Fortunately, the electrical conductivity through these overcoats can be finite, and devices can function with the overcoat over the conducting films. These dielectric coatings significantly affect the etching rate.

4.6.5. Patterned Dielectric Coatings

On occasion, it is necessary to produce a specified pattern in the thin film optical coating. One example of such a coating's use is to code the signal from a single gun vidicon television tube and produce a standard color signal output. A pattern that accomplishes this consists of fine bands (on the order of 100/cm) of a color separation (dichroic) coating separated by clear gaps. Superimposed on this pattern, but rotated by about 30 degrees, is a second similar pattern. The complete pattern is placed on a vidicon faceplate with a diameter of about 2 cm. Patterned photoresist masking was used to delineate the areas that were not to be coated. The coating is deposited at an elevated temperature; after the coating process is complete, the resist is stripped off, leaving the desired pattern. The resulting coatings can have high levels of durability, and photoresists are readily available that are compatible with the thin film coating process. Other patterns can be prepared in this manner. Thus, very fine patterns can be produced, and the performance of the resulting coatings can be made equivalent to those deposited on larger areas (Bartolomei, 1976).

4.6.6. Coatings for High-Power Laser Beam Reflectors

In these coatings, the absorption coefficient must be as low as possible, and the laser damage threshold as high as possible. This type of design is a subset of the previously mentioned high reflector types, but with some specifications peculiar to their use in high-power laser beams. In the visible and near infrared portions of the spectrum, silicon dioxide is currently the favored low index material because of its low absorption properties, whereas titanium dioxide is used as the high index material. The titanium dioxide does not have as low an absorption coefficient as the silicon dioxide, but it has the highest index of refraction in these spectral regions. The resulting high index ratio means that less material is required because a high reflectance level can be attained with fewer layer pairs. In addition, the actual physical thickness of a high index layer is less with the higher index for a given optical thickness, thus reducing the total amount of material in the coating. The materials used in other portions of the spectrum are selected on a case by case basis.

Three types of high reflector coatings are available to reflect high-power laser beams: (1) protected metals, (2) enhanced metals, and (3) all dielectric.

Protected metal coatings are simple and provide high reflectance over a broad band. Their durability, although better than unprotected metal layers, is not as good as that obtained with the other types of reflectors. They are good for continuous wave laser beams, but do not hold up well under high-power pulse irradiation. Because of the large bandwidth over which the reflectance is high, the choice of wavelength with which to align the optical system is also large.

Enhanced metal designs are more durable than protected metal coatings, but the wavelength region over which the reflectance is high is more limited. This approach is good for continuous wave lasers also, but fast (picosecond) single pulse lasers can cause mirror damage. These designs typically have between 2 and 5 periods of reflectance enhancing dielectric layers.

All dielectric reflectors rely entirely on the dielectric layers to provide the reflecting power of the mirror. These designs have very high reflectance, but the bandwidth is limited. Between 10 and 15 pairs of layers are used to attain the high reflectance. The coatings are typically very durable and perform well for both continuous wave and pulsed situations. Because the bandwidth of these mirrors is limited, the use of alignment lasers may be problematic if the alignment wavelength is far removed from the design wavelength. These coatings can also be used on metal or other nontransparent substrates. The penetration of the electric field is limited to the outer layers of the design so that the substrate does not contribute to the reflectance (or absorption).

4.6.7. Color Correction Coatings

Color correction reflectors have a high reflectance that is a function of the wavelength. The functional form of this curve is selected such that when the light of the incident beam is reflected by this mirror, the spectrum of the light in the reflected beam is altered to coincide with some predefined shape. For example, the visible spectrum of the light from an incandescent source can be converted, after reflection from such a mirror, to that of the solar spectrum.

A commercial use for this type of mirror is in color correction coatings on medical mirrors. The filament in the light source has a color temperature in the 2600 to 3100K range, and the color correction mirrors can change the apparent temperature to that of sunlight, which is approximately 6000K. The basic design of these filters is that of a cold mirror (see section 4.3.). These mirrors eliminate much of the near infrared energy emitted by the filament by virtue of the fact that they are highly transparent to wavelengths in this spectral range. Thus, these mirrors improve patient comfort, reduce biological tissue heat damage, and assure good color matching.

Good color correction eliminates metamerism, the apparent yet false matching of two different color samples under some lighting conditions. Dentists and other medical personnel that make color matches in an artificially lighted environment need this kind of color correction in their lighting systems. The use of a color-correction mirror is necessary whenever color matching is performed under artificial lighting conditions.

Similar color or spectral correction filters can function in a transmission mode as well. One use is in spectral trimming filters for silicon detectors. The response of the detector can be made to approximate the photopic eye response for use in photometers. Other uses include trimmers to shape the spectral response on exposure meters.

4.6.8. Switching Filters

We have mentioned in Chap. 1 how the term *optical constant* is a misnomer, as the values change significantly as a function of wavelength. In addition, the *constants* of some materials vary with temperature (at very high temperature, e.g., near the melting point, the indices of most materials change significantly). One such material, vanadium dioxide, has optical constants that change dramatically when the temperature exceeds a transition value. The transition is one that involves a change in the crystal structure, and the optical performance changes from a transmitting state to a reflective state. This effect can be used as a passive switch. The incident light heats the filter because there is some slight absorption in the transmitting state. As long as the power level remains below a critical level, nothing happens and the filter transmits most of the incident light. When the incoming light reaches a power density that heats the filter above a critical level, the filter switches to a reflective state and the transmission of light through it is blocked. Such a switch can be used to prevent damage to the optics behind the filter. When the light level decreases and the filter cools, the filter reverts to its original state and transmits the light once again. Such a scheme is at times superior to a shutter because there are no moving parts involved.

Additional thin films can be used with the switching material to increase its sensitivity, in the same way that the dark mirror structure is used to enhance the sensitivity of the optical data disk recording mechanism (see section 4.6.1.).

4.6.9. Emissivity Control

Coatings are available that can enhance or decrease the emissivity of surfaces. These coatings are useful especially when radiation is the primary mechanism of heat loss from a surface, and where conductive and convective heat transfer mechanisms cannot function effectively. Such conditions are found where a surface is surrounded by a vacuum, as may be found in space or in an evacuated container.

4.6.9.1. Emissivity Reducing Coatings. Emissivity reducing coatings have been used for many years in Dewars and other vacuum-insulated containers that are used to contain hot or cold substances; in these situations, the aim is to reduce the amount of heat transferred between two surfaces.

There are two major types of emissivity-reducing coatings. Opaque layers of high infrared reflectivity metals, such as gold, silver, copper, and aluminum, are used when visible transmission through a surface is unimportant. A transparent conductive coating, such as ITO, is used when visibility through the surface is required. In a vacuum environment, silver can be used as it will not tarnish. Silver has high reflectance in the visible as well as in the infrared, and is, therefore, the preferred material, though gold could also be used. The emittance of the surface facing the vacuum is the one that is to be controlled. The substrate in these instances is likely to be glass or a metal. These materials are opaque at the wavelengths where the blackbody curve peaks for most common temperatures at which temperatures and heat gains and losses need to be controlled. Thus, the coating should be applied on the vacuum side of the surface. Figure 4.43 shows the blackbody distribution for a number of temperatures. There are a number of slide rules available that allow the rapid calculation of various blackbody parameters.

A conductive coating, such as ITO, generally has a fairly high infrared reflectance as is shown in Fig. 4.42. As such, it can make a moderately good coating material to reduce the heat transfer between two surfaces, and yet retain visible transparency. Semitransparent metal films can also be used. In these metals, however, the infrared performance is improved over ITO, but at the price of a reduced visible transmission.

4.6.9.2. Emissivity Enhancing Coatings. An emissivity enhancing coating is used when it is desired that a surface emit more heat than it would otherwise, i.e., to couple more closely the surfaces thermally.

Spacecraft have surfaces designed to dissipate heat generated inside the craft

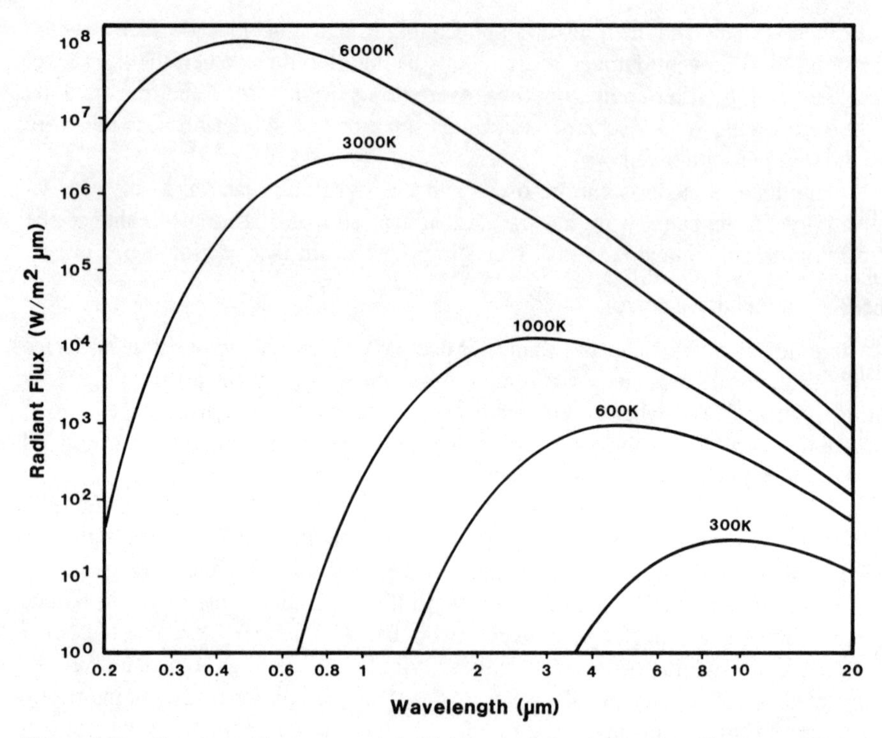

Figure 4.43. Spectral radiant exitance of a blackbody radiator. Curves are for temperatures of 300K (near room temperature), 600K, 1000K, 3000K (incandescent lamp filament temperature), and 6000K (approximate solar surface temperature).

by the electronics and other internal packages. Photovoltaic solar arrays are temperature sensitive and need to be kept cool to keep their solar to electrical efficiency as high as possible. For many years, thin fused silica (SiO_2) or glass sheets have been used to increase the emissivity of the silicon cells. The relatively high emissivity of fused silica, as well as the fact that it does not darken under particle irradiation, was responsible for its selection early in the space program for use in solar cell covers.

A thermal control surface for temperature control of the body of a spacecraft consists of a high reflector to reflect the solar radiation placed on the second (inner) surface of a fused silica sheet. A dielectric reflectance-enhancing stack over a silver layer is the choice reflector material because it keeps the solar absorptivity to a minimum. The uncoated (outer) surface of the fused silica serves as the emitting surface. This arrangement provides a surface from which the spacecraft thermal energy may be radiated even when it is in direct sunlight.

The hemispherical emissivity of silica is approximately 0.82 at a temperature of about 300K (Crabb, 1972; Jet Propulsion Laboratory, 1976). Note that it is necessary to specify a temperature when specifying an emittance value, since the normal emittance is defined as

$$\varepsilon(T) = \frac{\int A(\lambda)\, B(\lambda,T)\, d\lambda}{\int B(\lambda,T)\, d\lambda}$$

where $A(\lambda)$ is the spectral absorptance and
$B(\lambda,T)$ is the blackbody function for temperature T.

The integral has to be taken over the entire range of wavelengths, though from a practical standpoint, the blackbody curve will put reasonable limits on the range. A second integral, one over all angles needs to be calculated to obtain the hemispherical emittance. An angular weighting function is also required to take into account the directional nature of the emitting surface. In many thermal radiators, the substrate may be opaque in the infrared region of interest. Thus, in these instances, the absorptance is equal to $1 - R$, where R is the reflectance.

The emissivity of a surface may be increased by the application of coatings that reduce its reflectance in the range in which the blackbody curve has large values. For a surface at 300K, blackbody curve reaches a peak between 9 and 10 μm. The reststrahlen reflectance of silica also has peaks in this same spectral region, thus reducing the emittance of this material. These peaks can be reduced with appropriate coatings that form a dark mirror, which antireflects the reststrahlen

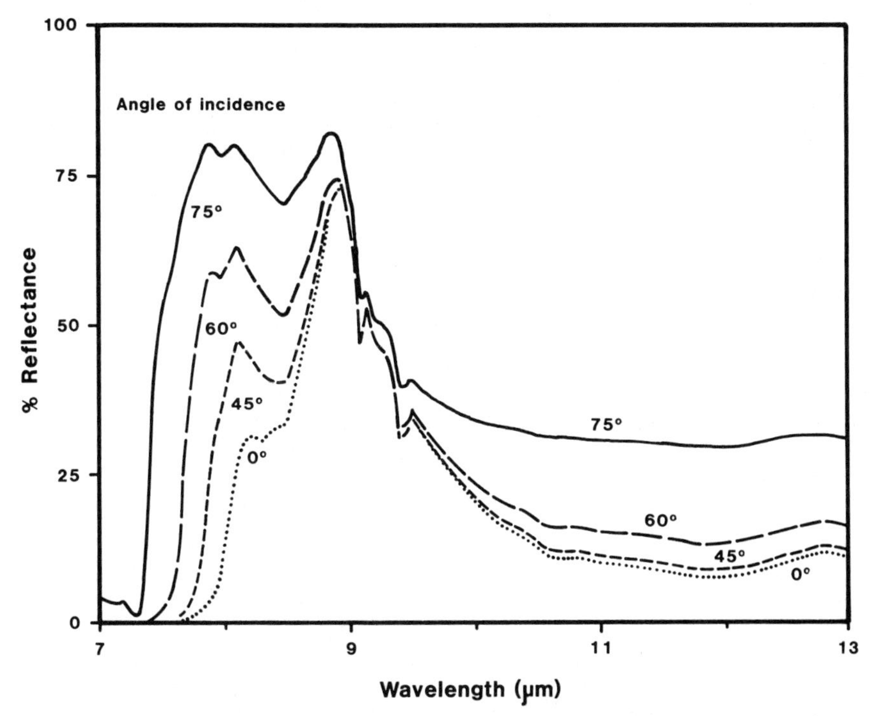

Figure 4.44. Reflectance of bare silica in the reststrahlen reflection region as a function of wavelength. The performance has been plotted for a number of angles of incidence (shown labeled in degrees). The reflectance increases monotonically with the angle of incidence at all wavelengths.

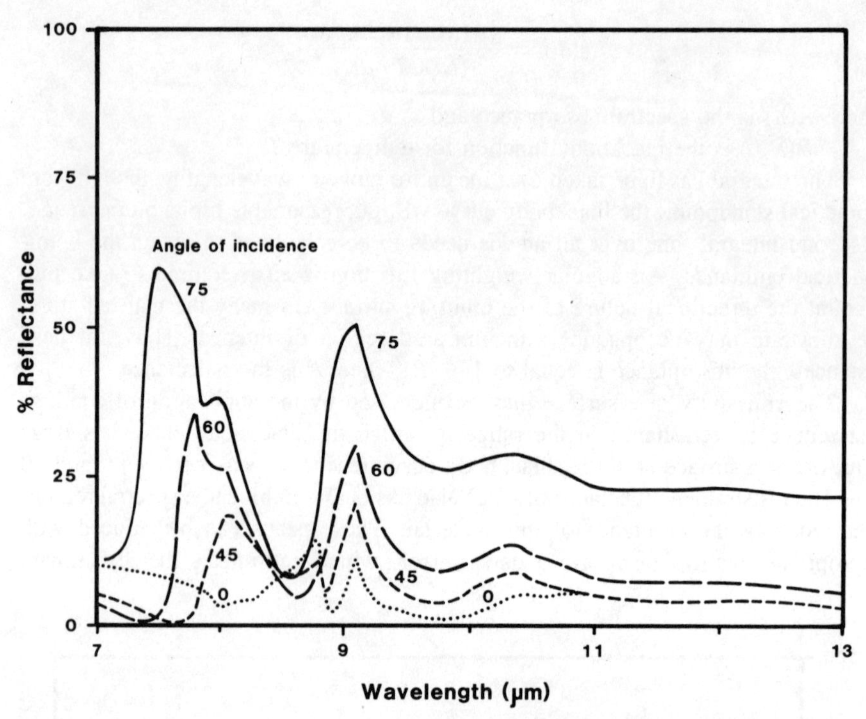

Figure 4.45. Reflectance of silica coated with an emissivity enhancing coating. The performance has been plotted for a number of angles of incidence (shown labeled in degrees). Most of the curves of the coated substrate lie well below the corresponding curve for the bare substrate (see Fig. 4.44).

reflection. In this way, the emissivity is increased by a reported 5 to 10 percent (Rancourt et al., 1984). Calculations show that the reststrahlen peak of the silicon dioxide widens considerably as the angle on incidence increases. The increase is considerably diminished with the coating, thus improving the high angle performance of these emitters as well. Figures 4.44 and 4.45 show the calculated reflectance of bare and coated silica for a number of angles of incidence.

Chapter 5

Filter Performance Measurements

This chapter deals with the functional characteristics, or performance, of typical thin film filters, and how one measures this performance. The measurement techniques must be up to the task that is required of them. For example, it is not acceptable to measure the transmittance of a narrow bandpass filter with a spectrophotometer whose resolution is less than that of the filter. This and other such problems are covered in this chapter. For further information, the reader is directed to the excellent book on photometry by Walsh (1958).

5.1. SPECTROPHOTOMETERS AND SPECTROPHOTOMETRY

Spectrophotometers are optical instruments that measure the intensity of light transmitted or reflected by objects as a function of wavelength. They are available in a variety of types and styles. Often, they are optimized for chemical studies, where the photometric accuracy, i.e., the precision with which the intensity is measured, is not necessarily critical. In many cases in optical filter work, however, there is as much interest in the photometric accuracy of a measuring instrument as there is in the wavelength accuracy.

The wavelength display of some spectrophotometers is calibrated in a scale the units of which are actually the reciprocal of wavelength. A common unit for reciprocal wavelength is the *wavenumber,* or reciprocal centimeter, i.e., the number of wavelengths that occur in the distance of 1 cm.

The wavelength accuracy of the dispersing mechanism can be verified (and calibrated, if necessary) by using the lines of atomic emission lamps, lasers, stable absorption filters such as a styrene film for infrared calibration, or other known standards. The photometric accuracy of an instrument cannot be verified quite so easily. The photometric linearity can be tested by using known neutral density filters in the optical path.

The accuracy requirements of the wavelength or wavenumber axis is a well-understood concept. Photometric accuracy, though often as important as wavelength accuracy, is frequently not tested and is accepted as correct. A few simple tests can be done to verify the linearity of the photometric axis, and a few operational habits ensure proper scaling.

The linearity of an instrument can be verified by placing known filters in the sample beam and noting if the output of the instrument is proportional to the transmission of the filter combinations. It is best to use filters whose spectral transmission curves are relatively flat as a function of wavelength to avoid possible complications from spectral shifts. By measuring the transmission of a number of individual filters and then measuring them in various combinations, the linearity of the instrument can be determined from a plot of the calculated versus the measured transmittances.

Once the instrument is known to be linear, it is necessary to know the two endpoints of the range to interpolate accurately. This is accomplished by ensuring that each measured point lies between two other known points: (1) there should be a lower value, either zero or some known lower transmission or reflection value and (2) there should be an upper value that is larger than the one being measured.

For a measurement of transmission, the upper value might be made with no sample in the beam, whereas the lower value might be made with a well-blocked sample beam. Specific reflectance measurement standards depend on whether the reflectance is very low, as in the case of an antireflection coating; medium, as in a beam splitter; or, high, as in an enhanced metal reflector.

The zero baseline can be obtained with a blocked sample beam. The upper value needs to be selected so that there is as much expansion of the output scale as possible.

For a low reflector, the flat face of a wedge of uncoated silica or a glass with a known index of refraction can be used to obtain a full scale reading in the neighborhood of 4 to 7 percent. A wedged standard is necessary to eliminate possible second surface effects. A sample reflectance of 0.5 percent will then produce a reading of one-eighth full scale with a zero baseline and a 4 percent reflectance standard. Similarly, a flat piece of silicon or germanium can be used to yield reflectances in the 30 percent range in the visible portion of the spectrum. Rhodium and other durable metal reflectors can provide 50 to 70 percent full-scale expansions.

The measurement of very highly reflecting surfaces can be problematic, especially if the reflectance exceeds that of a calibrated reference, such as freshly evaporated silver, the metal with the highest reflectance over a large part of the spectral ranges of interest in optics. In this case, it may be necessary to use extrapolation or rely on the linearity of the gain outside of a range that can be checked. On the other hand, as the extrapolation is over a small range if a known standard with a high reflectance is used, the error introduced in the measurement will likely be small.

The normalization of the performance value to 100 percent can present a problem because it depends on a number of parameters that cannot be precisely calibrated in a way similar to those used for wavelengths. In all spectrophotometers, the output is the result of a ratio: the intensity of the light through the instrument with the sample in place is divided by that of a calibration measurement. There are two ways in which this is done by spectrophotometers in current use, and the instruments can be classified accordingly: (1) single beam and (2) double beam.

5.1.1.1. Single-Beam Spectrophotometers. In the single-beam instrument, there is only a single optical beam through the instrument, as the name implies. The *baseline* scan is recorded for later division into the sample scan. The instrument must be stable enough so that its response or sensitivity does not change between the baseline scan and the sample scan. The components that must be stable include the lamp intensity, the detector response, and all other parameters associated with the instrument. The time frame during which these parameters must be constant can range from seconds to hours, depending on the time required to complete a scan, and the time between scans.

5.1.1.2. Double-Beam Spectrophotometers. Many instruments are of the dual-beam type, in which the light from the source is alternately directed into a reference beam and into a sample beam by a chopping wheel or other device. Each beam travels a different path. Theoretically, the two paths are identical in length, number of reflections, and so on, except for the sample's being in the sample beam. If such is the case, the ratio of the intensities of the two beams are proportional to the actual performance (reflectance or transmittance). The use of alternating beams reduces the errors that may be introduced by a fluctuation in the intensity of the light source, absorption in the light path, such as water absorption or mirror attenuation, and other similar common mode (common light path) effects as the two intensities are sampled a fraction of a second apart. Differences that are not common to the two paths and yet are not related to the filter's performance are the source of the measurement inaccuracies. An example of this is unequal conditions in the two optical paths, such as inadequate purging of the beam paths. If the path of one of the beams has more water vapor in it than the other, one can expect to see a difference in intensity between the two beams at wavelengths near the highly absorbing water band near 2.8 μm. Thus, spectrophotometers are frequently purged with dry nitrogen to ensure that the two paths are equal after a certain purge stabilization time. Some of the spectrophotometer measurement considerations noted here also apply to single-beam instruments.

There are other problems that are common to both types of instrument. These include: (1) substrate scatter that reduces the apparent transmission; (2) beam offset introduced by the substrate (the beam can be vignetted by apertures before it reaches the detector); and (3) in reflectance measurements where a reference reflector is used, the uncertainty in knowing its precise reflectance value.

The typical commercially available spectrophotometers were designed principally to measure transmission through a relatively small sample. To measure reflection, attachments have to be added by the user. These attachments lengthen the path in the two beams, so it is important that the original optical path be sufficiently flexible that this increase in length does not affect the basic performance of the instrument.

When the transmission of a sample is measured, it should be tilted a few degrees so that the optical axis of the instrument is not perpendicular to the surface of the sample. Otherwise, some of the light that reflects off the sample can travel back toward the source and upset the calibration of the instrument. This happens in some instruments when the energy reflected off the sample surface travels backward through the optical system and then enters into the reference beam system.

The dispersing element that is used in a given spectrophotometer is not important, as long as the overall instrument is adequate. Thus, prism, ruled or holographic gratings, and Fabry–Perot interferometer units are all acceptable from a fundamental viewpoint for general work. On the other hand, practical considerations such as stray light, scan time, ease of operation, or signal to noise ratio may sway the balance in favor of one type over the others.

5.1.1. Reflectance Measurements

The accurate measurement of reflectance poses a real challenge. Reflectance measurements are difficult to do at exactly normal incidence without a beam splitter or complex polarization schemes to divert the reflected beam to the detector and to prevent it from returning to the light source. The introduction of these additional intensity reducing optical elements into the system can reduce the sensitivity of the instrument, especially in low signal situations, such as in the measurement of antireflection coated surfaces. In addition, these auxiliary elements must be scatter-free to minimize their introducing noise into the measurements.

Because the true normal incidence geometry is so difficult, most measurements of reflectance are made at some small angle, such as 10 degrees. Usually, this does not introduce significant errors. At high angles of incidence, the polarization introduced in the optical path by the instrument, in addition to that arising from the sample, is significant and can distort the results, as discussed in section 5.1.5.

In a dual-beam instrument, to keep the path in the reference beam the same as in the sample beam, a standard reflector occupies a position in the reference beam that is identical to that of the sample in the measurement beam. The choice of reference reflector is arbitrary, though it is desirable that there be no steep variations in reflectance with wavelength, and that the reflected intensities in the reference and the sample beams be approximately equal. When a low reflectance is being measured, a low reflector should be placed in the reference beam to keep the intensities in the two beams approximately balanced. This reduces the possibility of stray light from an intense reference beam from interfering with the sample beam.

It also makes the taking of the ratio of the two intensities more accurate as the divider is working at close to unity ratios. The reflectance of the standard in the reference beam should be at least as high as that of the sample.

A problem in the measurement of reflectance is that a *reference,* the reflectance of which is well characterized, is usually needed. In either a single-beam or a double-beam instrument, a *baseline* scan is measured prior to making an actual measurement. In transmission, this is easily done by leaving the sample out of the measurement chamber. In measuring the reflectance, a mirror must occupy the sample position during the baseline scan. This reference mirror is then removed and the sample inserted in its place for the actual sample measurement. The ratio of the sample results to the baseline results gives the ratio of the two reflectances without any contributing factors from the instrument. The result of the ratio is multiplied by the reflectance of the reference mirror to get the sample reflectance. Any wavelength variation in the reflectance of the reference mirror must be taken into account. The problem is thus reduced to that of knowing well the reflectance of the reference mirror.

In the case of low reflectance values, the reflectance of polished samples can be used. A polished glass surface can be characterized on the basis of its refractive index. In addition, such a surface is hard and will not easily scratch. Other reference reflectors that can be used in the visible and the infrared are the polished surfaces of silicon and germanium. High reflectance references are more difficult. Typically, high reflectors can be made from freshly evaporated aluminum and silver for the visible, and silver and gold for the infrared. These reflectors are all relatively soft and can be damaged easily. They should be checked frequently. Rhodium metal films are fairly reflective and quite durable. In any case, the reference mirrors should be approved by appropriate quality assurance personnel.

Professor John Strong (1938) originated a technique to overcome the previously mentioned problem of needing a reference for reflectance measurements. His optical layout is shown in Fig. 5.1. It is sometimes referred to as a V–W configuration as the optical paths in the two positions look like these two letters. It should be noted that this arrangement, because it has two reflections off the sample surface, measures the *square* of the reflectance, and therefore, it is quite useful for mirrors and beam splitters with high reflectance. When the reflectance is less than about 30 percent, the squared value drops below 10 percent, and accuracy can be quickly lost at these lower values. The fact that one mirror has to be moved from the sample to the reference positions can pose a problem in the alignment of the optical path. Some instruments use separate fixed mirrors to redirect the beam, thus avoiding the alignment problems; the assumption is made that the two mirrors are identical. Another drawback that is sometimes encountered in this configuration is that the sample must be large enough to intercept and reflect the two beams that need to impinge on the sample. In addition, the larger the sample, the greater the need to ensure that the uniformity of the coating is adequate to ensure good results.

When the reflectance of a sample is low, the problem encountered is frequently one of low signal levels, combined with possible scattered light from the surface being measured, or from other sources such as stray light in the monochromator.

Accurate reflectance measurements of samples with very high reflectance levels

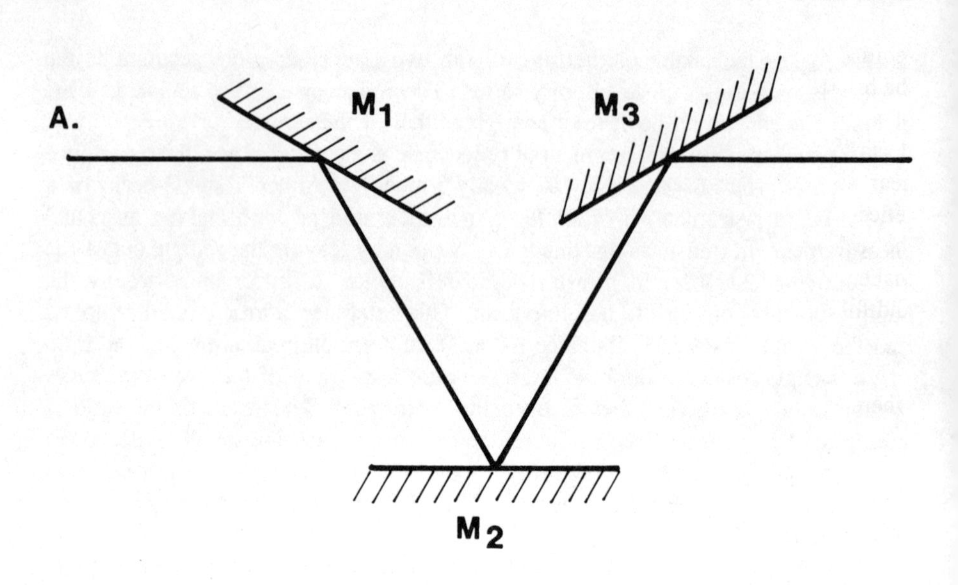

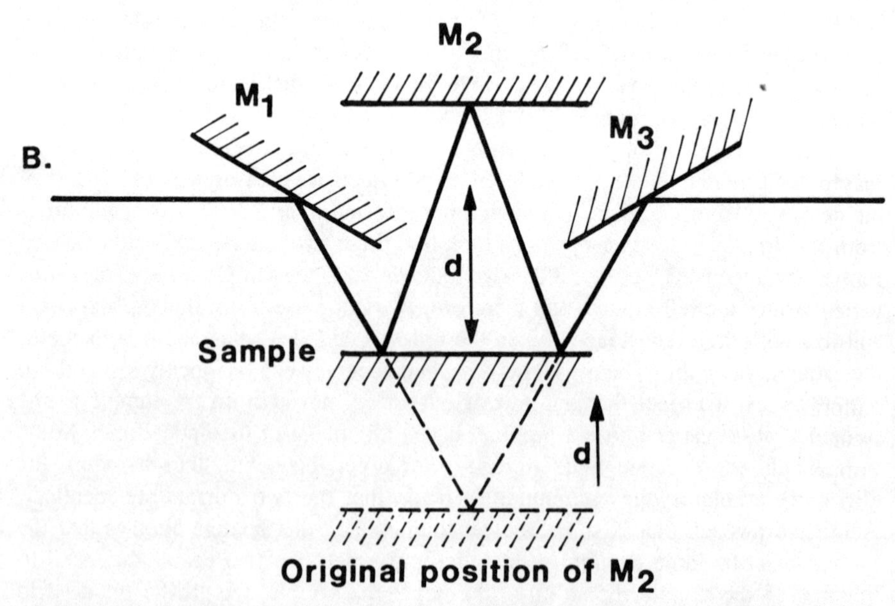

Figure 5.1. Diagram of Prof. John Strong's V–W reflectance attachment for the measurement of the reflectance of a sample without having to rely on a standard. Note that the beam reflects twice off the substrate. This yields R^2 as a result of dividing the intensity value from arrangement B by that from arrangement A. Note that this arrangement may be inaccurate f the reflectance of the sample is low (R^2 is too small).

require the determination of a small difference in intensity between the sample and the reference reflected intensities, which are large. Sometimes this is a small fraction of a percent difference between the mirror and 100 percent. In addition, the reflectance of the reference needs to be known precisely. For values of reflectance near unity, it is possible to design a measurement system such that the light is reflected many times off the sample surface. To first order, this technique increases the sensitivity of the measurement linearly with the number of reflections the beam makes on the sample, as can be seen by expanding the difference between unity and the reflectance losses:

$$R^n = (1-A)^n = 1 - nA + \ldots$$

where R is the value of the high reflectance, A represents the reflectance losses, and n is the number of reflections the sample beam makes on the sample surface.

The losses are assumed to be small (the reflectance is high) so that higher order terms can be neglected. By increasing the number of reflections off the test surface, we can increase the sensitivity. In such a situation, it is necessary to be certain that the auxiliary optics that are required to get the multiple bounces from the sample do not introduce undesirable effects of their own.

5.1.2. Beam Offset

The principle of most double-beam spectrophotometers is that both the reference and the sample beams traverse equivalent paths at all times during a measurement; i.e., the path lengths and the mirror reflectances are the same, and the beams impinge on the detector at the same point. A thick glass substrate or window inserted into one of the beams offsets it if it is not inserted perpendicular to the optical axis. The effect of offset can be subtle and difficult to detect. For example, the offset may cause vignetting of the beam, or cause the beam to strike a different part of the detector. Detectors can have variable sensitivity over their active surface, both from the standpoint of wavelength as well as position (Birth and DeWitt, 1971; Ballik, 1971; Sommer, 1973). In other words, the spectral sensitivity curve shape varies as a function of position. For precision work, these factors merit careful consideration.

A number of methods have been devised to reduce the effect of areal nonuniformities in detectors, including the use of integrating spheres. These approaches incur a degradation of sensitivity as the walls of the spheres always absorb a certain amount of light. Another method makes use of a scattering lens that diffuses as well as focuses the light onto a detector (Anthon, 1982).

5.1.3. Extra Surface Effects

The presence of extra surfaces in the optical path is another cause of measurement difficulties. For example, in the measurement of the reflectance of one surface of a thin glass plate that has been coated with an AR coating, the second surface (not AR-coated) may reflect about 4 percent of the light incident on it. The light reflected from the uncoated surface dominates that reflected from the treated surface. The exact value of the spurious component is difficult to determine because the light reflected from the second surface travels twice the thickness of the substrate, and will be offset from the nominal beam as well.

There are two frequently used methods to eliminate bothersome second surface reflectances. A wedge-shaped prism whose index is the same as the substrate can be placed behind the uncoated surface and an index-matching fluid introduced between the substrate and a wedge face. The result is to eliminate the optical interface. The transmitted beam is reflected by a second prism face. The deviation introduced in the beam, along with the extra path length in the prism, is usually sufficient to deflect the beam out of the optical train and the collection cone of the detector. The wedge must not be a right angle prism, as the deviation will be zero for this configuration and the beam will be returned to the spectrophotometer.

Another technique to reduce the second surface reflection is the use of a substrate whose second surface has been roughened, either chemically or mechanically. The treatment should reduce the specular reflectance of the surface to a negligible value.

The thicker the substrate, the less of a problem the second surface reflectance is as the beam reflected by the second surface will be displaced in proportion to the thickness of the substrate. The larger the deviation, the less the likelihood of this offending beam's finding its way to the detector. If the second surface is nonplanar, then the second surface beam will be defocused before it can get to the detector, thus further reducing the magnitude of the problem.

When an integrating sphere or other wide acceptance angle detector collects the light that is reflected by the sample, reduction of the second surface effect can only be done by absorbing the unwanted light. A paint that has an index equal to the substrate index and that is completely absorbing at the wavelengths of interest works nicely. Another similar method of addressing this problem is to attach an absorbing material, e.g., a piece of black glass, to the back of the substrate with an index matching oil.

When infrared substrates are used, the second surface reflection may be higher than that of optical glasses used in the visible. Index matching fluids are not generally available in this spectral region, therefore, second surfaces have to be treated to eliminate the specular beam. This has to be done carefully for the following reasons.

1. The substrates are frequently fragile. In the infrared, semiconductor crystals are frequently used as substrates since they may have very good transparency. In addition, these substrates are frequently thin, and this contributes to the fragility.

2. Scattering is a strong function of wavelength. The scattered light intensity can decrease as fast as the fourth power of the wavelength, depending on the type of scattering mechanism involved. As the wavelength of interest increases, it becomes more difficult to attenuate the second surface reflection by roughening it.

Some special purpose instruments have been designed to distinguish between the first and the second surface reflectances.

In the previous discussion, it is assumed that the light reflected from the second surface is adding in an incoherent manner to the light from the first surface. This may not be the case in some infrared, as well as visible, measurements. The two surfaces of the substrate can serve as the mirrors in a Fabry–Perot cavity with the result's being a so-called channel spectrum. This appears on a spectrophotometric scan as an oscillation in the reflectance. On a wavenumber scale, the oscillation period is constant and related to the thickness of the substrate. The amplitude of the oscillation in the reflectance depends on the reflectance of the two surfaces: the closer the reflectances are to being equal, the greater the amplitude of the oscillations. If the *f*/number (*f*/#) of the spectrophotometer is similar to that of the optical system in which the filter will be used, then the oscillations will also be present in the final system (assuming the filter is not optically bonded to some other component). If the second surface of the filter has an antireflection treatment, there should be minimal interference between the beam from the second surface and the front surface one.

5.1.4. Effect of the Refractive Index of the Incident Medium

Filters that are designed for use in an incident medium other than air may have a different performance in air, and vice versa. Examples of this are antireflection coatings applied to silicon photovoltaic cells. One- and two-layer antireflection coatings are used for this purpose. In air, the performance looks perhaps slightly better for the single-layer design. When the index of the incident medium is taken into account, the average reflectance of the single-layer design increases, although

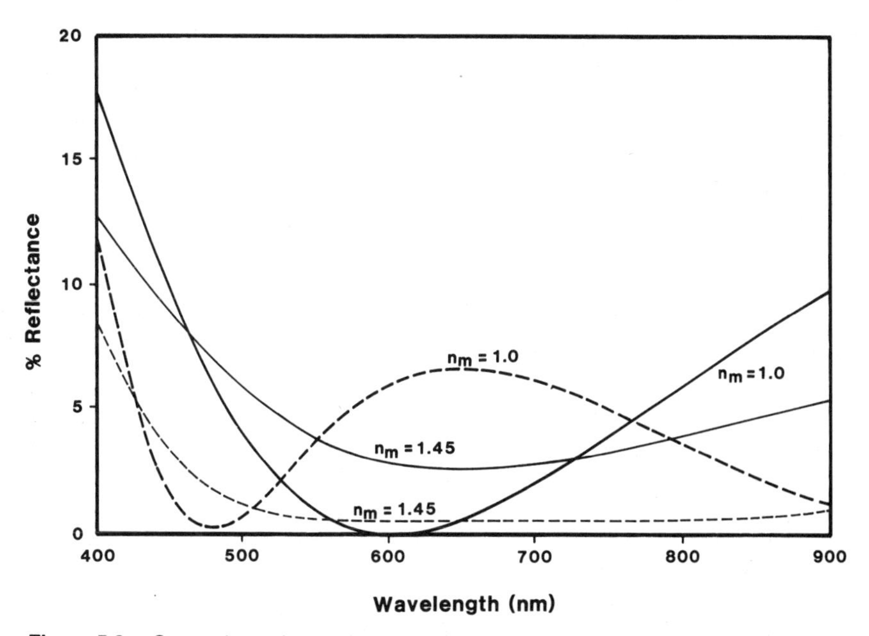

Figure 5.2. Comparison of the reflectance of a single layer (*solid lines*) and a double layer (*dashed lines*) antireflection coating applied to a silicon substrate in incident media of 1.0 and 1.5. Note that after immersion the reflectance of the single layer design increases, whereas that of the two layer one decreases.

the performance of the two-layer design improves. Figure 5.2 shows the individual performance curves. If the silicon cell is to be encapsulated in a protective organic material, the reflectance should also be measured immersed in a medium with an appropriate index of refraction.

To measure accurately the reflectance of the silicon/encapsulant interface, care has to be taken to consider the first surface reflectance at the air/encapsulant interface, as mentioned in the previous section.

Special purpose instrumentation is required when measuring the reflectance of an immersed sample at a high angle of incidence because it is difficult to introduce and extract a beam into the immersing medium at high angles.

5.1.5. Polarization

Unrecognized polarization can cause insidious problems in spectrophotometry. When a sample is placed in a dual beam spectrophotometer at or near normal incidence, any intrinsic polarization effects cancel because of the ratioing that takes place between the two beams.

When the sample is placed at a nonnormal angle with respect to the optical axis of the spectrophotometer, there will be some polarization effects introduced into the system since thin film filters behave differently in p as compared with s plane polarized light. The polarization state of the beams in the typical spectrophotometer without polarizers is unknown, and it is likely that the intensity of the two polarization components will be different. The reason for this is that spectrophotometers have many optical components that cause the light with which they interact to become partially polarized. Reflection diffraction gratings, for example, reflect light that is highly polarized. Prisms and mirrors that are used at nonnormal incidence can also introduce some polarization to the beam. Thus, in a spectrophotometer that has no specific polarizer, the interaction of the sample beam with the filter may lead to unpredictable results.

One solution to this problem is to install a polarizer in each beam of the spectrophotometer when measuring samples at nonnormal angles of incidence. For angles of incidence less than 30 degrees, however, this may not be necessary.

Sheet polarizers can be used for the visible portion of the spectrum, but these can introduce problems of their own because for the most part they are not very good optical elements:

1. They can introduce aberrations into the wavefront unless they are laminated between glass plates
2. Their spectral range is limited
3. They do not produce complete polarization of the transmitted light
4. They are absorbing, and thus reduce the amount of light available for the measurement

Their advantages include:

1. They are relatively inexpensive
2. They are compact and simple to use when compared with other techniques

For precision work over broad wavelength ranges, calcite and other crystals are used to make prism polarizers such as Glan–Taylor and Glan–Thompson. Some of these crystals are fragile and sensitive to moisture, and the assembled prism can be large (several centimeters long).

In the infrared portion of the spectrum, wire grids are frequently used as polarizers in addition to some crystals.

Polarizers are characterized by the amount of light they transmit (how much less than the theoretical 50 percent) and the attenuation, suppression, or extinction ratio of the two planes of polarization. The extinction ratio indicates how well the polarizers discriminate between the two planes of polarization. With this information, one can estimate the error that will be introduced in a measurement. This should be evaluated as a function of the wavelength as the transmission and efficiency of polarizers is not constant. In some portions of the infrared spectrum, the energy reaching the detector is quite low in the best of situations. Adding a polarizer that cuts the intensity by at least a factor of 2 makes the situation even worse. The combination of instrumental polarization and polarizer orientation may reduce the energy getting to the detector to an extremely low level. In some instances, a depolarizer or polarization rotator can be of some help.

5.1.6. Measurement Cone

The *f*/# of the cone of light incident on a sample in a spectrophotometer gives rise to a range of angles of incidence on the sample surface. The relation between the *f*/# and the maximum cone angle is given by

$$\theta_{max} = \tan^{-1} 1/(2 \cdot f/\#),$$

where θ_{max} is the maximum angle for the cone defined by the *f*/#. The values corresponding to this are included in Table 5.1. Table 5.2 lists the approximate *f*/# of some spectrophotometers commonly in use today.

An *f*/8 cone, for example, has a range of angles from 0 to 3.6 degrees, which is relatively small as compared with the angular sensitivity of most interference filters. The results of a performance measurement with such an instrument would not be skewed except, perhaps, for the narrowest of bandpass filters. An *f*/4 in-

Table 5.1 *F*/Number and Cone Angles

f/#	Half cone angle (degrees)
1	26.6
2	14.0
4	7.1
8	3.6
16	1.8

Table 5.2 Approximate *f*/number of Common Spectrophotometers

Spectrophotometer	f/#
Beckman DK2, DK-2A	10
Beckman IR4, IR-12	10
Cary 14 & 17	8
Cary 90	8
Cary 210, 2200, 2300	8 to 10
Gamma 180	5
Nicolet FTIR 6000	5
Nicolet FTIR MX1	5
Perkin Elmer 180	10
Perkin Elmer 983G	4

Note: Many instruments have slightly different f/#s in the vertical and the horizontal planes.

strument, on the other hand, has a maximum cone angle of 7.1 degrees. This could be the source of some differences between the theoretically evaluated (collimated light) and the measured spectra. As previously mentioned in this chapter, reflectance and transmission are almost never measured at exactly 0 degrees angle of incidence. If a 5-degree tilt of the sample is used for making a transmission measurement, and this is added to the 7-degree cone angle for an *f*/4 instrument, then the angle of incidence on the sample varies over the range from 0 to 14 degrees. This could be significant for some very narrow band (less than 0.5%) filters (see Fig. 3.15). In addition to the cone angle, the light in a spectrophotometer also has a spread of wavelengths caused by the finite slit width; this has to be added to the cone angle to get the total effect of the instrumentation on the measured result.

In narrow bandpass interference filters, the wavelength of the center of the pass band varies with the angle of incidence caused by the blue shift as discussed in section 3.6. The measured peak transmittance of a narrow bandpass filter illuminated by a cone of light differs from one that is illuminated by collimated light. The finite cone of light integrates the performance over a range of angles, and the effect of this is to reduce the total indicated transmitted intensity as compared with that of a collimated beam. Other types of filters whose measurement may be affected by this shift of performance with angle are (1) edge filters, where the slope of the edge may appear to be less steep than it actually is and (2) high performance AR coatings, e.g., V-coats at the minimum reflectance point, where the low reflection is not as low as expected based on normal incidence calculations with a 0-degree cone angle.

Reflectance attachments for spectrophotometers are frequently designed for 10-degree angle of incidence or greater. It is rare to find an instrument that can measure the reflectance at angles much closer than this to the normal. Thus, in reflection, one can expect that the angles may range from 7 to 13 degrees for an *f*/8 instrument.

Some commercially available spectrophotometers have different *f*/# in the vertical and the horizontal planes. This complicates an evaluation of the effect of the *f*/#.

In summary, the measuring of interference filters in a focused beam is useful because in many instances that is the way that they are going to be used. In certain measurement situations, it is simply necessary to account properly for the variations in the angle of incidence.

5.1.7. Nonspecular Samples

All optical elements have some level of scatter. Some of the scatter is caused by bulk effects in the substrate, and we are concerned about this type of scatter in this discussion. Some surfaces, for example, those used for laser gyroscope mirrors, are nearly perfect in this regard, with surface scatter of visible light measured in parts per million of the incident intensity. At the other extreme, some samples have surfaces that are intentionally made scattering. For example, the reflection from the new generation of frosted fused silica solar cell covers is entirely diffuse and has no specular component.

Specular generally means that the light beam retains its original distribution pattern after encountering an optical element. *Diffuse* means that some of the light in a specular beam is removed from the beam by the optical element and is redirected (scattered) in a direction other than the nominal one predicted by the laws of reflection and refraction for the macroscopic surface. The macroscopic scatter can be caused by reflection or refraction on a microscopic scale, or by true scatter from pointlike sites on the surface of the optical element. Note that the terms of specular and diffuse apply to both reflection and transmission. The effect of scatter is to reduce the intensity of the specular beam without reducing the total light in the system. Absorption, on the other hand, reduces the intensity of the specular beam as well as reducing the total light in the system.

In an ordinary spectrophotometer that measures the intensity of a specular beam, it is not possible to identify whether an observed intensity reduction is caused by absorption or scattering—they must be lumped together as one loss mechanism. Integrating spheres can be used to measure total integrated reflectance and transmittance as it can collect the energy scattered out of the specular beam as well as that which is specularly reflected from the sample. With such a technique, the only light unaccounted for is absorption. The many reflections that the light undergoes inside an integrating sphere before reaching the detector limits the sensitivity, thus making only relatively high levels of coating absorption in a scattering sample measurable by this technique.

Appropriately configured integrating spheres can be used to measure absorptance (Edwards et al., 1961). First, the sphere is calibrated for 100 percent with the sample removed from the sample beam. Then the sample coating (on a transparent substrate) is suspended so that it is at the center of the sphere and intercepts the sample beam. Now both the total transmitted and reflected beams are collected by the sphere, and the difference between the no sample and with sample measurements is the absorption in the sample.*

*Gier Dunkle Instruments, Inc., 1718 21st Street, Santa Monica, Calif. 90404, make such an attachment, Model SP220, for the Beckman DK-2 spectrophotometer.

A narrow band interference coating functions poorly if used in a beam that has a large diffuse component. The in-band light scattered at the higher angles of incidence is not passed by the filter, whereas some out of band light arriving at nonnormal angles of incidence light is transmitted. Note that light with wavelengths other than the nominal filter is transmitted because of the angle shift to shorter wavelengths. Similar situations can arise with edge filters that are used to discriminate between closely spaced wavelengths. If a component in an optical system is scattering, then narrow bandpass or other wavelength sensitive interference filters should be located so that the beam strikes the filter before it reaches the scattering element.

Scattered light is a problem in antireflection coatings (transmitting systems). The scattered light appears as a haze on the surface when viewed by eye (scatter in other portions of the spectrum may be detected with appropriate instrumentation). This is especially noxious when a bright light source is located off the optical axis of the optical system, and a dim object is being viewed through the system. A similar situation occurs in a reflective system when a mirror is used to observe a faint source and extraneous light floods the primary or other mirror, which may then scatter light into the detector.

In precision filters, it is necessary to use the highest quality substrates for peak performance. Furthermore, the coating should not increase the scatter already present in the substrate.

5.1.8. Nonplanar Surfaces

A nonplanar surface, for example, a concave or a convex lens surface, is very difficult to measure accurately without special tooling or fixtures custom designed for the purpose. The reason for this is that typical general purpose spectrophotometers have optical systems that assume the sample is flat and will not introduce any power or other aberration into the optical system. If the surface is not flat, the sample-introduced aberrations can change the optical system between the reference beam and the sample beam. This change in the optical system manifests itself as a change in the transmission or reflection of the part being measured, but which, in fact, is not related to the performance of the surface being measured.

It is easier to measure the transmission of a nonplanar sample, e.g., a flexible sheet, than to obtain its reflection. Relatively little aberration is introduced in a collimated beam by a thin sheet with parallel surfaces, and this is how a transmission measurement is made. In reflection, however, the nonplanar surface has a first order effect on the wavefront so that deviations from a flat surface can cause a large loss in the accuracy of the measurement. A spectrophotometer can be used to obtain a qualitative sense of the spectral shape of a nonplanar sample, but as some of the light may be lost after it has been distorted by the sample, the peak value may not be accurate. An integrating sphere may be helpful in desensitizing the detector to beam alignments and aberrations.

Some instruments have been available in the past that made spot reflectance measurements. These were capable of measuring curved parts more accurately than standard instruments. Currently, there is only one company making a spectrophotometer capable of measuring the reflectance of small, nonplanar spots.*

5.1.9. Sample Heating

Care should be exercised during the measurement of narrow bandpass or precision edge filters to ensure that they are measured at the temperature at which they will be used. There are two situations here:

1. The spectrophotometer can warm a sample while measuring it
2. The filter is intended for use at a temperature other than ambient

As mentioned in section 3.7, the performance of most filters is a function of temperature. Many antireflection coatings and other filters with broad, flat spectral shapes are not temperature sensitive because a small wavelength shift in their performance would probably go unnoticed. In a narrow bandpass filter, on the other hand, the spectral position of the pass band may be critically located to accuracies of better than 1 percent. In these instances, the shift introduced by a change of temperature can be significant.

Radiant heating of a sample by the light source in a spectrophotometer may occur:

1. If a filter is absorbing over the spectral range of the light source in the spectrophotometer
2. If the wavelength selective device in the instrument is located between the sample and the detector

It is common in infrared instruments to place the sample between the light source and the wavelength separator so that some visible light arrives at the sample to locate the beam; in visible spectrophotometers, the sample is placed after the wavelength separator. Another source of sample heating is the spectrophotometer light source via conduction and convection mechanisms. These sources of heat need to be evaluated when measuring heat sensitive filters.

Filters that are used at cryogenic temperatures should be measured at their operating temperature. A cryogenic sample holder is required to make these measurements (Fig. 5.3). Care should be exercised to prevent the window from frosting during the measurement process, otherwise unreliable values are obtained. The filter appears to change with time as the frost on the windows accumulates. The sample should be well cooled before measurements are started. After the measurements, the sample should be examined to ascertain that the low temperatures did not damage it.

5.1.10. Measuring Absorption by Calorimetry

Calorimetry is a very accurate way of measuring the absorption in a coating. This technique is accurate because it quantifies the absorption by measuring the tem-

*Model 1001 Spectrophotometer is available from Zygo Corporation, P.O. Box 448, Laurel Brook Road, Middlefield, Conn. 06455.

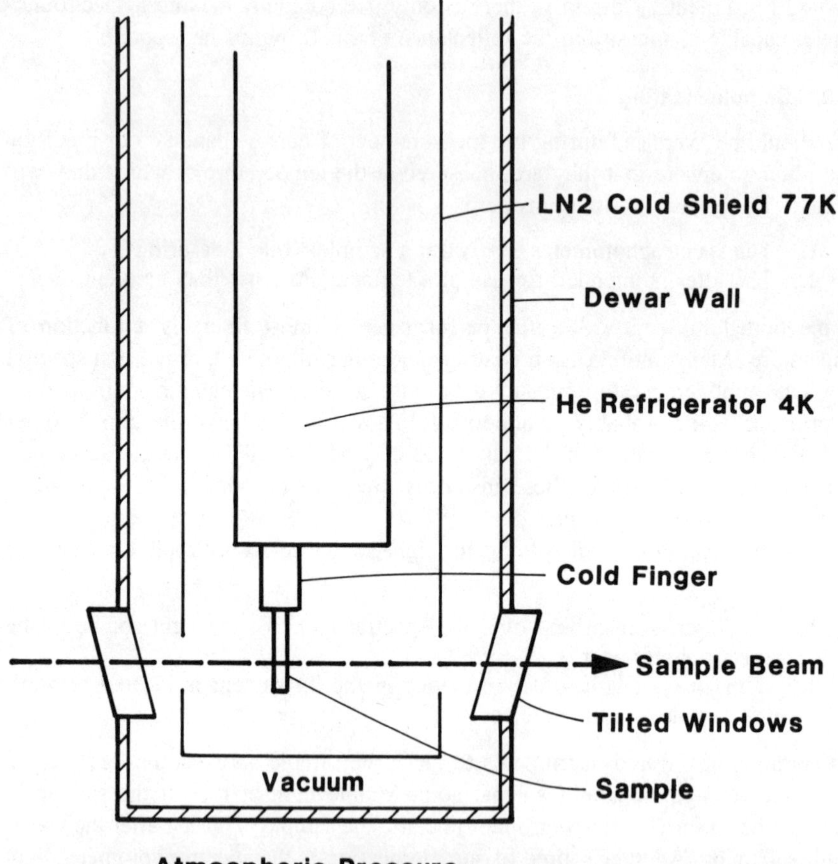

Figure 5.3. Diagram of a cryogenic sample holder for a spectrophotometer. The windows' tilts (shown exaggerated) prevent multiple reflections within and between them from contributing to the performance spectrum. Note that the windows must be kept warm or purged on the outside with a dry gas to prevent condensation of moisture on them. The windows need to be transparent over the spectral range being measured, and they may require antireflection coatings.

perature rise in a sample illuminated by a laser beam. The temperature rise is, to first order, proportional to the power in the incident beam (assuming, of course, that the power level is not so great as to damage the part being measured). High sensitivity is achieved by using low thermal mass and low specific heat substrates, and by designing special fixtures to hold the sample in thermal isolation. With proper instrumentation, light scatter from the substrate is not a problem. Ultrasmall thermocouples measure the temperature rise of the sample. Mounting the sample in a vacuum chamber evacuated to a pressure of 133 mbar (100 torr) eliminates any heat losses from the sample by conduction to the air and by convection. Figure 5.4 shows a schematic diagram of a calorimeter. In addition to the thermal con-

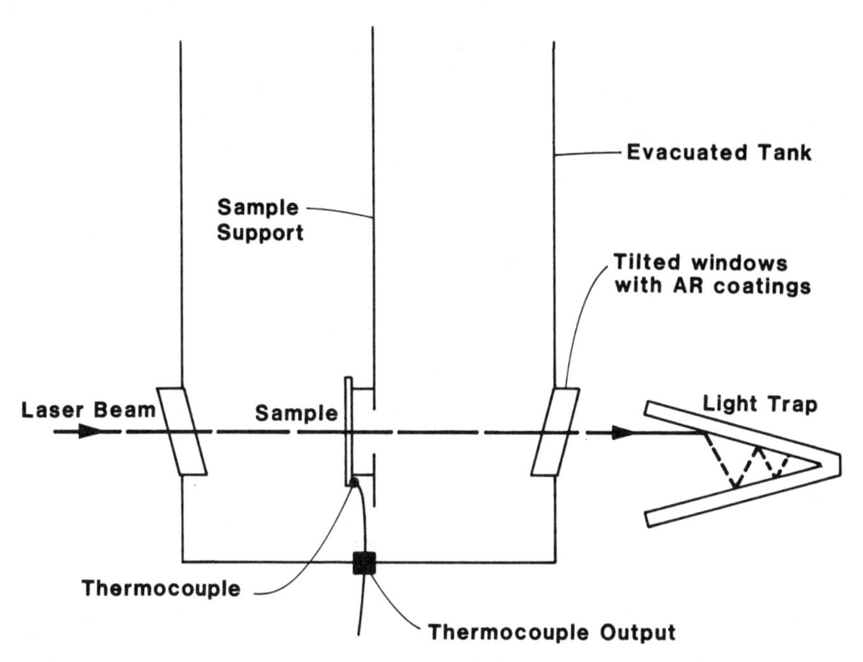

Figure 5.4. Diagram of a laser calorimeter. The incident laser power can be in the 1- to 10-watt range. The chamber is evacuated at least to a pressure at which the heat loses by air conduction and convection become insignificant in comparison to other experimental errors.

siderations, care needs to be exercised to assure that the transmitted beam (for transparent substrates) is directed properly to a beam dump, lest the reflected beam return to the sample area, additionally heat the sample, and cause systematic errors. A stray beam such as this could also heat the measuring thermocouple. The typical procedure relies on measuring the heating and cooling rate curves of the sample, and deriving the absorption from these data. Calorimeters can measure absorptances down to 2×10^{-6} (Allen et al., 1978).

5.1.11. Colorimetry

The perceived color of the light reflected or transmitted by a filter can be important in some applications. Colorimetry quantifies the color of a light beam and gives the results in standard terms. A common means of providing this is based on the Commission Internationale de l'Eclairage (C.I.E.) chromaticity diagram, Fig. 5.5. This chart is based on data gathered in the late 1920s and codified in 1931. Recent improvements to this chart include the 1976 CIE-UCS system (CIE, 1978), which displays an improved uniformity of color differences over its predecessor (Keller, 1983). This latter chart is shown in Fig. 5.6.

Two instrumental methods are used to evaluate color. In one system, the intensity of the light in the visible portion of the spectrum is measured. The color

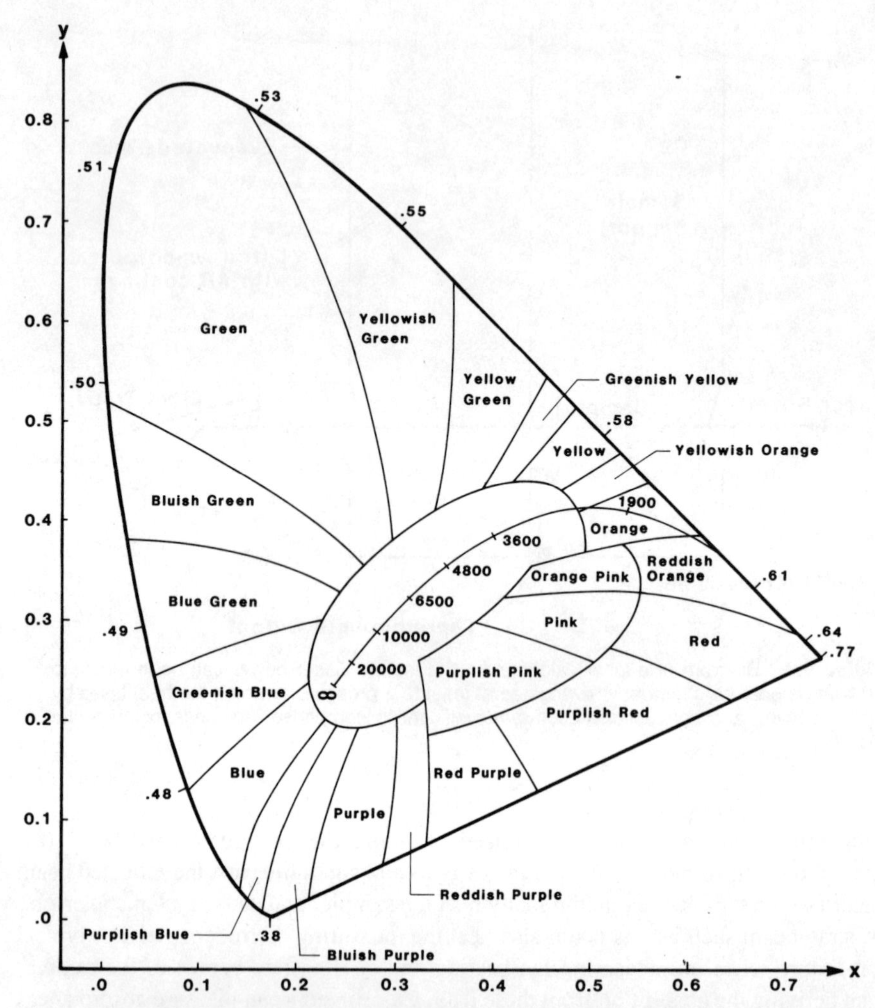

Figure 5.5. CIE color coordinate diagram (1931) (adapted from a Photo Research chart). This is the X–Y coordinate system with the color regions bounded by Kelley (1943). The solid line with numbered tics is the locus of the coordinates of the color of a blackbody emitter at the indicated temperatures. The wavelength in micrometers of the pure colors at the periphery of the diagram are indicated.

of the beam is quantified by calculating the tristimulus values according to the following equations:

$$X = \int \phi(\lambda) x(\lambda) d\lambda$$
$$Y = \int \phi(\lambda) y(\lambda) d\lambda$$
$$Z = \int \phi(\lambda) z(\lambda) d\lambda$$

where ϕ is the spectral radiant flux and x, y, and z are the red, green, and blue spectral tristimulus functions, respectively.

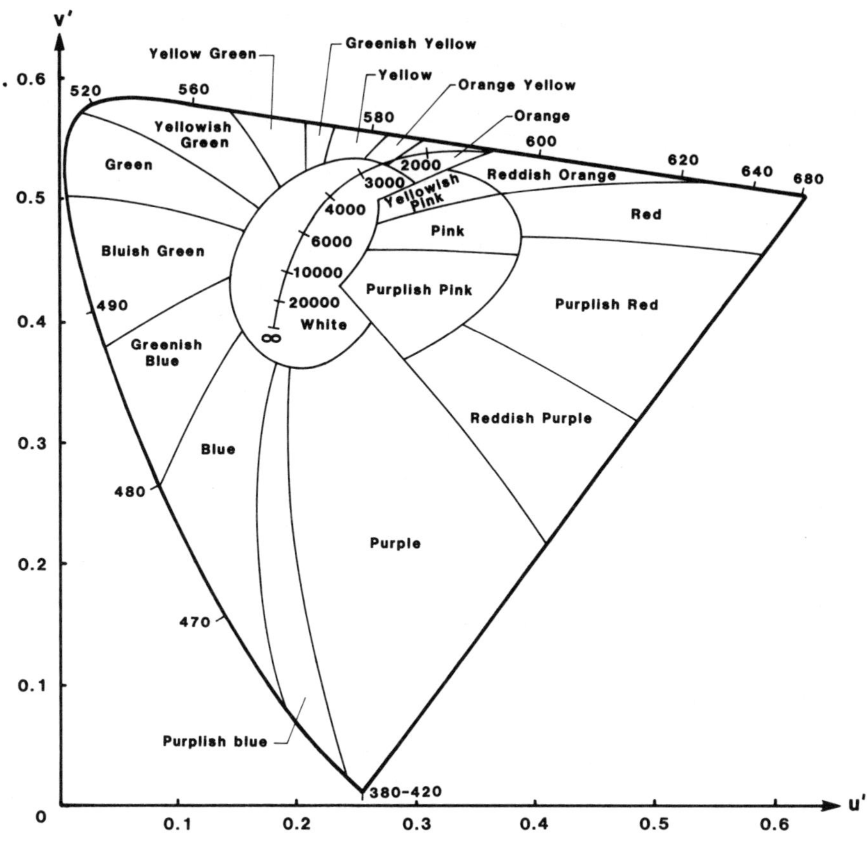

Figure 5.6. CIE (1976) color coordinate system (Keller, 1983) (adapted from a Photo Research chart). This is the u'–v' system. On this representation, equal distances approximate equal distinguishabilities of colors. The solid line with numbered tics is the locus of the coordinates of the color of a blackbody emitter at the indicated temperatures. The wavelength in nanometers of the pure colors at the periphery of the diagram are indicated.

The 1931 CIE chart displays normalized forms x and y of the tristimulus values:

$$x = \frac{X}{X + Y + Z}$$

$$y = \frac{Y}{X + Y + Z}$$

In the 1976 form of the chart, a different transformation is used:

$$u' = \frac{4X}{-2X + 12Y + 3} \quad \text{and}$$

$$v' = \frac{9Y}{-2X + 12Y + 3},$$

and the values of u' and v' are plotted on the latter chart.

The second method of determining chromaticity coordinates makes use of calibrated transmission filters to do these calculations optically. The filters' transmission pass bands match the three functions *x, y,* and *z*. One or more detectors convert the filtered light to electrical signals. Simple circuits can then be used to provide an indication of the color coordinates. In this approach, the calibration can be difficult and subject to drift. This is not the case with the technique that uses a dual-beam spectrophotometer. In addition, the spectral curve from a spectrophotometer supplies added information not available from the integrated quantities. On the other hand, the filter approach can be packaged in a small volume and taken out of a laboratory and into the field.

5.2. ENVIRONMENTAL PERFORMANCE AND MEASUREMENTS

An optical filter must be robust enough to survive in the environment in which it will be used, along with the conditions encountered along the way to that final location, e.g., shipping and storage. This is a difficult area to define ahead of time, as these conditions are subject to change with time. In addition, it is difficult to guarantee that a filter will survive for *x* years based on 30 days' testing, for example. Because the mechanisms for degradation are not known precisely, but rather only in qualitative terms, no theoretical models have been devised that will predict the lifetime of optical filters in storage or in use. This section describes the methods currently available to address these needs from a pragmatic point of view.

Pragmatic and functional tests have been devised over the years to evaluate filters more or less objectively. These tests have been codified in various governmental specifications; some of these are included in the Appendix B for reference purposes.

From an optical point of view, the *lifetime* of a typical coating is essentially unlimited because it is a passive component in the same way that a lens does not *wear out*. In some instances, however, the components of the filter may be attacked chemically or mechanically by environmental agents. An example of the former is sulfur attacking a silver coating, while the wiping of a coated surface with an abrasive material is an example of the latter. Obviously, it is not possible to predict how long a coating will last under all conditions as there are so many variables that could affect the reaction rate. Parameters, such as temperature and humidity, are synergistic, and they are not typically well controlled. The only way to evaluate completely the suitability of a coating is to get a sample of it, expose it to the environment, and observe its degradation. If it degrades too much or too rapidly, a different design should be investigated.

The tests mentioned in this section can be done singly on separate coated substrates or sequentially on one coated substrate. An advantage of separate substrates for each test is that the tests can proceed in parallel, thus cutting down on the time required to complete the tests. This can be significant, as some tests can take a week or a month to complete. When this is the case, i.e., the testing time is long and the available time is short, one can use a combination approach by making the assumption that the parts will pass the testing. Such an assumption is reasonable because the environmental durability of a coating type rarely changes given a well developed and controlled manufacturing process. In such a scenario, some witnesses undergo a cursory testing to ascertain in as short a time as possible that the basic process is sound, whereas others are subjected to the rigorous proof testing over a longer time period. In this way, it is possible to monitor a production process and receive timely feedback of information to control the process, without waiting for the definitive testing. On the other hand, these approaches can use up many witnesses and valuable space in the coating chamber.

Sequential testing is more rigorous than testing separate substrates. For example, a sample may pass a test such as adhesion prior to its exposure to humidity, and then fail the same adhesion test after the humidity exposure. Care should be exercised in deciding which tests are required before one specifies the environmental testing, because this could increase the difficulty of manufacturing the filter. Worse yet, in some instances, is the fact that it may not be possible to build a filter that meets the specified environmental durability constraints, or its optical performance may be reduced to meet them. If these requirements are not operationally necessary in the first place, then the result is an unnecessary reduction in performance.

In the evaluation of the results of an environmental test, one should be aware that the level of reflectance of the coating being tested has an effect on the appearance. Most tests rely on visual examination of the part after the test is complete to determine whether or not damage has occurred. A coating whose reflectance in visible light is very low will allow the observation of very low levels of scattered light, whereas the same physical damage in a coating whose reflectance in the visible is high may not be visible against the high reflectance background.

5.2.1. Use of Witnesses in Environmental Testing

In the previous discussion, we have referred to the testing of witness parts, rather than the testing of actual coatings and substrates that may actually be used in some application. There are a number of reasons why one should consider testing witness parts rather than the actual deliverable parts.

One or more witness parts must be used when:

1. The actual part is too large to be tested directly
2. The actual part is too expensive to risk in testing
3. The test is for informational purposes only
4. The shape or finish of the coated parts do not lend themselves to simple measurements
5. The testing is destructive

The measurement of the optical parameters, e.g., the reflectance, of a coating is generally a nondestructive test, and no damage is done to the coating. High-power laser damage testing is an exception to this. Environmental testing, on the other hand, must sometimes damage the substrate, as in abrasion testing that continues until damage is observed. Environmental tests can also leave a residue on the surfaces. Thus, a strong case can be made that an actual part should never be tested. In practice, most environmental testing is done on witness parts.

There are some precautions that must be taken to ensure that the witness is a true representation of the actual part. The use of a witness is acceptable if the results are really what would be obtained if the actual part were measured. This can usually be achieved by coating the witness(es) in the same chamber and at the same time as the actual part is being coated. Temperature, angle of deposition, and coating thickness are some of the coating parameters that must be the same on the witness and on the part. One must exercise care in the comparison of the actual and the witness parts to be certain that the coatings on them are indeed similar in performance.

5.2.2. Adhesion Testing With Self-Adhesive Tape

In the vast majority of uses, coatings should not become separated from their supporting substrate during their lifetime, nor should they fail at the interfaces between layers. Adhesion tests are performed to validate the integrity of a coating and to demonstrate the absence of a vulnerability to stress failures. A number of tests have been devised to determine if the coating adheres well to the substrate. In this section, we discuss only the test based on the use of self-adhesive tape. Other adhesion tests are discussed later in section 5.2.6.

The most familiar adhesion test is the Scotch Tape® test, or simply, *tape test,* in which a specified pressure sensitive adhesive tape is pressed onto the coated surface and then peeled off again. The test procedure specifies that if no coating material is removed by this test, then the coating has passed the test. The coating and the substrate should be clean and dry before this test is performed.

An interesting phenomenon occasionally occurs when certain films, such as metallic lead, are tested this way. The adhesive sticks much more tenaciously to the lead coating than to its cellophane backing. As a consequence, when the tape is pulled off the coating, the adhesive remains behind on the coating. The adhesive residue may be cleaned off with a solvent, and the part examined for damage. If there is no damage, the part passes the test (but the tape fails?).

The standard tape test is nondestructive if the coating passes the test. On the other hand, a failure is inherently destructive because some of the coating has to be removed by the tape.

A more challenging and rigorous adhesion test was adopted by the paint industry, to test the adhesion of paint to a surface. It is occasionally used to test optical coatings, though this is by no means a standard in this industry. It is inherently destructive. It consists of scribing with a razor blade, diamond knife, or other sharp tool a crosshatched pattern on the coated surface to be tested. Pressure-sensitive adhesive tape is applied to the scribed area and then peeled off. If any of the coating is found on the tape after it has been removed, the coating has failed the test. Naturally, because this is a destructive test, it should be performed on a witness substrate.

5.2.3. Abrasion Testing

Exposed coated surfaces frequently need cleaning, as do all other exposed optical surfaces. The typical user may wipe the optical surface with a cloth; sometimes, the cloth may be moistened, at other times, it may not. If the coating is soft, this treatment will likely damage it. The abrasion test attempts to evaluate and to quantify the durability of the coating to this type of abuse.

Two tests are in common use for measuring the abrasion resistance of coatings: the cheesecloth and the eraser tests. The tests are always performed on clean areas of the coating.

5.2.3.1. Cheesecloth Test. In the cheesecloth test, a pad of cheesecloth approximately 9.5 mm (3/8 inch) in diameter and 12.5 mm (0.5 in.) thick, is pressed against a coated surface with a force of 4.4. newtons (1 lb of force). The pad is moved back and forth 50 times (25 cycles or round trips) over a given coated area. The substrate is then examined under specified lighting conditions (see Appendix B) to determine if any damage has been caused by the test. This examination is frequently a source of disagreement as the examination conditions are critical to achieving uniform and repeatable results. The specifications state that *no signs of deterioration* will be seen; in marginal cases, the decision on the presence or absence of sleeks and scratches is somewhat subjective. The interpretation of these tests is a skill that improves with experience.

5.2.3.2. Eraser Test. The second abrasion test in common use is the eraser rub test. In this test, a well-specified pencil eraser (type and hardness of the rubber and the amount of pumice in it are specified) is pressed against a coated surface with a force of about 9 to 11 newtons (2 to 2.5 lbs. of force) and rubbed 40 times over a given coated area. As with the cheesecloth test, the coating is then examined for damage. This test is much more difficult to pass than the cheesecloth test as the pumice is a much harder material.

5.2.4. Moisture Exposure

A number of humidity tests are extant, and they vary in several ways. In some, the temperature and humidity are kept constant, whereas in others, one or both of these parameters vary as a function of time. In the tests in constant conditions, the specifications give the number of hours of exposure time the sample undergoes. In time-varying tests, the numer of cycles the witness must withstand is the criterion specified.

The goal of all these tests is to accelerate the aging process so that some meaningful data can be obtained in a short amount of time. In the ideal case, the coating systems behave in a linear manner, i.e., the amount of time in a humidity chamber is proportional to a certain amount of time in an actual or field environment. This is not necessarily true in reality. More likely, the degradation mechanism that takes place in highly accelerated testing is not the same one that causes the longer term degradation. The tests are used despite these problems because they are the best techniques currently known to accelerate the aging process simply and repeatably. In practice, the testing has evolved such that when a sample can pass a given test, one can be reasonably confident that the part will last the anticipated lifetime in some environment. There is no firm scaling factor for any of these environmental tests. The typical approach is that if a part, for example, passes a humidity test such the 24- or 48-hour test, then the coating is reasonably durable. The same can be said of the other tests.

5.2.4.1. Humidity Testing. Humidity tests consist of placing a witness part in a chamber whose temperature and humidity are controlled. Common time and humidity conditions are specified in Appendix B. The witnesses are placed in the chamber at specific times during cycled test sequences so that the possibility of condensation of water on the coatings is minimized.

An example of an effect that occasionally takes place in a humidity chamber in which both the temperature and the humidity level are cycled is that condensation sometimes does place, either intentionally or not. The liquid water on the surface may cause chemical reactions that would not happen if the part were kept free of liquid water. The fact that this water is essentially distilled increases its reactivity significantly. This reaction can cause the part to degrade, and thus give the appearance of a shorter lifetime than would be found in actual practice where the optic is kept dry.

5.2.4.2. Solubility Testing. Coatings may be required to withstand a number of solubility tests to demonstrate that they are durable enough for certain applications. A number of standard tests are available, for example, as given in Mil-C-48497 (see Appendix B). Some tests require that coatings withstand immersions in trichloroethylene, acetone, and ethyl alcohol for relatively short times to meet some specifications. Optional tests include a 24-hour soak in a saline solution or in distilled water. All of these tests are performed at or near room temperature.

5.2.5. Salt Fog Exposure

Filters that are to be used near saltwater should be tested for their resistance to degradation by it. A standard test consists of placing the witness parts into a chamber in which a continuous salt fog bathes them (Mil-STD-810D, Method 509.2). The composition of the artificial seawater that is used to make up the fog is also defined in the test specifications. This is the only test, other than the solubility ones, where liquid water intentionally comes in contact with the substrate for any length of time.

In general, a dielectric multilayer filter that survives for a given period of time in a normal humidity chamber will survive the same length of time in the salt fog test. Metal films, on the other hand, are much more susceptible to the salt fog and its reactive ions of chlorine and sodium. Filters with metallic layers in them must be screened carefully if they are to be used in or near saltwater environments.

5.2.6. Other Tests

Additional tests are available to evaluate coated parts. Practically all thin film interference filters undergo at least some visual examination for cosmetic defects before they are placed in service. At the very least, they are inspected when they are removed from the coating chamber, and again when they are installed in service. Microscopic examination is one test that is extremely sensitive yet not frequently used, possibly because it is not as quantifiable as other tests. Many times, defects that are not measurable are readily seen by eye.

5.2.6.1. Visual Inspections. One area where visual examination is the typical method used is in the examination of parts for edge chips, scratches, and cracks. Of course, this inspection must be done on the actual parts instead of witnesses. The chips must be less than a specified size, depending on their depth, and their number is limited as well. Standards for scratches and digs (pits) are available* which allow an inspector to compare visually with the standard the size (width, length, and depth) of a scratch or a sleek he or she has observed. The inspector who does the visual examination must have a certain talent and lots of experience in looking at samples to be repeatable, especially when the parts are only marginally sleeked.

5.2.6.2. Falling Sand Test. This test is useful in situations where the optical element will be exposed to an abrasive environment, e.g., spectacles and sunglasses (on a sandy beach), and coated architectural glass. In this test, sand in a hopper at a fixed height falls on a slanted substrate. The rate at which the sand falls is controlled, as is the test time. The sample is examined at intervals to see if the sand has eroded, abraded, or otherwise damaged the coating. This tests both the hardness and the adhesion of the coating. A similar test is formalized in Mil-C-14806, para. 3.11.5.

5.2.6.3. Other Adhesive Tests. Adhesion of a coating to the substrate is of prime importance to everyone who uses thin film filters. A number of tests have been devised to obtain a quantitative value for the strength of the bond between a coating and the substrate. A few of these are discussed here. The reader interested in more details about adhesion testing than is given here is referred to Mital (1978).

Topple Test. A true measure of adhesive strength between a coating and the substrate is to attempt to pull the coating off the substrate. This is what the adhesive tape test attempts to do qualitatively. There is no measure of how well the coating stuck to the substrate—it is a go/no go type of test.

An instrument is commercially available that measures the tangential force required to make a metal post fall over. From the geometry of the test fixture, it is possible to convert the tangential force to a normal force that is related to the adhesive forces between the substrate and the film, or between some internal film interfaces, if the failure does not occur at the substrate/coating interface. The post is attached to the coating with an organic adhesive and allowed to cure. This bond must be stronger than any of the bonds being tested; otherwise, it will be the one that fails. The tangential force needed to make the post topple is recorded and then converted to a normal force. This test is useful in that it provides a quantitative

*Procurement and/or calibration of scratch and dig standards can be obtained from: Armament Research and Development Center, U.S. Army Munitions and Chemical Command, AMSMC-QAF-I, Dover, N.J. 07801-5001.

value of the adhesive forces. It is assumed that in applying the adhesive to the coating there is no chemical reaction between the adhesive and the coating. For some epoxies, this may be difficult to determine. This is a relatively new test and it is not in widespread use to certify production coatings at this time.

Another test that determines the bond strength between a coating and the substrate is the measurement of the normal force required to pull off a stud securely cemented with an epoxy adhesive to a coated surface.* A mechanism applies an increasing force normal to the surface and provides an indication of the force at which the bond fails. The stated accuracy of this method is within plus or minus 10 percent. The primary cause of the variation is caused by the error in determining the size of the contact area between the stud and the coating.

Indentation Tests. This test was first described by Heavens (1950) and later expanded by Benjamin and Weaver (1959). It involves drawing a smooth point across the coated part. The load on the point when the film is removed is a relative measure of the adhesion of the film to the substrate. The absolute value of the shearing forces can be calculated from the load, the radius of the point, and the indentation hardness of the substrate. This method was used by Benjamin and Weaver (1961) to determine the adhesive strengths of a number of metal films on a glass substrate.

5.2.6.4. Combined Adhesion and Abrasion. A test that evaluates the adhesion as well as the abrasion resistance of a coating is the Taber† abraser test. A rubber wheel loaded with an abrasive rolls in a circular path on the surface to be tested. The motion is complex as the width of the wheel is such that some dragging or sliding takes place along with the rolling. This test is not widely used in evaluating optical thin films at this time.

No satisfactory correlation has been made among the results of the various abrasion tests.

5.2.6.5. Combined Temperature, Humidity, and UV Light. A humidity test sequence that has been found to be very vigorous in attacking coatings is the combination of a cycled temperature and humidity environment with simultaneous exposure to ultraviolet light. Some reactions, which otherwise may not be energetically feasible, can take place when the ultraviolet photons contribute their energy to the chemical reaction. At first glance, this type of test may appear too severe; however, especially in situations where the filters, in particular those with metal layers, are exposed to sunlight, this test may be valid. It has found use in the testing of paints and other similar coatings. No formal test procedure and specifications exist for use of this type of test for optical coatings.

*An instrument which performs this test is called the "Sebastian Adherence Tester," and is available from Quad Group, 331 Palm Avenue, Santa Barbara, CA 93101.

†Model 503 is made by Teledyne Taber, North Tonawanda, N.Y. 14120.

5.2.7. Cycled Temperature Testing

In some applications, filters are exposed to wide ranges of temperature. Prior to their being placed in service, they should be tested to ensure that they will survive the anticipated temperature excursions. The most common upper temperature limit encountered is 100°C. There are a number lower temperature limits in common specifications, including 0°C, −40°C (dry ice), 77K (liquid nitrogen), and even 4K. Cyclic temperature testing is frequently required of space satellite components and those installed in unpressurized portions of high altitude aircraft. The number of test cycles should be in excess of the anticipated operational number, but there is no hard and firm rule for the safety factor to use. It is probably safe to say that the first temperature excursion is the most difficult, and that if the filter survives the first cycle it will likely survive the succeeding ones.

5.2.8. Combined (Sequential) Testing

Some test procedures can produce hidden damage that will cause subsequent tests on the part to fail. If the tests were done on separate substrates or in the reverse order, the conclusions drawn from the results might be different. An example of this is doing an eraser test on a high reflector prior to a humidity test. The eraser test may cause minor damage to the coating, but as the coating is a high reflector, the damage goes unnoticed and the test is deemed a "pass." In the subsequent humidity testing, the moisture enters through the damage sites and attacks the inner layers. This could not happen if the surface layers were intact. Thus, it is easy to see that sequential testing can lead to a more difficult test sequence than individual or parallel testing of different samples. In addition, it is also obvious that the ordering of the test sequence may have an effect on the results.

A general rule concerning the order of testing is that the quick tests be done first to produce results as quickly as possible for process control. Tests, such as abrasion and adhesion, fit this criterion in that they can be done in a few minutes. If these tests fail, the information can be given to the production operation in a timely manner. The longer duration tests, such as humidity exposure, can then follow. Finally, the comprehensive product testing to verify to the customer's satisfaction that he or she is getting a quality product can be done.

5.2.9. When to Test

The age of the sample is sometimes important to the testing process. The wavelength shift of pass band and edge filters due to moisture absorption in the coating has previously been mentioned in section 3.7. Similar relationships can arise in environmental testing. For example, atmospheric moisture can lubricate a surface so that its coefficient of friction is not as high as one that is freshly removed from the coating chamber. An eraser abrasion test made on such a part will, therefore, vary with the age of the coating, as well as on the relative humidity. The rate of change will be the largest immediately after the sample's exposure to the atmosphere, and it will decrease to a low value after a few hours, depending on the relative humidity.

The adhesion of aluminum films to a glass substrate increases with time (Benjamin and Weaver, 1961). The conclusion one draws from all of this is that a standard period of time should pass after removing the parts from the coating chamber before testing to obtain the maximum amount of test to test repeatability.

Another factor that enters into testing is the cleanliness of the parts being examined. Abrasive particulate matter on the surface during eraser or cheesecloth testing may affect the test results. Similarly, particles between the adhesive tape and a surface could reduce the adhesive forces between the tape and the coating, thus invalidating the results. It is a standard practice to clean a surface before testing it. The solvents used for this purpose should be carefully chosen, as some of these solvents have been known to cause problems of their own. It was discovered some years ago that when chlorinated hydrocarbon liquids were used to clean parts before coating, the parts failed humidity testing. Apparently, some hydrogen chloride residue remained on the surface after the cleaning step, and this subsequently turned to hydrochloric acid when exposed to moisture. This acid then caused deterioration of the films.

Filters are sometimes subjected to moderately high temperatures (>100°C) during the manufacturing process. This may occur intentionally and be done by the filter manufacturer to densify the coating or desorb moisture. It can also be done incidentally, in that it is a part of the manufacturing process after the filter has been incorporated into another subassembly or component. The filter can shift its performance as a function of wavelength as a result of this heating. Also, the optical performance can charge because the films in the filter change their oxidation state. Generally, this means that the coating becomes clearer and less absorbing after heating. In other cases, however, the coatings become absorbing as oxygen is driven out from the layers, and then cannot get back in after the heat is removed. In other situations, the coating materials are simply unstable at elevated temperatures. Witnesses for filters that will be heated in use must also be heated prior to environmental testing to simulate the life cycle of an actual part. This is especially important if the filters undergo large temperature excursions during their subsequent processing or in their use.

5.2.10. Formal Specifications

Both military and civilian agencies have formulated standards by which thin films can be evaluated. The most common ones in the United States are the military specifications. The designations of these generally start with "Mil" and are, therefore, colloquially called "Mil specs." It is understood that the West German standards organization is preparing a set of specifications for thin films for international civilian use.

The original military specifications were for magnesium fluoride antireflection coatings and for aluminum high reflector coatings. These specifications were subsequently adapted and applied to other types of thin film coatings, to which they were more or less appropriate. Of late, a new set of Mil specs has been issued that deals more comprehensively with the specific needs of modern optical coatings, and interference filters in general. The designations of the new specifications are

Mil-C-14806 Coating, Reflection Reducing, for Instrument Cover Glasses and Lighting Wedges
Mil-C-48497 Coating, Single or Multilayer, Interference: Durability Requirements for
Mil-F-48616 Filter (Coatings), Infrared Interference: General Specification for

Because these are of general interest, they are reproduced in their entirety in Appendix B. The previous specifications (now obsolete for interference filter applications, though still referenced) are

Mil-C-675 Coating of Glass Optical Elements (Anti-Reflection)
Mil-M-13508 Mirror, Front Surfaced Aluminized: For Optical Elements

A standard testing specification is

Mil-STD-810 Environmental Test Methods and Engineering Guidelines.

All manner of testing military components and assemblies are described in this large manual. The majority of these tests are not applicable to optical coatings. Specific sections of this standard, however, are referenced in other specifications.

5.3. AGING

Optical filters are passive devices, and as such, should have an unlimited lifetime when properly used. There is nothing to wear out, and nothing in the filter is consumed. On the other hand, nonoptically related factors can sometimes put a limit on the shelf life and the working life of a filter. Of course, mechanical shocks can break some of these filters, whereas too high an energy level, e.g., too powerful a laser pulse, can damage them. Though most coating materials are inorganic, some molds and mildews can grow on just about any solid surface, including glass. Thus, an optical filter, like any other optical component, should be kept in a clean, dry environment for long-term storage.

5.3.1. Shelf Life

Many of the materials that are used to form coatings are refractory fluorides and oxides that are insensitive to temperature variations. On the other hand, humidity is a problem with some of these materials. For example, chiolite and cryolite (sodium aluminum fluorides with different ratios of sodium and aluminum ions, nominally approximately Na_3AlF_6) are two materials used in the earliest vapor deposited coatings. They were used because they are easily evaporated from resistively heated sources, and because they have low refractive indices, which make them among the best choices for single layer antireflection coatings. Unfortunately, these fluorides are soluble in water. When coatings with these materials in them are exposed to humid environments, they may undergo changes as the ambient humidity diffuses into the filters. The moisture may simply dissolve the material and remove it. Note that a layer 1 quarter wave in optical thickness in the visible portion of the spectrum consists of only 100 to 200 atomic layers. In terms of mass density of the antireflection coating, this translates to about 300 ng/cm^2 (3×10^{-7} g/cm^2). Very little moisture is needed to dissolve away this small amount of material. Fortunately, many of the materials currently used by film manufacturers are nonsoluble compounds, so the problem of moisture's dissolving the film materials is minimized. In some spectral ranges, however, the manufacturer might still have to use somewhat soluble materials to meet a specification.

Although the use of nonsoluble oxides minimizes the problem caused by moisture, it is by no means completely solved. Condensed moisture is akin to distilled water, and this can be a surprisingly strong etchant for thin films. Only a small number of atomic layers make up a layer, therefore, it is not difficult to see why freshly condensed moisture, dissolving only a few molecules from the surface at a time, will eventually change the optical performance. Thus, a coated optical element exposed to condensing moisture may have a shorter life than one that is kept free of moisture, or even one that is immersed in less pure water.

Because humidity is such a deleterious component of the environment, various means have been designed to avoid the problem of humidity degradation. The most successful one is the hermetic seal, wherein the moisture is prevented from entering the region where the films are located. The seals can take the form of a solder or other glass to metal contact, a glass frit seal where a layer of glass is used to prevent the diffusion of water into the filter, and other similar arrangements. Inorganic materials are generally less permeable to moisture than organic ones. A rule of thumb is that water permeates a given thickness of organic material an order of magnitude faster than an equivalent thickness of inorganic material. Therefore, inorganic sealants are preferred over organic compounds for hermetic seals. Examples of coatings that need to be sealed are ones that have thin metal layers, such as Fabry–Perot cavities with silver reflectors, and optical data disk coatings that have a thin metal layer as the sensitive recording medium. High-quality filters of these types are generally supplied sealed.

Most coatings have intrinsic stresses in the individual layers, even though the net stress in the total coating may be low. As a result of aging, some of these stresses may relax because of the rearrangement of the atoms or molecules in individual films. The result of such an annealing over time could be to upset the original stress balance in the design. A low stress coating is preferable for long-term stability.

5.3.2. Material Interactions and Interdiffusion

In some uses, filters are exposed to high temperatures. This environment can make several degradation mechanisms energetically allowable. This can decrease the lifetime of a filter.

5.3.2.1. Oxidation and Reduction. A metallic coating can be prone to oxidation when exposed to high temperatures in air or other oxidizing atmospheres. Nonoxide compounds can likewise be converted to oxides under appropriate conditions of heat and oxygen. For example, a coating consisting of magnesium fluoride and zinc sulfide can degrade over a period time at an elevated temperature ($>100°C$)

in air. Several chemical reactions can occur, including the conversion of the sulfur, zinc, and magnesium to their respective oxides. Some of the reaction products are volatile, and this obviously leads to a change in filter performance. Conversion of a sulfide or a fluoride to an oxide is generally accompanied by a change in refractive index, and a corresponding change in filter performance. Similarly, oxide formation can occur with other compounds under appropriate conditions.

Some molecules, for example, titanium dioxide, can lose a portion of their oxygen complement at elevated temperatures. A chemically stable suboxide may form that is stable and that does not return to the fully oxidized stoichiometric state following removal of the heat source. This may occur because the suboxide is intrinsically stable or because the oxygen that was released diffused out of the coating, or was driven out of the thin film structure by a pressure gradient. As the temperature is reduced, the oxygen does not reenter the coating at a sufficient rate during the cooling off period. Once the system is back at room temperature, the reaction kinetics are insufficiently rapid to cause further oxidation to take place.

5.3.2.2. Agglomeration. In metal mirrors, notably silver ones, hillocks can form when the film is heated in the presence of oxygen (Presland et al., 1972). The hillocks scatter light, thus destroying the high specular reflector properties of the surface, while at the same time adding to the veiling glare in the system. If optically acceptable, overcoats can decrease the tendency for hillock formation and at the same time improve the overall environmental stability of the design.

5.3.2.3. Interdiffusion. The interdiffusion of adjacent thin film layers is another phenomenon caused by high temperatures. This effect is not ordinarily observed in films kept at room temperature. Interdiffusion consists of the intermingling of the atoms or molecules of the materials in adjacent layers. A stack composed of two or more different materials can be degraded as the two materials interdiffuse. It is also possible for the diffusion to take place between the stack and the substrate. In the ultimate situation, the coating changes from a system of alternating homogeneous, discrete layers, to a homogeneous and monolithic single layer. At intermediate stages, the coating appears as a single inhomogeneous mass. The optical properties change and the filter no longer functions as intended. On the bright side of this phenomenon, a small amount interdiffusion at layer interfaces lowers the stress levels and improves the interlayer adhesion in marginal cases.

It is difficult to predict how quickly interdiffusion effects degrade a coating. The process depends on the materials, the temperature, the time at temperature, and other factors. If such a situation in the usage of the filter is anticipated, it should be brought to the attention of the supplier so that he or she can take appropriate action to evaluate whether a problem exists or not, and to suggest appropriate tests that may be performed to verify that indeed the problem does not exist in the delivered product. In addition, the manufacturer's choice of materials could be influenced by their tendency (or lack thereof) to interdiffuse if this were indicated to be a potential problem.

5.3.3. Biological Effects

Because most filters are constructed from inorganic materials, the general rule is that one need not be concerned with biological degradation. This is correct in most instances. In some situations, however, the degradation mode of a filter can be biological, specifically fungal, in nature. Fungi (mold and mildew) live on just about any surface, including optical filters. Their secretions can cause etching of the surface, and eventually, complete failure of the coating. In an operating environment that is hot and damp, the optical surfaces should perhaps be treated with an antifungal agent. Once again, the filter supplier should be apprised of the environment so that some suggestions can be made as to what antifugal treatments the coatings may and may not survive.

5.4. COSMETIC AND OTHER VISUAL DEFICIENCIES

In this section, we discuss features that may be seen on the surface of an interference filter. Some defects may be apparent only when a part is examined under the proper lighting conditions. Other features in a coating cannot be detected with standard instrumentation, such as spectrophotometers, yet they are readily apparent with the naked eye. In either case, the fact that a surface defect can be observed does not necessarily mean that the part is unsatisfactory or otherwise inadequate to serve the intended function. It indicates that the human eye is very sensitive to certain types of optical patterns and color changes, and these may or may not be relevant to the inspection of a given sample. We give examples of some of these situations.

5.4.1. Sleeks, Digs, and Scratches

A sleek is a very small scratch whose width cannot be resolved with the naked eye. By comparison to a sleek, a scratch is relatively large. The United States military specification designated Mil-O-13830, revision L, specifies how to evaluate a scratch and give it a numeric value. The number given to a scratch is 10 times the width of the scratch in micrometers (more or less). These dimensions are not precise and are more a rule of thumb for evaluating the cosmetic appearance of a scratch (Young, 1985, 1986; Young et al., 1985).

A *dig* is a scratch whose length is equal to its width, and it is evaluated in the same way as a scratch.

The design of the inspection station used to examine parts for sleeks, digs, and scratches is very important in determining the level of sensitivity that can be attained. The arrangement of the illumination and a dark background influences the visibility of the light scattered from the defect. A dark background reduces the light reflected by a high reflector, while it provides a dark background against which to view an antireflecting coating.

Scratches and digs may originate from a number of causes.

1. Improperly prepared substrates: The raw material for the substrate needs to be of high quality, homogeneous, and bubble-free. The optical surface must be properly ground and polished, or otherwise given a smooth surface.

2. Improper handling: Substrates are subjected to a large number of handling operations, including incoming quality inspections, cleaning, loading into and unloading from the coating chamber, and final inspection and packaging. Unless carefully done, each of these steps may degrade a part.

Tooling is typically configured so that the parts are supported around the periphery of the coated area. Improper care in tooling design, poor fabrication of the tooling, or differential thermal expansion between the tooling and the substrate material may all lead to scratching of the substrates during coating.

In this discussion, we ignore the obvious problems that can occur in improper substrate handling and packaging.

5.4.2. Blotches and Stains

Coated surfaces sometimes have areas in which the reflected color of the coating is different from other areas. These have a number of names including *discolorations, blotches,* or *stains*. Generally, these are quite faint and can be seen only in reflection and against a dark background. Even when the stain is quite severe and is very apparent to the eye, it is frequently very difficult to measure the difference between the two areas with a spectrophotometer. The reason this is so difficult is that the human eye is extremely sensitive to color differences in adjacent areas. The typical stain has quite sharp edges to the discolored area, thus making it easy to see the stain. From a functional point of view, such a stain will typically not degrade the optical performance of the filter, though it might be an indication of the care taken when the substrate was cleaned. On the other hand, it should be noted that there are some optical glasses that inherently stain easily.

The appearance of a stain is sometimes observed only after a part has been coated. Even an opaque coating does not always hide a substrate defect. Stains that are invisible on the bare substrate affect the nucleation and growth of the film, causing some areas to become delineated and apparent only after coating.

5.4.3. Replication of Substrate Defects

A thin film will not *bury* defects that exist in the substrate prior to its being coated. In fact, in many instances, the coating exacerbates the situation as substrate defects perturb the otherwise planar process of nucleation and growth. Inevitably, the amount of light scatter increases after the surface is coated.

5.4.3.1. High Reflectors with Opaque Metal Layer. An opaque metal layer often replicates and enhances substrate defects. An improperly cleaned substrate causes the metal layer to be hazy. If certain areas are dirty, whereas others are clean, then some areas will have a highly specular reflectance, whereas other areas will be much more diffusely reflecting. These areas of differing specular and diffuse reflectance are readily apparent to the naked eye.

5.4.3.2. Low Reflectance Surfaces. Surfaces that have a low reflectance readily expose scattering caused by defects. Because they reflect little light, the ratio of scattered intensity to reflected background intensity can be quite high for these surfaces.

Scattering from low reflectance transparent substrates may be observed by viewing a brightly illuminated surface against a dark background. Any scattered light becomes a bright spot on a dark background, and this is readily discerned. Any dark, opaque substrate is its own dark background, and it is only necessary to illuminate the surface with a bright light.

A fundamental axiom in the construction of thin film filters is that a thin film replicates the contour of the surface being coated. This varies in degree for the various coating materials. For example, a screen of wire mesh with a spatial period of approximately 2 mm was coated with 2 to 3 cm of aluminum oxide over a long period of time. The pattern on the surface of the deposit exactly matched that of the underside where the screen was located. This was a vivid demonstration of the extent of replication that is possible. The structural morphology in many growing thin film layers can explain this kind of behavior. The growth is generally columnar, with the columns perpendicular to the surface of the substrate if the evaporant is also arriving normally. If the evaporant is arriving at some nonnormal angle, the angle of growth follows a rule that states that the tangent of the angle of growth is one-half the tangent of the angle of material arrival (Dirks and Leamy, 1977; Pulker, 1984, p. 327)

It is possible that a coating can mask surface scratches and sleeks. A germanium substrate with polishing sleeks was coated with an antireflection coating containing zinc sulfide. Microscopic examination of the coated surface showed no signs of the sleeks, yet they were readily visible on the substrate surface. Similarly, electron microscopy has revealed that some coating materials can smooth out the roughness of previously deposited layers.

5.4.4. Coating Voids and Other Defects

Coating voids, called pinholes when they are small, are areas of the substrate where there is little or no coating. A void may extend all the way to the substrate, or may only involve a limited number of layers. These defects have a number of causes.

Particulate matter on the substrate when the coating process starts can cause poor adhesion in a small area. The coating can suffer local adhesive failure as the inevitable stresses build up, either during the vacuum coating process, after the coating is complete, or during the cool down and vacuum chamber venting to atmospheric pressure. The result is an area without its full complement of the coating.

Another source of particulate material is the coating source itself. For an evaporation source, the particles can come from the source, having been ejected from the molten pool of material. This phenomenon is frequently the reason a manufacturer chooses one evaporation source type, e.g., electron gun source or covered or uncovered resistance source, over another.

In sputtering, a momentary arc between the target and some charged surface or the plasma can cause material to be blown off the target in a small area. This material can travel in all directions, so that sputtering with the substrate above the target will not eliminate this problem. When sputtering takes place with the substrate below the target, the heavy deposits that build up around the target can flake off and fall on the substrate. This configuration also is subject to contamination of the substrate from material from any of the walls of the vacuum chamber. As stresses build up, the materials on the wall may snap off with enough lateral velocity to reach the substrate.

Typically, pinholes occupy a relatively small area of the total coated surface. They are frequently apparent because they have a large contrast ratio with the adjacent coated areas. The significance of pinholes in a filter is highly dependent on the application. In some, voids are functionally insignificant, whereas in other instances, they are intolerable. For example, in a coating that will be etched with a fine pattern, a void could obliterate a part of the pattern, or cause a line to have a break in it, perhaps causing a portion of a large circuit to fail. Small openings in a blocking filter, which must have a great deal of attenuation, can allow an objectionable amount of undesired radiation to pass through the filter.

The observation of pinholes can be a tricky affair, especially when large areas need to be examined. Visual examination of a small coated area with a microscope can readily detect a void, either by the difference in intensity or by a difference in color between the defect area and the normal coating. If the substrate is transparent and the coating is opaque, the light can be incident from one side of the coating and the detector can observe from the opposite side. If the coating is an antireflection coating, then one can search for the pinholes by looking at the reflected light. A difficult situation is encountered when the coating is semitransparent, as happens in a single film of indium tin oxide on a glass substrate. Then the contrast between the coated areas and the pinholes is poor in both reflected and transmitted light. Thus, the scale-up from scanning a small sample with a microscope to scanning a large substrate surface area for pinholes is a real challenge.

Inclusions are trapped particulates. They can be caused by particles that are ejected from the coating material source or from dust or other particles that are incorporated into the film as it grows. Figure 5.7 shows a scanning electron microscope photograph of an infrared film with a particle that initiated a conical defect zone. As the film grew, the disturbed area increased in size, and the amplitude of the disturbance remained unchanged. The particle continues to make its presence felt even after it has been buried to a depth many times its diameter. The size of the disturbance increased in diameter so that the slope error became less, but the overall volume of the disturbance increased. If the particle detached from the underlying material, either due to the buildup of stresses in the coating with increased thickness, or perhaps due to differential expansion from the heating of the film from the coating source, a void many times the diameter of the particle will form. Even if the inclusion has the same index of refraction as the adjacent material, a scattering site is formed because of the displacement it causes to the other layers.

5.4.5. Cleaning Coated Parts

Thin film filters require occasional cleaning to remove dust and other environmental dirt and grime that may deposit on them. Cleaning them can be simple or difficult, depending on the type of coating and on the substrate. For example, unprotected metal layers, such as silver, gold, or aluminum, can be cleaned only with great

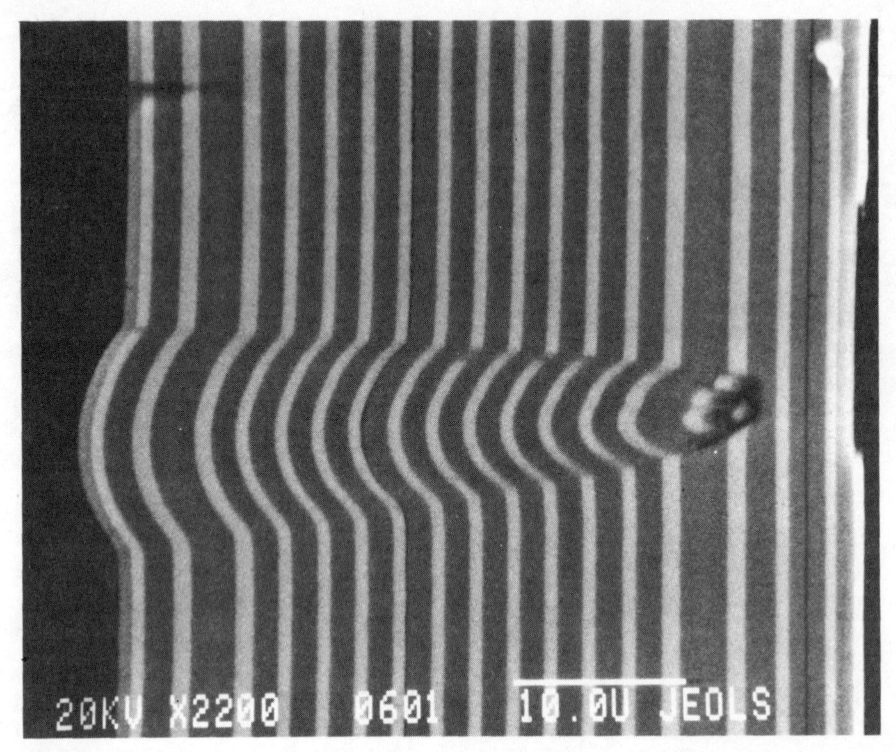

Figure 5.7. Scanning electron microscope (SEM) photo of a conical defect arising from a point defect. The white bar is 10 μm long. As more layers are added to the stack, the disturbance caused by the defect increases in lateral extent, but the defect height remains constant. (Courtesy of Phil Swab of Research Dept., O.C.L.I.)

difficulty because the metal surface is soft and easily damaged. On the other hand, a hard coating of magnesium fluoride on a glass lens is nearly as durable as the bare substrate. One would not attempt to clean a finely polished glass lens with ordinary techniques such as might be used on a window pane, and a similar caution is warranted for coated optics.

Loosely attached particles can be removed from a coated surface by blowing them off with a gas jet. This is described in the section of cleaning soft coatings.

5.4.5.1. Hard Substrates—Hard Coatings. For our purposes here, a hard substrate is one with a hardness similar to that of glass. When both the coating as well as the substrate are hard, it is straightforward to proceed to clean the coated surface with a solvent and a lens tissue. It is important that no abrasive material be picked up by the tissue and dragged across the surface, otherwise a scratch or sleek will result. The solvent could be a soap or detergent solution, an organic solvent such as isopropanol, or some commercially prepared mixture. It is assumed here that the coating is indeed inorganic, deposited in some vacuum process, and is water insoluble. Strong acids, such as nitric and hydrofluoric, and strong bases should be avoided as they can etch some of the thin film layers as well as damage some of the optical components. Chlorinated hydrocarbon solvents should be avoided when metallic layers, especially silver, are present in the coating because the chlorine can attack them.

The solvent or other cleaning solution should not be allowed to dry on the surface. There are a number of reasons for this:

1. The solution may have picked up and suspended or dissolved contaminants that were on the surface. Upon evaporation of the solvent, the contaminants will be concentrated into the remaining droplets. Thus, the residue from the entire surface will be concentrated into a few dirty spots on an otherwise very clean surface.

2. For organic solvents that are initially water-free, the evaporation of the solvent from the surface being cleaned will reduce the temperature locally. This may cause the moisture in the air to condense in the vicinity of the droplet. Finally, as the water dries, a mark will be left at the site.

3. If an organic cleaning solvent has been exposed to the atmosphere for a period of time, it can absorb some of the atmospheric moisture. This contaminates the solvent; the water will be left behind on the surface when the solvent is used for cleaning, again causing a water mark.

There are several ways of overcoming the staining problems just mentioned:

1. Do not allow the solvent to dry on the surface. A jet of filtered, dry nitrogen can be used to blow the solvent off the surface before it has a chance to dry or evaporate.

2. A clean paper tissue dragged over the surface will absorb and remove the solvent, so that no droplets are left behind on the critical surface.

3. The solvent selected should have a low volatility so that it will not readily evaporate. Fluorinated hydrocarbons are one class of such liquids. These solvents will not dry readily on the surface, nor will they cool the surface and induce moisture to condense on it. When the cleaning process is complete, the solvent can be removed by any number of ways, including those mentioned above.

Although the preceding remarks are directed at coated optics, they are applicable to uncoated surfaces as well. The treatment of substrates before coating is important to the coating manufacturer. Water marks are just about impossible to see before a surface is coated. After the coating has been deposited on the surface, however, the stained areas are readily seen. The coating materials nucleate differently in the stained areas, and this leads to different properties of the coating. In turn, these differences lead to the scattering of light, and this is usually a cause for concern.

5.4.5.2. Hard Coatings on Soft Substrates, and Soft Coatings. Although it is possible to apply a hard coating onto a soft substrate, the lack of a solid support for the hard coating will cause it to crack and fail when a sharp point force is applied to it. Short of a self-supporting film, attempts to harden completely a soft surface are doomed to failure ("You can't armor plate a marshmallow" is a classic phrase in the thin film industry).

Soft coatings are problematic regardless of whether the substrate is hard or soft.

The ultimate soft coatings are probably the metallic coatings with no overcoats, such as gold, silver, and, to a lesser extent, aluminum. As previously mentioned, there is no way that these can be cleaned with a contacting method. Particles can be blown off the surface with a gas jet, either manually supplied as from a rubber bulb, or from some other source such as a compressed/liquefied gas in a can, a gas cylinder, or a gas manifold. When using any gas source to blow off a surface, it is necessary to ensure that the gas itself is clean and does not contaminate the surface. If a rubber bulb is used, the bulb should be inspected to ensure that it is clean. For liquefied gases in aerosol cans, it is difficult to ensure that none of the liquid itself comes out of the nozzle; if this happens, it will cause spotting on the surface. Finally, when the gas comes through pipes and hoses, either from a compressed gas cylinder or from a central gas supply, it is essential to have a final filter near the nozzle to remove particles that may be in the gas lines. Filters are available that can remove particles as small as 1 μm in size.

While blowing off a surface, care should be exercised so as not to touch the surface with the nozzle. This is particularly important with a manually operated rubber bulb, as motion is required to squeeze the bulb, and this could result in a movement of the nozzle toward the surface.

A surface charge of static electricity attracts and retains dust. If the relative humidity is low, the surfaces can acquire and hold significant static charges from the cleaning process. The neutralization of surface charge reduces the amount of dust that collects on a surface. The presence of ions in the environment accomplishes this, as does a moderate to high ambient humidity level. An accessory that is useful in areas of low humidity is an electrostatic ion generator. It can be installed near the tip of the nozzle used to direct the gas that blows off a surface. The ion generator ionizes some of the gas flowing through the nozzle, and these in turn neutralize the surface charge. In another form, the generator can be part of the room air filtration system and can flood an entire room with ions.

A tissue with a solvent can be used to clean surfaces. This is a contact method, and can remove particles that are more adherent than those that are removed with a simple gas blast. The force of gravity and the liquid forces are the only ones acting between the tissue and the surface. The forces applied to the surface by the wet tissue are small, and therefore, this technique will in general suffice for plastic and other fragile surfaces. The wet tissue is dragged in overlapping paths over the entire surface without applying a normal force. The solvent should be removed before it has a chance to dry (Stowers and Patton, 1978).

Perhaps the ultimate challenge for cleaning substrates are the soluble materials such as sodium and potassium chloride. Because of their nature, they must be handled in a room in which the humidity is kept at a low value. In addition, these materials are excellent insulators, and so acquire and hold an electrostatic charge very easily.

5.4.6. Coating Thickness Uniformity

In most instances, one desires a coating whose performance is uniform over the entire useful aperture. The need for thickness uniformity arises from the fact that optical filters are generally sensitive to the thicknesses of most of the layers that comprise it. On flat substrates, this translates to a coating in which the thickness of all layers is uniform over the useful aperture.

In the simplest situation of a small flat part, the color of the light reflected by the coating can give a qualitative indication of the variation in the thickness of a film on a surface. It is fairly easy to measure the reflectance or transmission of flat parts with a spectrophotometer to determine quantitatively the thickness uniformity of a coating. On the other hand, if the part has a curved surface, such as a lens or a dome might have, then a typical spectrophotometer cannot be used to get a good photometric value.

Good photometry is not necessary in some cases to determine to first order the uniformity of a coating system. For instance, if the performance of a design has a feature with a steeply rising slope, the wavelength position of the half maximum point can be plotted as a function of position. This technique assumes that the shape of the curve remains constant as the thickness changes. This might not be the case if the ratio of the thicknesses of the materials that are used in the coating changes. This is detectable by comparing the shapes of all of the curves measured to verify that indeed they are identical except for a wavelength shift.

Some instruments are capable of measuring accurately the reflectance of surfaces with optical power (see section 5.1.8.). They are very useful in measuring curved surfaces with unknown coatings and for which there are no flat witness pieces. Of course, special purpose instruments can also be built.

On complex shapes with steep surfaces, such as fast lenses or domes, the analysis of uniformity is more complicated than that of flats because the angle of incidence varies as a function of position over the surface. To be more quantitative, one needs to measure the reflectance (or transmittance) at several points on the surface at the appropriate angle of incidence. The angle of incidence should not be neglected on steep parts as this can have a significant effect on the performance (see section 1.3.). The uniformity is sometimes deliberately adjusted to compensate for the angle shift on large steep-surfaced optics.

When the parts are small enough to be placed into the sample compartment of a spectrophotometer, the task of measuring a number of points is limited by the beam size that gives adequate signal to noise ratio (assuming the focusing problem is somehow solved).

When the parts are too large to fit in an instrument, the manufacturer has three choices.

1. Cut the part into pieces small enough to fit on the equipment available
2. Design and build larger measurement equipment that can handle the large part
3. Install small witness pieces in the coating equipment to stimulate a larger part, and measure each witness individually after it has been coated

The assumption is generally made that the witness pieces will fairly represent the coating that is placed on the actual part.

When many small parts are being coated at one time, the uniformity over an individual part may be quite good. On the other hand, the thickness may vary from part to part within a single lot, as the various parts sample different portion of the uniformity curve.

5.4.7. Microscopic Examination

Microscopic examination is frequently a very revealing technique for evaluating coating-related problems. There are several reasons why this technique is so valuable:

1. Many coating failures can be traced to microscopic phenomena
2. The eye can analyze patterns and features that may give a clue to the origin of the difficulty
3. Being sensitive to spatial color changes, the eye can detect subtle clues in the coating
4. The instant feedback of being able to observe the coating first-hand can lead to successes that cannot be realized in any other way

For example, the reason a coating is scattering might be identified in a matter of moments with a microscope: stress cracking can be seen and the type of stress identified, or perhaps there are numerous defects that are causing the problem. We discuss these examinations below.

The microscope is an excellent method for investigating adhesive failures in coatings. Microscopic color variations in the coating will frequently indicate whether the failure has occurred at the film/substrate interface, or within the film stack itself. If the failure occurred at the substrate interface, the problem may be one of substrate cleaning, whereas a failure within a stack could indicate that there has been a buildup of stresses to the point where the stress overcame the adhesive forces. Another possibility is that some process abnormality, for example, a leak in a vacuum chamber occurred during the coating run, and this had the effect of weakening the adhesive strength at that point in the design.

The microscopic patterns observed with tensile stress failures are radically different from those obtained with compressive stress failures. In compressive failure, the film delaminates from the substructure and buckles, as shown in Fig. 5.8. In tensile failures, the film also fails in a distinctive pattern: a series of cracks, as shown in Fig. 5.9. The key to the interpretation of this figure is that the lines only intersect at right angles. Although some of the lines may be somewhat parallel, at the points of intersection, the two cracks meet at right angles. The explanation for this phenomenon is that the crack relieves the stress locally in the direction perpendicular to the crack direction. The stress vectors that are oriented parallel to the crack are, to first order, unchanged by the rupturing of the film. The further one moves away from the crack the greater the stresses in all directions. Thus, two cracks can propagate in roughly parallel directions. As one of the cracks turns to intersect its neighbor, it finds the stresses in its neighbor's vicinity to be closer and closer to a one-dimensional stress. At the edge of the neighboring cracks, the stress is entirely one dimensional, therefore, the new crack intersects at right angles to the previous one. Depending on the level of the stresses and the amount of relief that is afforded by the cracking, the point at which the new (second) crack turns

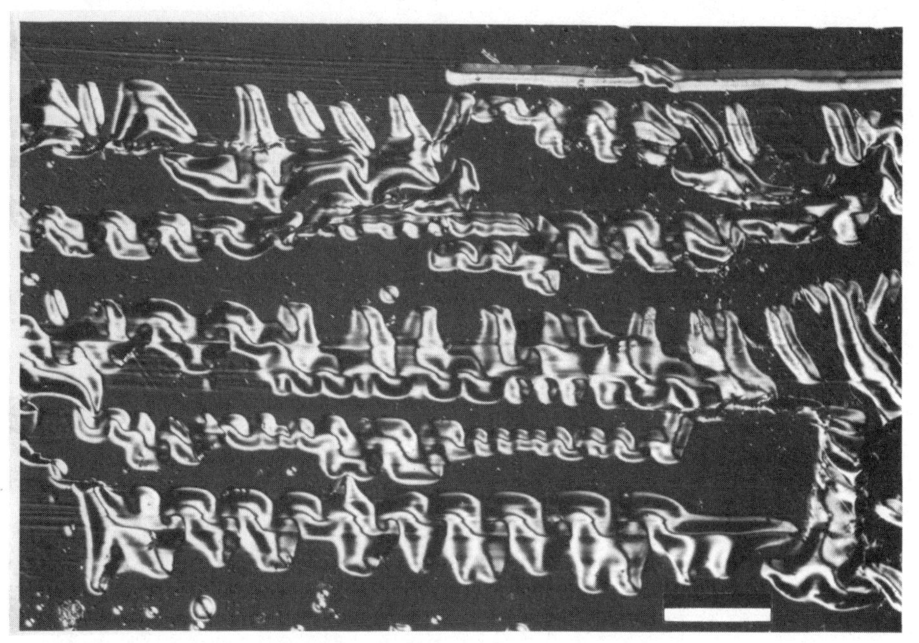

Figure 5.8. Photo of compressive failure as seen through an optical microscope. The bar is 100 μm long. The film has delaminated from the substructure and buckled upwards. (Courtesy of T. Tuttle Hart of Research Dept., O.C.L.I.)

to align itself in a perpendicular direction to the old (first) crack is a measure of the stress level remaining in the coating. The higher the stress level, the more cracks that form in the film. The orientation of the cracks also indicates whether there is a preferred stress direction in the coating.

A microscope can also help determine the type and the source of dirt and debris on or in a coating. These particles can be located in three places:

1. Under the coating, which means that they were there before the coating was applied
2. In the coating, which means that they were a result of the coating process
3. On the surface of the coating

The particles in the third category can usually be removed by carefully cleaning the coating, as outlined in a previous section. Microscopic examination reveals what needs to be removed, and what is left after the cleaning process is complete. Of course, when such work is being done, a clean bench or other clean area should be used to ensure that extraneous contaminants are not confusing the issue.

Lint and other loose surface dirt are readily apparent under microscopic examination, as are particles imbedded in the film by source spatter during the coating process. Pinholes can be identified in a transmitting filter by alternating between reflected and transmitted illumination: the pinhole area changes color or brightness with the change in lighting condition. A pinhole in an opaque filter on a transparent

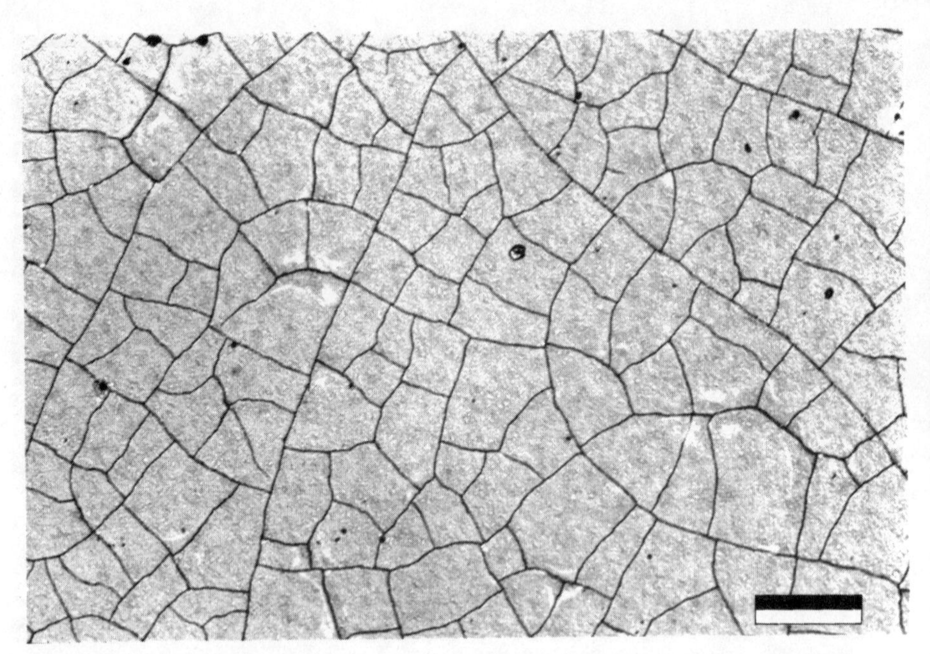

Figure 5.9. Photo of tensile failure as seen through an optical microscope. The bar is 100 μm long. Note that at the intersection of all lines, the two cracks meet at right angles. (Courtesy of T. Tuttle Hart of Research Dept., O.C.L.I.)

substrate appears bright when examined in transmitted light. A pinhole in a filter on an opaque substrate may be difficult to identify unless the coating is highly colored and the area of the hole is dark.

If areas of the coating have failed and the failed material has been removed, it is frequently possible to determine at just what level in the coating the problem occurred as long as there are not too many layers in the stack. The interference colors of the coating change when layers are removed. This is true in visible and infrared designs. Colors can be seen in infrared designs, as the higher order effects in these coatings frequently give rise to spectacular colors in the microscope. It is a simple matter to count the different colors to see which layer from the top was the last layer left intact. Ultraviolet coatings, and other designs that have very thin layers, may not produce the vivid colors of coatings that are designed for the longer wavelengths, but one can still recognize the individual layers by the edges where the layers are discontinuous.

A number of features simplifies using a microscope for the examination of coated specimens. First of all, the objectives should be of the metallurgical type. Biological objectives are designed to function in conjunction with a cover glass and the specimen immersed in an index matching liquid. If the cover glass is not present, the image suffers from spherical aberration.

For the proper investigation of thin film coatings, both transmitted and reflected images need to be examined. It is helpful to have a microscope with an illumination system in which the light can be made to be incident on the substrate through the objective. Such illumination systems are standard on metallurgical microscopes.

Another useful microscope technique in the investigation of thin films is the use of dark field illumination. The objective illuminates the sample with a cone of light at relatively high angles of incidence. The specularly transmitted or reflected light cannot be collected by the objective as it is outside of its collection cone. Only the light scattered into the objective is seen. Features that scatter light such as dust, voids, edges, and other discontinuities are readily seen, whereas smooth features of the specimen appear dark. The typical dark field work is done in reflection, and this requires the use of special objectives as a well as special microscope optical system.

An interference or phase contrast, also called *Nomarski,* attachment to a microscope is a useful technique for investigating the surface contour of thin films. The method uses the interference of light to magnify the vertical features on the surface of the film. Features at different heights are seen as different colors or shades. Although it is difficult to quantify the extent of the roughness, one can get a good qualitative feeling for the extent of a problem. This technique produces colorful displays that can be easily evaluated by eye, but as the effects are caused by a gradation of colors, it is difficult to place an exact value on the height of a feature.

5.5. INSPECTION LOT SIZE

When the quantity of coated parts is small, it is not a major effort to scan or otherwise measure every part. When many parts are being produced either in one or multiple coating runs, it becomes prohibitively expensive to measure all parts. A typical approach in these cases is to qualify a vendor by measuring a large fraction of the initial lots of a large order. As the buyer and seller gain confidence in each other and in the fact that the process is under control, the requirements for testing can be reduced to a small sample of each coating run.

When 100 percent inspection is required and the quantities are large, it is important to automate the measurement process to reduce costs.

Chapter 6

Substrates and Materials

The type of substrate upon which a filter is deposited generally determines many of the characteristics of the filter, such as durability, stability, and cost. There are three major categories of substrates:

1. Glasses and similar crystalline materials
2. Organics and plastics
3. Metals

Substrates can also be classified as either rigid or flexible. The majority of high-quality optics are fabricated of rigid materials. A large volume of web-coated substrate is produced each year for decorative and other uses.

In this chapter, we discuss in more detail the influences the various substrate materials and forms can have on thin film coatings.

6.1. SUBSTRATES IN GENERAL

The primary driving factor in the selection of a specific material as a substrate for a thin film coating is the wavelength region over which it must function. Transparency may be the major concern, as coatings can modify only the surface-related characteristics of the material, such as the reflectance; they cannot alter the internal transmittance or absorptance of a substrate. Table 6.1 lists the ranges over which some common optical materials are transparent. This table should be used with care; the limits shown in it are not absolute as the absorption is a function of the thickness of the material being considered. The absorption frequently increases gradually as the wavelength approaches the limits. If the sample is thin enough, it if possible that the transmittance of a substrate outside the spectral range given in Table 6.1 may be adequate for some applications.

Substrates should be strong enough to survive the handling that they will undergo during the coating process. In a typical evaporation process, fixtures support the parts by contacting them on the coated side only at the edges or outside of the useful aperture. The fixtures are designed so that they will constrain the substrate from moving laterally. Typical tooling consists of a milled recess in a metal rack in which the part is situated as shown in Fig. 6.1. The supporting protrusion, the *jig lip,* is usually made as small as practicable to minimize the area of the coated substrate that is shadowed by it and is, therefore, not coated. This results in an uncoated area around the periphery of the element. If the coating must extend to all edges of the surface, then the fabrication process must include steps that coat

Table 6.1 Common Substrate Materials Used for Optical Filters

Material	Short wave limit (μm)	Long-wave limit (μm)
SiO_2	0.12	4.5
Al_2O_3	0.14	6.5
MgF_2	0.11	7.5
CaF_2	0.13	12.0
GaAs	1.0	15.0
Si	1.2	15.0
BaF_2	0.25	15.0
Ge	1.8	23.0
C (diamond)	0.25	80.0+
ZnSe (CVD)*	0.5	22.0
IRTRAN – 1 (MgF_2)	0.5	7.5
IRTRAN – 2 (ZnS)	0.4	14.5
IRTRAN – 3 (CaF_2)	0.13†	11.5
IRTRAN – 4 (ZnSe)	0.5	22.0
IRTRAN – 5 (MgO)	0.23†	9.0
IRTRAN – 6 (CdTe)	0.9	33.0

Notes: Transmission limits include surface reflections and are for a T = 10 percent level and for a sample thickness of 2 mm. IRTRAN data from Kodak Publication U-72, unless otherwise indicated. IRTRAN is an Eastman Kodak Company trademark.

*CVD Incorporated brochure.

†Lichtenstein, 1979.

From Wolfe, 1965, p. 327.

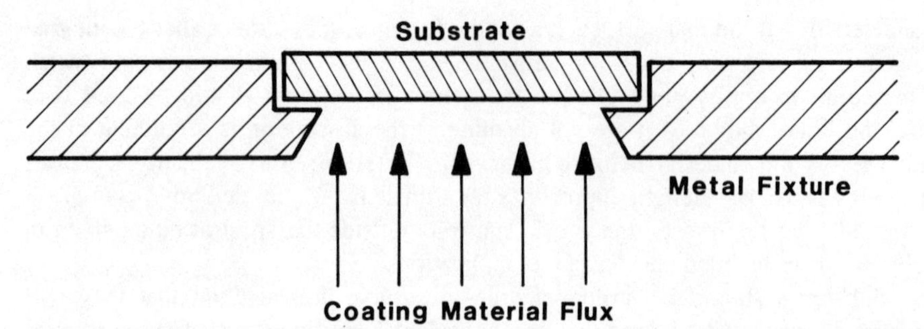

Figure 6.1. Cross-section of a typical coating fixture supporting a substrate. The substrate is held in place by protrusions from a rack that usually rotates so that the substrate is coated with a uniform deposit.

an oversize part that is subsequently trimmed to size. In a sputtering process in which the part is supported above a source, a similar uncoated area exists on the substrates. On the other hand, some sputtering systems allow the substrate to be below the source, thus making the supporting structure very simple and allowing for complete coverage of the surface. Substrates are frequently covered to protect those surfaces that are not to be coated.

Coatings can be deposited on substrates that are at ambient or cooler temperatures. Elevated process temperatures, however, usually lead to harder, environmentally more durable filters. The range of process temperatures used is frequently 100° to 200° (or higher). A number of factors related to the substrate limit the upper temperature that can be used: (1) glass transition temperature, (2) recrystallization temperature, or (3) melting temperature.

If one of these temperatures is exceeded, the substrate may be degraded by the coating process.

The amount of time required to heat and cool a part is a concern in a production environment. In general, the cooling rate is more critical than the heating rate for glasses because the cooling process places the surface in tension. In a vacuum coating chamber with multiple rotational motions, the only way to effect heat transfer to the substrates is by radiation, and it is difficult to accelerate the process effectively. The challenge for the coating supplier is to adjust the process so as not to damage the substrate.

Most optical coatings in the past have been applied to rigid substrates, and these have usually been of a glassy nature. Recent advances have enabled the application of optical coatings of interference quality to nonrigid substrates. In particular, webs of sheet materials have been coated with optical grade metal and dielectric layers to make interference filters. The substrate materials can be metal foils, such as aluminum and stainless steel, or organic polymers, such as polyethylene terephthalate (PET) and fluorinated hydrocarbons. Glass webs are not flexible enough to be run through these coating machines at this time.

6.2. SUBSTRATES TYPES

6.2.1. Glasses

Glasses are the most common substrate materials for thin film filters. One reason for this is that most optical elements are glass, though plastic lenses are fast becoming popular in high volume units such as cameras. In both low and high-quality products, most optical elements are coated to reduce surface reflectance losses, ghost images, and veiling glare. We also include in the category of glasses some crystalline materials, such as silicon, germanium, and calcium fluoride, which, from a coating standpoint, behave quite similarly to glasses. These crystalline materials can be heated to typical coating temperatures with no degradation of their optical properties.

The optical designer has a number of trade-offs to consider when selecting optical glasses. In addition, some thought needs to be given to how these elements will be cleaned and coated once fabricated by an optical shop.

The substrate material preferably should not darken under ultraviolet, electron, or ionized particle irradiation, as a number of coating processes generate such conditions in the vacuum chamber. In evaporative coating processes, a glow discharge cleaning step is often used, and the glow plasma in which the substrate is bathed generates ultraviolet light as well as energetic ionized particles and electrons. A similar plasma may exist during the entire sputtering coating process. If materials must be used that have poor radiation resistance, the filter manufacturer can tailor the coating process as necessary.

Polished substrates need to be cleaned at least twice before they are coated. The optician who polishes the surface must remove the polishing compound immediately after the last polishing step, lest a residue remain on the surface. A similar cleaning step is required immediately after diamond-turned optics are completed. Otherwise, a tenacious film of the polishing compound or the diamond turning lubricant forms, and it cannot be removed at a later time (Buckmelter et al., 1975). This is true of both conventionally polished and of single point diamond turned (SPDT) surfaces. The layer that is left behind can then be removed only by additional polishing. Thus, it is necessary for the optical shop that is preparing the surface to be certain that they clean off the surface as best they can and as soon after they are finished with the optical element. Because they are familiar with the properties of the materials used in fabricating the element, they are in a good position to know how to remove this surface contaminant. Also, these materials become more adherent as they dry out and age on the surface; thus, time is of the essence in their removal.

Surface cleanliness is a matter of vital importance. Organic and other residues from the surface finishing process can leave a layer on the surface. This residue can act as a release layer for the coating, reducing the adhesive forces between the substrate and the coating, and possibly causing the coating to separate from the substrate.

A final cleaning of the substrate, done shortly before coating, prevents as much as possible:

1. Contaminants' reducing the adhesion between the first layer and the substrate; and,
2. Particles' remaining on the surface and leading to scatter and pinholes in the coating.

Other symptoms of poor cleaning are streaks and wipe marks on the surface. Unfortunately, these may become visible only after coating, necessitating removal of the coating, perhaps by repolishing, and then recoating. Thus, it is vitally important to have good communication between the user and the coating vendor. The latter must understand the history of the surface so that it may be cleaned properly.

Some glasses stain more easily than others (see Schott and other glass company catalogs for these ratings). It is difficult to clean some of these glasses as aggressively as one would like, lest the cleaning process succeed in cleaning off the contaminants but stain the glass at the same time. On the other hand, the coating may well improve the durability of a surface to further degradation from staining by protecting the glass from the environment. This improvement in resistance to staining cannot be guaranteed, but it is to be expected that the additional surface material reduces the substrate's exposure to the active elements in the environment and thereby protects the surface.

The Schott glass catalog recommends that their glasses not be heated above a temperature which is 200°C below the glass' transition temperature ($T_g - 200$). At higher temperatures, the optical figure may deform and the refractive index may change. Table 6.2 lists some of the glass transition temperatures for selected glass

Table 6.2 Approximate Limit Temperatures for Various Substrate Materials

Material	Temperature (°C)
Plastic (acrylic)	90–105§·#
Plastic (polycarbonate)	150§·#
Plastic (polyimide)	310–365§·#
Plastic (polyester)	255§·#
BK7	365†
Germanium	936*
BaF_2	1280*
CaF_2	1360*
silicon	1420*
SiO_2 (fused silica)	1710†
Al_2O_3	2030*

Notes: *Melting temperature. †Softening temperature. ‡Maximum safe temperature ($T_g - 200$).
§Glass transition temperature (T_g).
#Agranoff (1985).
From Wolfe, 1965, p. 330.

types. It should be noted from this table that there are some glasses for which even a 200°C coating temperature may be marginal.

One of the considerations the coating manufacturer has to take to account is the expansion coefficient of the substrate material. If the coating process involves heating, enough room must be allowed for the expansion of the substrate in the supporting fixtures. The thermal expansion of the tooling and the substrates must be matched through the thermal excursion of the coating process. Table 6.3 gives the expansion coefficients of some typical substrates and tooling materials.

The fabrication of glass substrates into optical components is an art that has been developed over many centuries, and we will not go into those details here. We touch upon the dimensional sizing of substrates, however, as this is frequently something the coating manufacturer is called upon to do.

Optically flat glass sheets can be made flat in two ways:

1. The most expensive of these is the traditional grinding and polishing as done in an optical shop but on a larger scale.

2. Flat glass can be produced directly from the liquid phase by floating it out on a pool of molten tin. The liquid tin/liquid glass interface is very flat, as is the glass/air interface. The two surfaces have slightly different physical and chemical properties, and these differences may be important in some applications. The tin and air sides of a float glass sample can be identified by directing the light from a lamp that emits in the near-UV on the float glass sample: the tin side fluoresces while the air side doesn't.

Flat glass can be cut to size by scribing, either manually or with a machine, a line on it with a carbide or a diamond tool and then breaking the glass along the line. The accuracy that can be obtained with this technique is on the order of +/− 125

Table 6.3 Expansion Coefficients of Commonly Used Filter Substrates

Material	Expansion Coefficient ($\times 10^{-6}$/°C)
Aluminum	25*
BK7 glass	7.4†
SiO_2	0.5
PET	65‡
Si	4.2
Ge	5.7
Al_2O_3	5.8 (ave.)
CaF_2	24

Note: These values are for the temperature range normally experienced by optical systems. *From Weast, 1974. †From Ohara Glass Catalog, 1974. ‡Agranoff, 1985

From Wolfe, 1965, unless noted.

μm (5 mils). This technique works well with most glasses, but it cannot be used with fused silica or with sapphire substrates as the break line does not necessarily follow the scribed line for these materials.

Other methods of cutting glass and other hard substrates include (1) loose abrasive saws that use either carbide or industrial diamonds as the cutting agent and (2) diamond-tipped saws. Inside diameter diamond saws, which work so well with semiconductors such as silicon, perform poorly when used to cut fused silica because of heat buildup as the silica is a poor heat conductor.

6.2.2. Plastics and Organic Substrates

At the present time, organic materials are not very useful as transmitting optical elements in the ultraviolet. The energy of UV photons can break the chemical bonds in most plastics, and this causes them to become absorbing. This damage manifests itself as a yellowing of the material. In addition, many plastics oxidize and become brittle and crack and craze with age. Chemical additives can delay these reactions, but it is difficult to stop them completely.

Organic plastics have limited uses in the infrared portion of the spectrum as a multitude of vibrational bands resulting from carbon–hydrogen bonds occur in that part of the spectrum. As such, they can make good wavelength calibration samples, and sheets of polystyrene are frequently used for this purpose.

Plastic substrates are more difficult to handle than glass substrates for several reasons. First, they are softer, and therefore, more susceptible to damage than glass. Next, they can be a source of contamination in a vacuum chamber. Certain plastics contain low molecular weight plasticizing components. When these plastics are placed in a low pressure environment, the low molecular weight constituents may diffuse out of the bulk material, especially if the temperature increases as it would during a coating run. Apart from the fact that the loss of these volatile components may well change the properties of the plastic, their loss in the vacuum chamber has the effect of increasing the gas load the pumping system must handle. These gases may change the actual process gas composition of the coating chamber, and this may affect the deposited layer. Thus, plastics that do not contain any low molecular weight components, including unreacted monomers and other precursors, are preferable to those that do contain these components (from a coating process standpoint). Finally, most plastics absorb a large amount of water from the air; when the pressure is reduced, they release it again. This can similarly change the gas environment during the coating process. Table 6.4 lists some examples of plastic materials that can be used as substrates without detrimental effects to the coating process provided care is exercised in selecting process parameters. Most of these plastics are rigid, or flexible in thin sheets.

For plastic substrates, the maximum allowed process temperature is either the softening temperature, or a transition or crystallization temperature. The limiting temperature of multicomponent plastics may be determined by the outgasing of some of their constituents, such as plasticizers. The temperature limits on plastics is difficult to determine, and is most often done on a trial and error basis.

Typically, the inorganic materials used in coatings produce more durable de-

Table 6.4 Plastics Substrates with Good Coating Characteristics

Material	Comments
PET	Available in web form for roll coating
Polyimide (Kapton*)	Available in orange or black versions
Polycarbonate (Lexan†)	Tough but soft
Acrylic (Plexiglas‡)	Low softening temperature; polished surfaces are very soft

*Kapton is a trademark of the Dupont Corporation. †Lexan is a trademark of the General Electric Company. ‡Trademark of the Rohm and Hass Co.

posits at higher process temperatures, so it is normally desirable to heat the coating chamber. Because many useful plastics cannot withstand very high temperatures, the coatings deposited on them are not as durable as similar coatings deposited on glass substrates at a higher process temperature. Table 6.2 indicates the maximum operational temperature for several plastics.

A thin sheet of polyethylene plastic can serve as an interference film material in the very far-infrared portion of the spectrum. The plastic is impregnated with various materials that have desirable far-infrared properties, and then sheets with alternating properties are laid up and the entire assembly is treated to make it a single massive structure. The layered structure may be pressed in various ways to reduce the overall (and individual layer) thickness. The result is an interference stack for the far-infrared portion of the spectrum that has remarkably good properties (McCarthy, 1967). This approach is feasible in the far-infrared portion of the spectrum since the wavelength is so large. A wavelength of 50 μm (2 mil) is greater than the thickness of many commercially available wrapping films made of polyethylene.

6.2.3. Metals

Metal substrates obviously cannot be used for transmitting optics. They can only serve as mirrors or absorbers. Because metals are generally good thermal conductors by comparison to dielectrics, they can function in situations where the heat load on the substrate is high, e.g., high-power laser mirrors. Metal substrates can be configured with cooling coils or integral channels through which a coolant can flow to remove heat. Molybdenum and beryllium are two materials that are often used for mirror substrates. Other metals can also be used to make optical components.

Beryllium is valuable in those instances where lightweight and low thermal gradients are needed. Molybdenum is useful as it has good thermal characteristics and can be machined and welded straightforwardly. These substrates are not typically coated directly. For beryllium substrates, a nickel-based plating (Kanigen) is applied to the surface prior to figuring. This plating has better polishing characteristics than the bare beryllium metal.

Large molybdenum mirrors can be plated with copper by a number of techniques, including electroplating. The copper coating is then turned on a lathe with a single-point diamond tool. The copper is much easier to turn than the molybdenum since it is much softer. The highly reflecting coating is subsequently applied to the figured (turned) surface.

Because a numerically controlled diamond turning machine can produce an arbitrary surface, it is useful in producing off-axis optics. These find use in instruments such as X-ray telescopes. Such surfaces require coatings with extremely thin layers to function properly in these wavelength regions. (See Chap. 7 for more information on these new coatings.)

6.3. SAFETY GLASS

There are two common ways to control glass as it shatters in order to limit the damage that can be caused by large shards: lamination and tempering. In lamination, a layer of plastic bonds two conforming sheets of untempered glass together. When such a glass panel breaks, the plastic interlayer holds the fragments together. Tempered glass, on the other hand, has a high level of controlled internal stresses. When tempered glass breaks, the result is that the internal stresses cause the pane to break into small, relatively harmless pieces.

6.3.1. Lamination

The plastic used to bond glass sheets is frequently polyvinyl butyral (PVB), and it keeps the pieces of broken glass together, thus minimizing the potential for injury or other damage from glass shards. The plastic bonding layer is fused to the two glass surfaces by placing the unbonded assembly in an autoclave. High isostatic pressures force the two glass sheets against the intervening plastic. Heat causes the plastic to melt slightly and bond to the two glass sheets. The PVB plastic layer may be colored for functional or cosmetic purposes.

PVB plastic can absorb a relatively large amount of water from the atmosphere. If a bonded panel is to be coated, the absorbed water can be released in the coating chamber during the coating process and degrade the coating. It is, therefore, generally recommended that the coating be applied before the lamination of the substrates, if possible. If a coated surface is laminated, the coating must be able to survive the autoclave bonding process temperatures, which may reach 125°C.

Placing a coating on the interior surface of a bonded assembly gives it additional environmental protection; e.g., it cannot be scratched, abraded, or otherwise damaged.

6.3.2. Tempering

A properly performed tempering process increases the breaking strength of glass, even though the tempered glass has a great deal more internal stresses than annealed glass. Tempered glass cannot be cut after it has been treated; thus, it must be cut to size before the tempering treatment is performed. Two types of tempering processes are in common use: thermal and chemical.

In the thermal tempering process, a hot piece of glass is rapidly cooled so that the surface develops a compressive stress, while the interior of the sheet is under a tensile stress at room temperature. This process works well on glass sheets that are over 3 mm (1/8 inch) thick; the necessary thermal gradients cannot be set up when the glass sheet is much thinner than this. The thermal tempering process may be done either before or after the coating has been applied to the surface. On the other hand, it should be verified that the coating can survive the tempering process before a commitment is made to temper coated parts.

In chemical tempering, an ion exchange process replaces the sodium ions near the surface of the glass with larger ones, such as potassium. The increase in surface volume causes compressive stresses at the surface, as does heat tempering. Chemical tempering can be applied to even very thin sheets of glass. The limitations in this process revolve around the issue of size, as the entire glass sheet must be immersed in a vat of molten salt. These parts must consequently be cut to shape and then tempered.

The coating affects the ion diffusion rate, and in some cases can reduce it markedly. This change depends on the materials in the coating and their total thicknesses and crystalline form. It is difficult to generalize a tempering process with the sort of variables involved, therefore, it is generally agreed that the coating process is the last step in a manufacturing process involving tempered parts.

6.4. SUBSTRATE SPECIFICATIONS

Optical glasses are very tough materials as long as there are no local defects that can cause stress concentrations that exceed the tensile strength of the glass. Glass fails mechanically from tensile stresses, not compressive ones. In some circumstances compressive stresses can give rise to tensile ones. This phenomenon can be seen in the conical shape of the hole formed in a glass pane where a small hard object strikes it. The break occurs where the glass is in tension during the impact.

A piece of glass with an edge crack breaks more easily than one that has a smooth finish on its edges. Stresses, such as may be generated from uneven heating from heaters in a vacuum chamber or from a coating source, concentrate at the edge cracks. The stress at the tip of the crack causes it to propagate very quickly to the interior portion of the part, and perhaps causes catastrophic failure of the substrate. Thus, the edges of glass parts are generally carefully inspected for cracks prior to coating them.

The United States Armed Forces have established specifications for the sizes of chips and scratches in glass optics. (Section 5.4. describes the relevant specifications.) These defects can scatter light, which is undesirable in a high performance optical system. Bennett (1978) discusses how the digs and scratches influence the amount of light scattered by the substrate. Scheele (1977) has measured the forward and the backward scatter at some common laser wavelengths of some typical infrared substrate materials. Tables 6.5 and 6.6 show some of the data from his publication.

The surface finish of a substrate is at times a problem for the coating manufacturer. A surface that is acceptable from an optical perspective may not perform well as a substrate for thin films. A surface can have nucleation properties that

Table 6.5 Total Forward Scattering

Sample	Thickness (mm)	Wavelength (μm) 0.633	1.06	3.39	10.6
IRTRAN 1	3.0		0.14	0.027	
	6.0		0.17	0.43	
IRTRAN 2	2.0		0.089	0.012	0.0007
	6.0		0.14	0.024	0.0013
IRTRAN 4	2.0		0.045	0.028	0.0025
	6.0		0.15	0.11	0.015
Al_2O_3	3.0	4.5×10^{-5}	9.4×10^{-5}		
CaF_2	10.0	0.016	0.0044		
Ge	4.6				4.7×10^{-4}
ZnSe	5.2	0.011	0.024	0.0065	9.0×10^{-4}

From Scheele, 1977

cause films to grow in a manner that causes scatter. These surface conditions are frequently invisible, and show up only after the thin film has been applied. Coating manufacturers are in the best position to advise those unfamiliar with these problems.

6.5. BLOCKING PROPERTIES

When wide wavelength bands need to be blocked by a filter, it is frequently not possible or economical to accomplish this solely with interference stacks. It is sometimes possible to block the transmission over certain ranges by a suitable choice of substrate material.

All materials have transmission as well as absorption bands at various places in the electromagnetic spectrum. Table 6.1 gives the approximate ranges over which a few materials are useful as optically transmitting substrates. The ideal substrate

Table 6.6 Total Backward Scattering

Sample	Thickness (mm)	Wavelength (μm) 0.633	1.06	3.39	10.6
IRTRAN 1	3.0		0.046	0.0078	
	6.0		0.066	0.0068	
IRTRAN 2	2.0		0.037	0.0077	0.00053
	6.0		0.045	0.014	0.0011
IRTRAN 4	2.0		0.017	0.013	0.0017
	6.0		0.043	0.037	0.0063
Al_2O_3	3.0	1.6×10^{-5}	4.6×10^{-5}		
CaF_2	10.0	0.0086	0.0021		
Ge	4.6				7.2×10^{-5}
ZnSe	5.2	0.0024	0.0014	5.6×10^{-4}	1.5×10^{-4}

From Scheele, 1977

material has high transmittance to the wavelengths of interest and high absorptance of light energy at those wavelengths that must be blocked.

Absorption filters have no angle shift such as there is in interference filters. A shift in the transmission level can take place in an absorption filter since the angle on incidence increases as the path length within the absorbing substrate increases in proportion to the secant of the propagation angle within the substrate. Thus, the attenuation increases (transmission decreases) with increasing angle of incidence. (In calculating the transmission through an absorbing substrate, the surface Fresnel losses must also be included to get a precise result.)

The attenuation of a light beam is a negative exponential function of the thickness, and an appropriately absorbing substrate can be very effective at blocking unwanted wavelengths. It may be quite difficult to get the same level of attenuation with a reasonable number of thin film layers. An absorption coefficient of about 140/cm allows a transmission of less than one part in a million through a substrate 1 mm thick. If this same material is used in a thin film design in a layer 1 μm thick, the transmission is of the order of 99 percent (ignoring any interference effects).

When absorption is used to block energy, the incident energy level should not be so high that it excessively heats the substrate. If the substrate could overheat, a reflective coating could be added to the blocking filter to reduce a portion of the heat load. Otherwise, active cooling of the absorber is required.

The blocking properties of a substrate may be temperature dependent. For example, germanium changes its optical properties when the temperature approaches 77K. In addition, the transmission properties of germanium and other semiconductors can decrease as the temperature is increased. Care should be taken to be certain that substrates that are to be used as blockers, or as transmitters, for that matter, at temperatures other than room temperature are carefully checked for their blocking or transmitting capabilities at the operating temperature.

The use of absorbing blocking filters is frequently a wise choice from a technical, as well as from an economic, standpoint. The limit on this type of filter is that the useful blocking bands are fixed by available materials and are not as adjustable as are interference filters.

6.6. RESTSTRAHL BANDS

Crystalline materials have spectral regions in which the reflectance increases to high values. Relatively thin sections of the material are essentially opaque in these regions. These bands are often found in the infrared portions of the spectrum. When broadband infrared light is reflected by one of these materials, only light with wavelengths corresponding to these bands remains after several reflections. Thus the bands became known as *reststrahlen* reflection bands, from the German word for *remainder* or *residual* ray (Kittel, 1968, p. 154; Jenkins and White, 1957, p. 454).

These bands exist in the spectrum of both crystalline and amorphous phases of a material, though the structure in the reflection peaks is sharper and more distinct in the former. These bands are a manifestation of lattice or interatomic bond resonances. The detailed structure depends on the long and short range order of the material. A good example of this is silicon dioxide; crystal quartz has a much different reflectance spectrum than does amorphous fused silica, yet they both have the same chemical composition.

The index of refraction undergoes rapid changes as a function of wavelength in a reststrahl region. The reflectance at wavelengths in the reststrahl band can also vary quite rapidly as the angle of incidence varies. Figure 4.44 shows the theoretically calculated reflectance for a number of angles of incidence for fused silica.

The reststrahl reflectance can sometimes be used to advantage in designing a filter. In other instances, it is a hindrance, as when a highly absorbing coating is desired. Thus, it is wise to know as much about a material as possible when making a selection.

6.7. ISLAND FORMATION IN THIN FILMS

In the formation of a thin film, the atoms or the molecules that condense on the substrate surface may have a finite surface mobility. They may move about on the surface prior to becoming affixed to a particular site. The range of the particles depends on the relative abundance of nucleation sites, the surface mobility of the condensate, the kinetic energy of the molecule or the atom, and so forth. In the initial stages of film formation, the substrate surface may have a number of nucleating sites on it. The arriving molecules preferentially attach themselves to these points. In metals especially, the atoms have good surface mobility and may move a significant distance to find a low energy site. In the former case, material builds up at the nucleating site at the expense of adjacent areas where there is relatively much less material. The succeeding molecules that arrive preferentially attach themselves to the molecules that are already condensed and therefore, there is a buildup of material in localized areas. The result of this process is the formation of islands of the film material. These microscopic islands are the subject of a great deal of current study, as their properties are not always what one might expect on the basis of bulk properties of the material. Effective media theories provide means of treating discontinuous island films as homogeneous films.

The optical properties of spherical particles immersed in a dielectric (massive) medium have been studied for many years. J. C. Maxwell-Garnett (1904) wrote one of the first treatises on the subject. A number of theories have been proposed that can approximate these films with equivalent refractive indices. Depending on the exact shape of the metallic particles, the resulting equivalent complex index of refraction can take on a variety of values. Thus, it is quite difficult to model accurately the behavior of these films, especially when the index of the surrounding medium varies.

This phenomenon is the basis of some practical coatings. Optical data disk designs have been patented that are based on the transformation of one type of island film into another (LaBudde et al., 1983). The sensitivity of very thin metal films to elevated temperatures is used to transform the index of the film. Initially, the metal islands are very small. The islands in a spot heated with a focused laser beam coalesce and form much larger spherical particles. The difference in index of refraction between these two regions allows the data bits to be read optically. A film composed of noncontiguous metal islands has little thermal conductivity in the plane of the film. This reduces the amount of power that needs to be deposited in the film to raise its temperature to the point where an irreversible change in its optical properties occurs.

The transmittance of visible light through a growing thin metal film is interesting. For example, the transmission of a glass substrate starts at approximately 96 percent and decreases as silver is deposited on it. At the point where the islands begin to coalesce with one another to form a continuous film, the transmission begins to increase, until, at the point where most of the film is a single continuous sheet, the transmission again begins to drop. It is at this point that the index of refraction of the film begins to become stable. The exact thickness at which this occurs varies with the nucleation properties of the underlayer. A similar behavior of the index is observed for noble metals. Metals, such as nickel and molybdenum, do not show this property quite as distinctly as the noble metals. This may be caused in part by the fact that the noble metals form more distinct islands.

6.8. THERMAL RUNAWAY IN SEMICONDUCTORS

Germanium, with its high index of refraction in the infrared, is a very useful material for thin films designs. It is generally not a good material to use in high-power laser systems. At room temperature, germanium has sufficient absorption that a high-power laser beam can heat it. Germanium, like all semiconductors, becomes more absorbing as its temperature increases. This results in more heating from the laser beam, and more absorption. This vicious cycle is termed *thermal runaway*. This avalanche process increases until the laser beam is turned off, the optic is damaged, or the transmission goes to zero. In some instances, this effect has been used as a passive switch, but it is generally a drawback in most optical systems. The use of germanium should be avoided when the temperature is expected to exceed 55°C (Savage, 1985). Cooled germanium substrates cannot tolerate power densities greater than 88 w/cm^2. Other semiconductor materials can show this effect also, but it is most evident with germanium as it has a small band gap. Materials with a wider gap, such as silicon, do not have this problem as their intrinsic absorption at room temperature is much smaller.

6.9. NUMBER OF MATERIALS IN MULTILAYER COATINGS

Although in theory there is no upper limit to the number of materials that may be used in a thin film design, most manufacturers try to limit designs to as few materials as possible. A trade-off is sometimes possible between the number of materials and the number of layers in a design, and the manufacturer will select the optimum one given the various constraints.

Designs with two materials are far easier to control because the thickness ratio of the two materials is a major parameter in the performance of a filter. Generally, the number of layers and the number of materials in a design are of little interest to the user, but of significant interest to the manufacturer.

6.10. TOXICITY

Manufacturers naturally prefer to produce, and users prefer to purchase filters with materials that are not toxic. At times, it is necessary to use less desirable toxic materials in a design because they are the only ones that allow the filter to meet the user's specifications. These materials can frequently be incorporated safely into a design with no adverse effect on the user or the environment.

A good example of a toxic material that can be safely used and is quite valuable from an optical point of view is thorium fluoride. This material is radioactive because it contains thorium, an emitter of soft alpha particles. Optically, it is an excellent material for thin film coatings as it has an optical transparency from the near ultraviolet to the far-infrared (350 nm to 12 μm). In designing a coating that incorporates thorium fluoride, some material other than thorium fluoride is generally used as the outer layer. With a sufficient thickness of material over the thorium fluoride, the thorium's alpha particles can be stopped before they leave the coating. Thorium fluoride is perhaps one of the *best* toxic materials available because it is easily detected. A Geiger counter with an appropriate window, or one of its more modern counterparts, can easily detect the presence of the material, whereas most other commonly used toxic materials require elaborate chemical analyses to determine their presence. Thus, one can easily check a coating for stray radiation, and can check the area around a coating machine for residual thorium fluoride without elaborate equipment.

Although a fairly large data base exists on the effects of various bulk materials on the human body, there is very little information available on these and other materials when they are in thin film form. Thus, one has to assume the worst case situation for the materials one uses in fabricating thin films. This undoubtedly leads to excessive caution in some instances to be on the safe side. Fortunately, the quantity of materials involved in a coating is relatively small in terms of total mass involved in a given filter.

Chapter 7

Future Prospects

The technology of fabricating optical thin films has been very dynamic over the years, and the prospects for the future are of more exciting advances to be expected. Many coating problems are being addressed in universities and industrial laboratories alike. Also, new technological advances, some of which are fallout from the space program, can be expected. In this section, we review a few of the current areas that hold some promise of changing the state of the coating art and some of the implications that these advances have for the users of these coatings.

7.1. ION BEAMS

The conditions under which a coating is deposited determine the quality of the thin film coating. The surface mobility of the arriving species is usually an important parameter in the physics of the deposition and nucleation process. At low temperatures, the arriving atom or molecule does not move very far from the initial point of contact with the substrate before it loses its kinetic energy. If the substrate is hot, the atom or molecule can migrate over a larger area before becoming immobile, thus increasing its chances of finding a nucleation site with the lowest possible energy state. Another important parameter in deposition physics is the kinetic energy of the arriving species: atoms with higher kinetic energy move about on the substrate surface more than those that arrive at low speed with less energy. This explanation accounts for some of the differences observed between sputtered and evaporated films. Sputtering processes impart much more energy to each atom or molecule that is knocked out of the target than do conventional evaporation processes in which the evaporant has the energy associated with the evaporation temperature. Unfortunately, sputtering processes frequently have a significantly lower deposition rate than do evaporative ones for many oxide materials, and some materials are not sputterable at all. Thus, a great deal of development effort is being expended to understand better the deposition physics to get the best of both worlds: high quality films formed at high deposition rates at low substrate temperatures.

In evaporative processes, the substrate is frequently heated to enhance the surface mobility of the evaporant (Lyon, 1946). Higher temperatures in the chamber can yield better, more durable, denser (higher index) films. Unfortunately, these higher temperatures are costly in terms of energy consumption, process time to get the substrate to the coating temperature and to cool it down again (without thermally fracturing it), and coating chambers and equipment that will handle these temperatures. In addition, some substrates degrade at high temperatures. In particular, most common organic (plastic) materials degrade at temperatures above 70° to 100°C. Some glass and metal precision optics may warp (on a submicrometer scale) at temperatures that yield durable films.

One would like to have a method whereby the energy that depositing species need to be able to move freely on the surface prior to becoming immobile is supplied by a means other than heat. Several schemes have been proposed to accomplish this. The most successful ones have involved the use of particle beams that flood the substrate surface with low energy ions (<1000 eV, and frequently <100 eV) or neutral atoms. The energy of these atoms should not be so high that they cause a loss of stoichiometry or other coating degradation. For example, titanium dioxide films are particularly susceptible to a loss of some oxygen, which in turn makes them absorbing in the visible and the UV portions of the spectrum. Finally, there should not be a net electrical charge in the particles arriving at the surface, lest the surface becomes charged and repels the activating beam.

The ideal beam source has a large flux of activating species with a low (by ion gun technology standards) energy per atom or ion. The problem with today's ion sources, such as many of those that have been developed under the auspices of the NASA space program, is that the particle flux that can be drawn out of them at low energies is limited. Thus a central theme in contemporary research on ion sources for thin film deposition use is to find a way to achieve high currents at low voltages. In particular, as Ohm's law applies, what is needed is a source with a very low impedance so that a large current can be obtained at low voltages.

Research on sputtering sources is also proceeding apace. It has been recently reported that magnesium fluoride has been successfully deposited with sputtering equipment. This is a significant breakthrough as magnesium fluoride is a low index material that is very useful in making wide band antireflection coatings in the visible portion of the spectrum. Prior to this announcement, fused silica was the lowest index material that was sputterable. The goal of sputter research is to enable the process to deposit materials at higher rates than previously possible. In addition, in some situations, it is desirable to sputter at lower pressures than allowed by conventional systems. This lower pressure operation reduces the gas that may be trapped in the film. Some low-pressure sputter sources are beginning to find their way into production environments.

Sputtering processes are prone to arcing between the plasma and the metal objects. This can result in a deposit of debris on the substrate. Once the physics of the process is better understood, we can expect significant advances in sputtering technology.

The results of this research will be more durable, more stable coatings and filters that can be deposited at lower temperatures (McNally et al., 1986). One finding is that ion beams can influence the stresses in coatings. Materials with high tensile stresses, in particular, will benefit from deposition in the presence of an ion beam. Thus, the kind of coatings that are today only possible on glass substrates could one day be found on plastics and other materials that cannot withstand high temperatures.

7.2. NONVACUUM PROCESSES

The processes used in the commercial manufacturing of optical thin films are predominantly based on vacuum technology. The sources of the deposited species are sputter targets or evaporation sources. Other technologies are now being developed that may play some role in the fabrication of the optical filters of the future.

One promising new coating technology consists of using organometallic compounds as the source of the metallic ion. After the compound has been applied to the surface of the substrate, the molecule is oxidized during a bake in an oxidizing atmosphere. The oxidized hydrocarbon components are all volatile, consisting for the most part of carbon dioxide and water. The metal ions form an oxide that remains as a film on the substrate.

There are two basic forms of this process. In the *sol-gel* technique, the molecule containing the metal ion is *suspended* in an organic solvent solution, whereas in the other technique, the solution consists of the organometallic molecule *dissolved* in an organic solvent. Either of these liquid solutions can be sprayed onto a substrate, but the more common usage involves a dip process, wherein an immersed substrate is withdrawn from the solution at a controlled rate. The thickness is determined by a number of parameters, including viscosity, temperature, withdrawal rate, angle of withdrawal, and surface tension of the solution.

The equipment required for the dip process is relatively simple when compared with the capital investment required for an equivalent vacuum-coating process. On the other hand, there is still a great deal of work ahead for the developers of this technology. A major difficulty at present is the high level of stresses in the film after baking.

There is a great deal to be gained in pursuing this technology. If the problems can be overcome, the cost of coatings could decrease significantly, and thus open new areas where thin film coatings might be used.

High-pressure (by vacuum-coating standards) chemical vapor deposition (CVD) is a standard coating process in the fabrication of semiconductors. This technology may one day be applied to the manufacture of large quantities of thin film filters.

7.3. INHOMOGENEOUS FILMS

Inhomogeneous films have been produced since the earliest days of the making of optical thin films. For the most part, they have not been desirable as their index profiles were unpredictable. In addition, with the exception of simple antireflection coatings, there were few workable systems for designing complex filters with inhomogeneous index layers incorporated into them.

Some recent breakthroughs have given inhomogeneous coatings new importance:

1. Recent improvements in computer algorithms have enabled the theorist to calculate the complex inhomogeneous index profiles required for novel filters (Dobrowolski, 1986; Southwell, 1985).
2. The rate control of coating sources is far superior today than it was just a few years ago.
3. Today's computers can evaluate the performance of a growing inhomogeneous film in real time. This enables the coating technician or system to monitor the index profile as the film grows, and make adjustments or corrections as necessary.

Inhomogeneous films have several advantages over discrete layers. The lack of discrete interfaces between layers eliminates one of the weakest mechanical points in a thin film stack, the interlayer boundaries. By eliminating these, the durability of the coating should be improved. In addition, the elimination of the interface may reduce some of the absorption in the coatings (see section 3.5.). This concept can be extended to discrete layer designs by grading or blending the interfaces between layers.

The use of a continuously variable index introduces additional degrees of freedom for the thin film designer to use in the design process. With inhomogeneous layer design techniques, it may become possible to design filters whose performances are far superior than can be achieved with discrete homogeneous layers. These advances may well be reflected by the emergence of inhomogeneous films in the commercial marketplace.

7.4. CODEPOSITION

Thin films can be made by codepositing two materials with different refractive indices. The index of the resulting deposit usually is intermediate between the two indices of the initial materials, and the specific value depends on the the composition of the coating. By varying the relative coating rates of the different sources, the index can be adjusted from one limit to the other. Inhomogeneous films can be obtained by varying the deposition rates according to a prescribed function of time or film thickness. This coating technique, being dynamic, requires good stability of the coating sources and a monitoring system adequate to the task of ensuring good control of the index profile.

Several problems must be overcome if a uniform index profile is to be obtained over a substantial substrate area. In evaporation, the two simultaneously operating sources cannot be located at precisely the same point, and different parts of the coating plane or surface will get different ratios of the two coating materials. A common technique to obtain homogeneous coevaporated films is to use two separated sources and rotate the substrate above them so that the substrate is exposed alternately to one and then the other source's maximum coating flux. Although the index profile has a high frequency modulation component, on the average, it closely approximates the desired shape. If the theoretical profile must be matched even more closely, then the evaporation rates of the two sources can be slowed to the point where the substrates get coated with a monoatomic layer or less during each revolution. Then, the two components will be very well intermixed. This latter approach is necessary to coat a filter with a carefully controlled index profile over a large area, if current technology is used. Thus, if coevaporation is to become commercially viable, new deposition techniques require development.

The geometry of a sputtering system lends itself to different solutions to the uniformity problem. One means of obtaining a mixed deposit is by sectioning the target so that some portions are of one composition, whereas others are from the second component. The ratio of the two areas determines the compositional ratio of the deposit. A second approach is to move the substrate between two targets in a manner similar to what is done in evaporation systems. The profile can be varied by changing the coating rate, and this is easily accomplished in sputtering by simply adjusting the power. It is not clear that an inhomogeneous film with an arbitrary index profile can be deposited over large areas in a sputtering process. One exception to this exists in reactive sputtering of pairs of materials in which there is a common constituent that comes from the target, and the other component comes from a variable gas mixture.

Some materials that may not be compatible in bulk form can be made to coexist in a film form by codeposition. The basic effect is a solid state condensation taking place at a relatively low temperature so that a chemical or a physical reaction cannot take place. The two materials may exist in a metastable state for a long time as energy is not available to overcome the potential barrier that prevents a reaction between them. The atoms are frozen into place as they condense, and they do not have the energy to undergo a phase separation. This approach to making new materials has not yet resulted in many materials being manufactured in this way. It is possible as technology develops and new requirements arise that materials of this type may become economically viable. Cermets are one class that may benefit from this type of technology.

7.4. X-RAY COATINGS

Researchers are pushing the frontiers of optics to ever shorter wavelengths (Underwood, 1986). Current research focuses on wavelengths in the 1 to 300 Å range.

As in conventional optics, these new spectral regions require high reflectors, antireflection coatings, and bandpass filters. In the past, X-ray imaging involved the use of mirrors at glancing angles of incidence. This approach limits the available aperture, results in large aberrations in the image, and limits the resolution of the system. At present, we are just beginning to understand how to design and construct multilayers that can function efficiently at these extremely short wavelengths.

There are many problems associated with the making of coatings that function in the extreme ultraviolet and the X-ray portion of the spectrum. The most severe problem is that of finding materials that have appropriate optical constants. At these wavelengths, most solid materials are absorbing, and the real parts of their refractive indices are close to unity. In addition, the energy of the X-rays is such that the entire electron cloud around an atom is involved in determining the optical response, so that changes in the chemical composition have little effect on the refractive index. Thus, the material choices are reduced to a consideration of the properties of the elements. Design schemes are being developed that take advantage of these physical constraints, or that work around them.

Another problem being addressed is that of making these filters. Some layers may have to be as thin as 3 atomic layers to form a quarter wave thick interference film. Monitoring such thin layers during the deposition process becomes crucial and may involve the installation of X-ray sources in the coating chamber (Spiller, 1981). In addition, stability of the films to both diffusion and chemical attack may be difficult to obtain.

Optical elements need to have surfaces that are smooth to a small fraction of a wavelength of the light with which they will function. At far-UV and X-ray wavelengths, the substrate surfaces must be far smoother than that needed for visual optics. At least two current technologies are available that can produce surface with the requisite low surface roughness. Semiconductor wafers have surface finishes in the 3-Å rms range, but unfortunately their surface figures are not generally good enough for imaging work. Laser gyroscope mirror substrates are prepared with good surface finishes as well as good figure control. The scatter is the coatings needs to be improved, and this is one area of study that should prove fruitful for all spectral ranges.

It is just a matter of time before production techniques for X-ray coatings will no longer be state of the art and become routine and commonplace. The result will be complex optical systems with multiple reflecting surfaces that yield quality extreme ultraviolet and X-ray images.

7.6. EQUIPMENT ADVANCES—AUTOMATIC COATING MACHINES

Significant advances in the area of coating equipment has been taking place in the past few years. Microprocessors have been added to coating chambers, and new process monitors make interfacing them to the actual deposition process even simpler. As a result, it is now possible to purchase a totally automated vacuum-coating system.

The state of the art has not yet progressed to the point where it is possible to buy a turnkey system that is guaranteed to produce a particular coating, but these will probably be available in a few years. The systems will be set up to manufacture one specific coating design, and it will be difficult for the novice to alter the process to coat successfully another type of design. The dedicated coating manufacturers will continue in business, and their customers will continue to return to them for most of their coating requests.

7.7. MATERIALS RESEARCH

The area of thin film technology that will advance the field the most in the next few years is undoubtedly materials science. Fundamental data in the thin film area is still lacking at the most basic levels. As an example, Ennos' paper (1966) on stresses is still the only readily available publication on the subject. The work was performed 20 years ago!

It is necessary to undertake fundamental studies of the basic physical parameters of matter in thin film form. Many of the physical properties of materials in thin film form are unknown: stresses, thermal conductivity in both the film plane and perpendicular to it, and Young's modulus and other mechanical constants. Even the refractive index of most compounds is poorly known except in a few wavelength regions. The advent of lasers, synchrotron light, and other light sources in unexplored portions of the spectrum will mean new requirements for coatings and coating materials in these wavelength regions. It will be necessary to know the optical constants of materials in these wavelength regions before adequate filters can be designed and constructed.

The chemistry of thin films is not the same as that of bulk materials. When the density of a material is 70 to 90 percent of that of the bulk, one can be assured that there is a large surface area within the film. The effect of this on the chemistry of the material is poorly known. Silver is a very good example of what can happen to a material when it is in thin film form. In bulk form, silver is relatively inert, save for the formation of the sulfide. In thin film form, it seems to participate in many different reactions in the presence of moisture. Thus, this seems to be a fertile field for study, and the results of these studies will be of great benefit to those working in this field.

Appendices

Appendix A

How to Specify Optical Filters

By far the most important thing one must do when one specifies a filter's performance is to determine one's absolute "needs" and to distinguish them from one's "wants" or desires. The needs relate to the functional aspects without which the filter and the rest of the optical system will not function properly. These needs do not necessarily have to be optical in nature: the tolerance on the size of the filter or the environmental parameters might be very important needs.

The wants, on the other hand, are additional features that add value to the system but are not critical to its fundamental operation. A bandpass filter may be adequate with a certain tolerance on its bandwidth, but the system's performance might be significantly better if it were made narrower: this would be classified as a want.

There are always gray areas that need to be resolved subjectively on an individual basis. Hopefully, these guesses are minor when compared with the total number of decisions required in the system. Usually, more information or a better understanding of the system in which the filter will be used can help to clear up ambiguities.

Cost considerations are always of paramount importance in commercial procurements. On the other hand, in high technology and military purchases, the primary considerations are likely to be optical and environmental performance.

To get the maximum benefit from a filter supplier's expertise, it is wise to involve that person in the design of the system. The function and the location of the filter may have an effect on the system's performance, and the supplier may be able to make some suggestions that will have the effect of reducing the total system cost.

The filter manufacturer should be brought into the design picture at the earliest possible stage. Filters are an integral part of an optical system, and if a system change needs to be made late in the design cycle, the costs of the design task may escalate. Polarization effects are one area where a small change can sometimes make a large difference in the difficulty of producing an acceptable filter. A simple rotation of the plane of incidence or of the filter can simplify the design by changing the plane of polarization with respect to the plane of incidence on the filter.

Coatings are not separable from the performance of the optical system in diffraction limited systems. Filters should not be left to the end of a project; an adequate filter may not be available off-the-shelf.

The following questions are typical of those that are likely to be asked by the representative from a coating house in response to a call for a filter. Of course, they may not all be appropriate in any given instance.

1. What type of filter is desired? What is the spectral range over which the coating must function? What performance (transmission, reflection, and absorption) is required at each point?
   - **a.** Questions Specific to Long-Wave Pass Filters
     What attenuation level is required over what wavelength range? At what wavelength is the edge located? How precisely must this be positioned? Is there a requirement on the steepness of the slope?
     What must be the minimum and average transmission in the high transmission region? Over what range does this specification apply?
   - **b.** Questions Specific to Short-Wave Pass Filters
     At what wavelength is the edge located? How precisely must this be positioned?
     What passband width is required?
     What minimum and average transmission levels are desired over this range?
     Is there a requirement on the steepness of the slope of the edge?
   - **c.** Questions Specific to Bandpass Filters
     What is the center wavelength?
     What is the desired bandwidth? How is this specification defined?
     What is the minimum or average transmission in the pass band?
     Is any blocking outside of the pass band required? How low must it be?
2. Will the filters be used at normal incidence?
   Will there be a range of angles of incidence on the filter?
   Will the light be collimated, or will there be a steeply convergent beam of light in the filter plane?
3. If the angle of incidence on the filter is nonnormal, is the incident light polarized?
4. What are the tolerances on all numbers specified?

5. In what kind of environment will the filter be used?
   What environmental tests should be performed on the filter to verify adequate durability?
6. Will the coating be exposed to high-power beams? Will the light be coherent or incoherent? Will this energy be in-band or out-of-band?
   Will the coating be exposed to other energy sources, such as ultraviolet light, ionizing radiation, or high energy particles?
7. If the filter has absorbing layers in it and is not symmetric, is there a requirement on the reflectance from the second surface (reverse reflectance)?
8. Is scattered light a problem in the optical system? If so, to what level must scatter be controlled in the filter? How should this be measured to verify that the desired level of performance has been attained?
9. Will the filter be exposed to elevated temperatures (>100°C) in storage or in use? If so, what level will the temperature reach?
   Will the filter be exposed to temperatures below 0°C?
   How long will the filter be exposed to these temperatures?
   How rapidly will the temperatures change?
10. What is the incident medium? What is its index of refraction?
11. Of what type of material(s) will the substrate be made? If glass, what type and what is its index of refraction? Similar questions will be asked for plastic and infrared materials.
12. What is the physical size and shape of the substrate? Will it be round, square, or some peculiar shape? What will be the aspect ratio (ratio of the largest dimension to the smallest)?
    Who will supply the substrate? If the filter manufacturer is to fabricate the substrate, who will supply the raw material for the substrate? What are the specifications for this?
13. Are there any special precautions that need to be taken with the substrates; e.g., is the substrate material prone to staining or sensitive to heat?
14. How many substrates will be required? Over what period of time? What will the lot size be? How many extra pieces can be supplied for testing purposes?
15. When will the customer want the first parts delivered?
    When will the coating manufacturer receive the first shipment of substrates (if customer supplied)?
16. Does the customer have any special cleaning requirements or recommendations? This is especially important if the coating manufacturer is unfamiliar with a specific substrate material.
17. Are there any requirements to assemble the filter into an assembly or subsystem?
18. Are there any other requirements or specifications that the filter must meet that might not be obvious to the supplier?

Appendix B

Selected United States Military Specifications that Relate to Thin Film Optical Filters

A number of United States Armed Forces specifications are used to qualify optical coatings. An early specification, Mil-C-675, was specifically designed for single layer antireflection coatings of magnesium fluoride on glass optical elements. In fact, it even calls out the range of thicknesses the coatings can have. In addition, however, it specified that the coated element should pass certain humidity, solubility, abrasion, and adhesion tests (Table B.1).

These environmental tests have been applied to a much broader class of coatings than originally intended by the authors of this specification.

Another specification, Mil-M-13508, was written specifically to deal with aluminum mirror coatings, which are typically less durable than a hard magnesium fluoride AR coating.

A third specification, Mil-STD-810, covers material testing in general and is not specific to optics. Some sections of this rather long document, however, codify tests that are suitable for performing accelerated degradation tests on optical coatings. Sections 507.2 ("cycled humidity") and 509.2 ("salt fog") are very rigorous tests that cause degradation in all but the most durable coatings.

Specifications -675 and -13508 are now obsolete. These have been replaced by three specifications that are more appropriate to the wide variety of modern coatings. The new documents are

Table B.1 Tests for Coated Elements

	C-675	M-13508	STD-810	C-14806	C-48497	F-48616
Revision	C	C	D	A.2	A	Ammend. 1
Date	Sept. 10 1982	Mar. 19 1973	July 19 1983	May 1 1974	Sept. 8 1980	Aug. 22 1980
Abrasion	Severe: 40 rub eraser	50 rub cheesecloth	none	20 rub eraser	50 rub cheesecloth	50 rub cheesecloth
	Moderate: 50 rub cheesecloth				Severe: 20 rub eraser optional	Severe: 20 rub eraser optional
Adhesion (tape)	fast	slow	none	none	fast	fast
Humidity						
				Method A		
Temp (°C)	48.9	48.9	30↔60	82↔160	48.9	48.9
Temp (°F)	120	120	86↔140	27.8↔71.1	120	120
% Rel. Hum.	95–100	95–100	85↔95	85↔95	95–100	95–100
Hours	24	24	24(cycle)	240	24	24
Conditions	Constant	Constant	Variable	Variable	Constant	Constant
				Method B		
Temp (°C)				48.9		
Temp (°F)				120		
% Rel. Hum.				95–100		
Hours				24		
Conditions				Constant		
				Method C		
Temp (°C)				82↔160		
Temp (°F)				27.8↔71.1		
% Rel. Hum.				85–95		
Hours				48		
Conditions				Variable		
Salt Fog/Salt Spray						
Time (hr)	24	24	48	48	none	24(option)
Temp. (°C)	48.9	35	35	35	?	?
Temp. (°F)	120	95	95	95	?	?
Solubility	(hr)					
Water	—	—	—	—	24(opt.)	24(opt.)
Saltwater	24	—	—	—	24(opt.)	24(opt.)
Other					10 min in each: trichloroethylene, acetone, and ethanol	

Table B.1 (*Continued*)

	M-13508	C-14806	C-48497	F-48616
Temperature				
(°C)	−62 to 71	−54 to 71	−62 to 71	−62 to 71
(°F)	−80 to 160	−65 to 160	−80 to 160	−80 to 160
Time at each temp. (hr.)	5	48	2	2
Shock (*dT*/*dt*)		<5 min to cover range Room T to 160, hold 4 hr, to −80, hold 4 hr, to 160, hold 4 hr (°F)	<4°F/min	<4°F/min
Dust		28 hr in fine sand in 2000 ft/min airstream		
Fungus		28 days in a mixed spore suspension		
Notes		Test sequence (order) and delay time between tests is specified These specifications include scratch and dig values		

Mil-C-14806 Coating, Reflection Reducing, for Instrument Cover Glasses and Lighting Wedges
Mil-C-48497 Coating, Single or Multilayer, Interference: Durability Requirements for
Mil-F48616 Filter (Coatings), Infrared Interference: General Specification for

This last specification also includes a number of supplements. These are intended to be specific to certain types of optical filters.

As a result of the availability of these specifications, which are directed at multilayer thin film interference technology, it is now possible to specify tests that are directly applicable to the product being tested. Mil-C-675 is directly replaced by Mil-C-14806, and therefore, the latter should be now used exclusively. Similarly, the other two specifications cover the general field and the cases of specific infrared coatings.

All of these specifications are subject to change at any time. The specifications given here were correct as of their release date. Revisions are designated by a suffix letter, starting with the letter "A." Users should assure themselves that they are referring to the latest edition of a specification, or know the changes that have been made to a given edition.

Some of the newer specifications explicitly define the order in which the test procedure is to be conducted on a coating sample. The change in procedure has two significant implications: (1) the test is much more severe than that when the same tests are run in parallel on separate substrates, and (2) the time that it takes to perform all of the tests is much longer than when tests are run in parallel, especially if there are some long duration humidity tests to be performed before abrasion and adhesion tests.

In most tests, the criterion for passing the test is that after the test is complete no degradation due to testing be observed. Any degradation visible under the specified examination conditions indicates that the coating has failed the test.

MIL-C-14806A
Incl. Amendment 2
1 May 1974
SUPERSEDING
MIL-C-14806
27 May 1968

MILITARY SPECIFICATION

COATING, REFLECTION REDUCING, FOR
INSTRUMENT COVER GLASSES AND LIGHTING WEDGES

This specification is mandatory for use by all Departments and Agencies of the Department of Defense.

1. SCOPE

1.1 This specification covers reflection reducing coatings applied to substrate material having indices of refraction within the range of 1.47 to 1.55 to be used as instrument cover glasses and lighting wedges herein referred to as optical elements.

2. APPLICABLE DOCUMENTS

2.1 The following documents, of the issue in effect on date of invitation for bids or request for proposal, form a part of this specification to the extent specified herein.

SPECIFICATIONS

Military

MIL-E-12397	Eraser, Rubber-Pumice for Testing Coated Optical Elements
MIL-O-13830	Optical Components for Fire Control Instruments; General Specification Governing the Manufacture, Assembly and Inspection of

STANDARDS

Military

MIL-STD-105	Sampling Procedures and Tables for Inspection by Attributes
MIL-STD-810	Environmental Test Methods
MIL-STD-1241	Optical Terms and Definitions

(Copies of specifications, standards, drawings, and publications required by suppliers in connection with specific procurement functions should be obtained from the procuring activity or as directed by the contracting officer).

FSC 6650

MIL-C-14806A

3. REQUIREMENTS

3.1 Optical terms and definitions.- Terms and definitions peculiar to the general field of optics as used herein are defined in MIL-STD-1241.

3.2 Coated area.- Unless otherwise specified, optical elements shall be coated over their entire effective aperture except for an allowable uncoated holding area as shown in the table below.

Diameter or Maximum Diagonal Dimension of Optical Element	Max Width of Uncoated Area
0 - 1.5"	0.030"
1.5 - 2.5"	0.040"
2.5 - 3.5"	0.050"
3.5 - 5.0"	0.060"
5.0 - 7.0"	0.100"

3.3 Cemented and bonded surfaces.- An optical element surface or portion thereof which is to be cemented or bonded to another surface or portion thereof shall not be coated on the area to be cemented or bonded.

3.4 Specular reflectance.- When applied to substrate materials having indices of refraction within the range of 1.47-1.55 the specular reflectance from each coated surface shall not exceed the following limits for energy incident on the surface at an angle within the range 0-15 degrees inclusive and at an angle of 30 degrees:

Wavelength Range (Nanometers)	Maximum Reflectance (%) (at any wavelength within specified spectral range)	
	0° to 15° incl.	30°
440-460	0.55	--
460-660	0.45	--
660-675	0.55	--
425-460	----	1.0
460-620	----	0.7
620-665	----	1.0

3.5 Light loss.- Within the wavelength range 425 to 700 nanometers, light loss (absorbance plus (+) diffuse reflectance) in the coating shall not exceed the following limits per surface unless otherwise specified (see 6.2):

Maximum average loss	0.5%
Maximum absolute loss	2.0%

3.6 Coating quality.- The coating shall be uniform in quality and condition, clean, smooth, and free from foreign materials, and from physical imperfections and optical imperfections as follows:

3.6.1 The coating shall show no evidence of flaking, peeling or blistering.

3.6.2 The coating shall not contain blemishes such as discolorations, stains, smears and streaks.

3.6.3 The coating shall show no evidence of a cloudy or hazy appearance.

3.6.4 Unless otherwise specified, the coating shall meet the following scratch and dig requirements. Coating defects shall be counted in addition to those allowed by the substrate specification.

Lighting wedges or light piping glass	-	40-20 per MIL-O-13830
Cover glass for lighted instruments	-	60-20 per MIL-O-13830
Cover glass for unlighted instruments	-	80-50 per MIL-O-13830

Coating spatter shall be counted as a dig or inclusion.

3.6.5 Pinholes in the coated surface shall not exceed twice the allowable dig size. Unless otherwise specified, pinhole distribution shall be ignored.

3.7 Abrasion resistance.- There shall be no visible damage to the coated surface when rubbed with an eraser conforming to MIL-E-12397 under a force of 2 to 2 1/2 lbs.

3.8 Humidity - "B".- Continuous exposure for 24 hours in an atmosphere controlled at 116°F to 124°F and 95 to 100% relative humidity.

3.9 Salt fog.- A 5% solution at a temperature of 95°F for 48 hours.

3.10 Humidity - "C".- Continuous exposure for 48 hours in a controlled atmosphere with a relative humidity ranging between 85 and 95% at temperatures ranging from 82°F to 160°F.

3.11 Environmental.- Coated optical elements shall withstand the following environmental exposures.

3.11.1 High temperature.- Temperature of 160°F for not less than 48 hours.

3.11.2 Low temperature.- Temperature of -65°F for not less than 48 hours.

3.11.3 Temperature shock.- Temperature of 160°F for 4 hours and transferred within 5 minutes into a temperature of -65°F for 4 hours and returned within 5 minutes to 160°F for 4 hours.

3.11.4 Humidity - "A".- Continuous exposure for 240 hours in a controlled atmosphere with a relative humidity ranging between 85 and 95% at temperatures ranging from 82°F to 160°F.

3.11.5 Dust.- A fine sand-laden air with a maximum velocity of 2000 feet per minute for 28 hours.

3.11.6 Fungus.- A mixed spore suspension for 28 days.

4. QUALITY ASSURANCE PROVISIONS

4.1 Responsibility for inspection.- Unless otherwise specified in the contract or purchase order, the supplier is responsible for the performance of all inspection requirements as specified herein. Except as otherwise specified, the supplier may utilize his own facilities or any other facilities suitable for the performance of the inspection requirements specified herein, unless disapproved by the Government. The Government reserves the right to perform any of the inspections set forth in the specification where such inspections are deemed necessary to assure supplies and services conform to prescribed requirements.

4.2 Certification.- Unless actual performance of the tests is specified in the contract (see 6.2), the supplier shall certify product conformance to the requirements of 3.11 for each shipping lot.

4.2.1 Certification testing.- When such testing is required, each of six test parts shall be subjected to the sequence of tests in Table I. Upon completion of the test sequence, the test parts shall be assembled into an appropriate instrument or test device and evaluated. The ability to read the instrument without a substantial reduction in clarity shall be considered evidence of successful completion of the testing. Test procedures are specified in MIL-STD-810.

4.2.2 Certification test samples.- Unless otherwise specified in the contract (see 6.2), certification test samples shall be crown glass of refractive index 1.523 and Abbe number of 58.6, and shall be of such a size and configuration that they can be assembled into an instrument or suitable test device for evaluation.

TABLE I - CERTIFICATION TESTS

Characteristic	Requirement	Test Method
Light Loss	3.5	4.4.5
Temperature - High	3.11.1	MIL-STD-810, Method 501-1
Temperature - Low	3.11.2	MIL-STD-810, Method 502-1
Temperature - Shock	3.11.3	MIL-STD-810, Method 503-1
Humidity "A"	3.11.4	MIL-STD-810, Method 507.1
Salt Fog	3.9	MIL-STD-810, Method 509.1
Dust	3.11.5	MIL-STD-810, Method 510.1
Fungus	3.11.6	MIL-STD-810, Method 508.1

4.3 Inspection provisions.

4.3.1 Final acceptance inspection.- The classification of defects in Table II constitutes the minimum inspection to be performed by the supplier prior to government acceptance or rejection by item or lot.

4.3.1.1 Acceptance and rejection.- Rejected lots shall be screened for all defective characteristics. Removal or correction of defective units and resubmittance of rejected lots shall be in accordance with "Acceptance and Rejection" as specified in MIL-STD-105.

4.3.1.2 Submission of product.- Unless otherwise specified by the contracting officer, inspection lot size, lot formation, and presentation of lots shall be in accordance with "Submission of Product" provisions in MIL-STD-105. A lot will be considered to be a shipping lot (see 6.5).

TABLE II - CLASSIFICATION OF DEFECTS

Level II of Table I with Sampling Plan Table IIA of MIL-STD-105

Class	Requirement	Test Paragraph
Critical: None		
Major: AQL 1.0% defective		
101. Coated Area	3.2	4.4.1
102. Cemented and Bonded Surfaces	3.3	4.4.2
103. Coating Quality	3.6	4.4.3.1, 4.4.3.3, 4.4.3.4, 4.4.3.5

4.3.2 Final acceptance test.- The classification of defects in Table III shall constitute the minimum tests to be performed by the supplier prior to government acceptance or rejection by item or lot.

4.3.2.1 Test samples.- Unless otherwise specified (see 6.2), two (2) test witness pieces of a similar refractive index and surface finish as the parts to be coated and coated in the same evaporation lot (see 6.4) as the parts, shall be subjected to each individual test of Table III. No two (2) test witness pieces need be subjected to more than one test. The test witness pieces shall be placed in the coating chamber such that they represent the whole evaporation lot.

4.3.2.2 Acceptance and rejection.- Failure of any test shall cause the evaporation lot to be rejected. Rejected lots shall be screened for all defective characteristics. Removal and correction of defective units and resubmittance of rejected lots shall be in accordance with "Acceptance and Rejection" as specified in MIL-STD-105.

TABLE III

Characteristic	Requirement	Test Paragraph
201. Humidity "B" and	3.8	4.4.6, 4.4.3.1, 4.4.3.2
Abrasion	3.7	4.4.7, 4.4.3.1
202. Reflectance	3.4	4.4.4
203. Salt Fog	3.9	4.4.8, 4.4.3.1
204. Humidity "C"	3.10	4.4.9, 4.4.3.3

4.4 Test methods and procedures

4.4.1 Coated area.- Coated area shall be visually examined to determine conformance to 3.2.

4.4.2 Cemented and bonded surfaces.- Coated area shall be visually examined to determine conformance to 3.3.

4.4.3 Coating quality.- Coated surfaces shall be examined for conformance to 3.6 as follows:

4.4.3.1 The coating shall be examined by both transmission and reflection for evidence of flaking, peeling or blistering. The test method is as shown in Figure I-b. Such evidence shall be cause for rejection.

4.4.3.2 The coating shall be examined by both transmission and reflection for evidence of discoloration, stains, streaks, or smears. The test method is as shown in Figure I-b. When such evidence is found, the blemished area shall be evaluated by specular reflectance. Parts exceeding the requirements of 3.4 shall be rejected.

4.4.3.3 The coatings shall be evaluated as shown in Figure I-c for evidence of cloudy or hazy appearance. Such evidence shall be cause for rejection.

4.4.3.4 The coating shall be examined by transmission as shown in Figures I-a and I-b for pinhole, and scratch and dig requirements. Parts exceeding the requirements specified in 3.6.4 and 3.6.5 shall be rejected.

4.4.3.5 The coating shall be examined by the transmission method for evidence of discoloration, stains, streaks and smears. The test method is as shown in Figure I-b. Such evidence shall be cause for rejection.

4.4.4 Specular reflectance.- To determine conformance to the requirements of 3.4, the test shall be conducted with a calibrated spectrophotometer or a low level reflectometer capable of measuring within the 400 to 700 nanometer wavelength range. The measuring equipment shall be suitably equipped to measure specular reflectance at the angles specified.

4.4.5 Light Loss.- To determine conformance to the requirements of 3.5, this test shall be conducted using a calibrated spectrophotometer capable of measuring within the 400 to 700 nanometer wavelength range. Several elements are scanned in series with the spectrophotometer. The elements may be skewed in the beam, precoating and post coating measurements are compared, reflection losses are extracted and the resulting absorption loss per surface determined.

4.4.6 Humidity - "B".- The test samples shall be placed into an environmentally controlled test chamber at a controlled temperature of 120 ± 4°F and 95% ± 5% relative humidity and held for 24 hours minimum. After completion of this test, the parts shall be evaluated for conformance to 3.6.1 and 3.6.2. Following this evaluation, the parts shall be subjected to the abrasion test of 4.4.7.

4.4.7 Abrasion.- The coating shall be tested by rubbing the coated surface with a standard eraser conforming to MIL-E-12397 mounted in the holding device. A force of between 2 and 2 1/2 lbs. shall be applied. Strokes of approximately one inch shall be made if the size of the element permits; smaller strokes for smaller elements. All strokes shall be made on one path for twenty strokes. After completion of this test, the parts shall be evaluated for conformance to 3.6.1.

4.4.8 Salt fog.- The coatings shall be tested as specified in MIL-STD-810, Method 509-I. After completion of this test, the parts shall be evaluated for conformance to 3.6.1.

4.4.9 Humidity - "C".- The coating shall be tested as specified in MIL-STD-810, Method 507-I, except that the test time shall be 48 hours (2 cycles). Upon completion of these tests, the parts shall be evaluated for conformance to 3.6.3.

5. PREPARATION FOR DELIVERY

5.1 Packaging shall be accomplished in a manner to ensure that the coated glasses during shipment and storage, will not be damaged by chipping, abrading or staining.

6. NOTES

6.1 Intended use.- The coatings covered by this specification are applied to the surfaces of optical elements primarily to reduce the intensity of reflected surface light and increase light transmission.

6.2 Ordering data.- Procurement documents should specify the following:

a. Title, number and date of this specification.

b. Performance of certification tests if required (see 4.2) and quantity of test parts required (if different than 4.2).

c. Type and size of certification test parts (if different than 4.2).

d. Light loss permitted (if different than 3.5).

e. Number of test samples (if different than 4.3.2.1).

6.3 Abrasion testing device.- A device conforming to Frankford Arsenal Drawing No. D7680606 may be used for the abrasion testing of 4.4.7.

6.4 Evaporation lot.- An "evaporation lot" shall be defined as the group of parts which has the coatings applied at the same time and in the same evaporation chamber.

6.5 Shipping lot. - A "shipping lot" shall consist of units manufactured under essentially the same conditions and at approximately the same time.

6.6 International standardization. - Certain provisions (paragraphs 3.4, 3.5, 3.6, 3.7, 3.8 and 3.9) of this specification are the subject of international standardization agreement (ASCC AIR STD 10/61, Coating, Reflection Reducing for Aircraft Station Displays). When amendment, revision, or cancellation of this publication is proposed which will affect or violate the international agreement concerned, the preparing activity will take appropriate reconciliation action through international standardization channels including departmental standardization offices, if required.

Custodians:

Army - MU
Navy - AS
Air Force - 11

Review Activities:

Army - MU, AV
Navy - AS, OS
Air Force - 11, 17, 26

User Activities:

Army - WC, MI
Navy - SH
Air Force - None

Preparing Activity:

Army - MU

Civilian Agencies
Interest-DM

Project No. 6650-0035

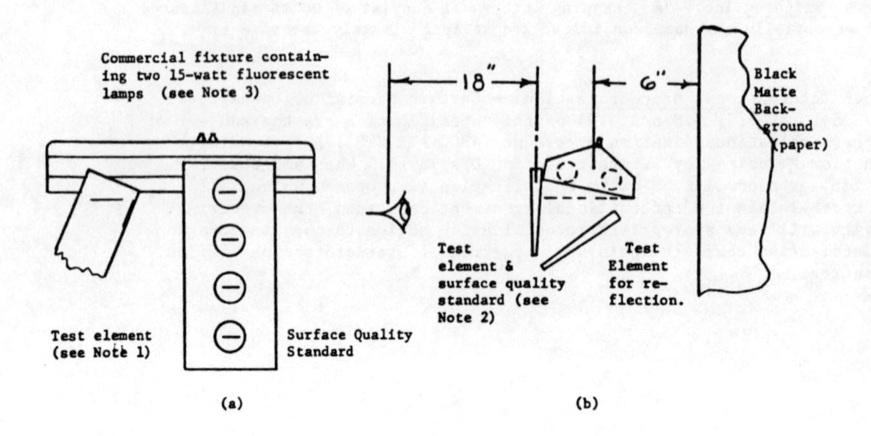

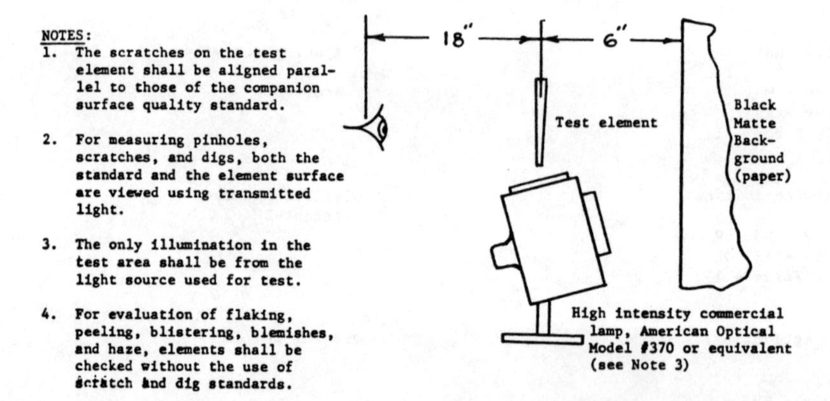

NOTES:

1. The scratches on the test element shall be aligned parallel to those of the companion surface quality standard.
2. For measuring pinholes, scratches, and digs, both the standard and the element surface are viewed using transmitted light.
3. The only illumination in the test area shall be from the light source used for test.
4. For evaluation of flaking, peeling, blistering, blemishes, and haze, elements shall be checked without the use of scratch and dig standards.

FIGURE 1

MIL-C-48497A
8 September 1980
SUPERSEDING
MIL-C-48497 (AR)
27 June 1974

MILITARY SPECIFICATION

COATING, SINGLE OR MULTILAYER, INTERFERENCE: DURABILITY REQUIREMENTS FOR

This specification is approved for use by all Departments and Agencies of the Department of Defense.

1. SCOPE

1.1 Scope. This specification establishes minimum quality and durability requirements for single layer and multilayer interference coatings that are primarily used within the protective confines of sealed optical systems. (See 6.1).

2. APPLICABLE DOCUMENTS

2.1 The following documents, of the issue in effect on date of invitation for bids or request for proposal, form a part of this specification to the extent specified herein.

SPECIFICATIONS

Federal

L-T-90	Tape, Pressure-Sensitive, Adhesive, (Cellophane and Cellulose Acetate)
O-A-51	Acetone, Technical
O-E-760	Ethyl Alcohol (ethanol), Denatured Alcohol, and Proprietary Solvent
O-T-634	Trichloroethylene, Technical
CCC-C-440	Cloth, Cheesecloth, Cotton Bleached and Unbleached

Military

MIL-E-12397	Eraser, Rubber Pumice, for Testing Coated Optical Elements
MIL-E-13830	Optical Components for Fire Control Instruments; General Specification Governing the Manufacture, Assembly and Inspection of
MIL-I-45607	Inspection Equipment, Acquisition, Maintenance and Disposition of

Beneficial comments (recommendations, additions, deletions), and any pertinent data which may be of use in improving this document, should be addressed to: Commander, US Army Armament Research and Development Command, ATTN: DRDAR-TST-S, Dover, New Jersey 07801, by using the self-addressed Standardization Document Improvement Proposal (DD Form 1426), appearing at the end of this document, or by letter.

☆U.S. GOVERNMENT PRINTING OFFICE: 1980—703-023/4247 FSC 6650

STANDARDS

Military

MIL-STD-105	Sampling Procedures and Tables for Inspection by Attributes
MIL-STD-109	Quality Assurance Terms and Definitions
MIL-STD-1241	Optical Terms and Definitions

DRAWINGS

U. S. Army, Armament Research & Development Command

7641866	Surface Quality Standards for Optical Elements
7680606	Coating, Eraser Abrasion Tester

(Copies of specifications, standards, drawings and publications required by suppliers in connection with specific procurement functions should be obtained from the procuring activity or as directed by the contracting officer.)

3. REQUIREMENTS

3.1 Optical terms and definitions. Terms and definitions peculiar to the general field of optics as used herein are defined in MIL-STD-1241.

3.2 Coated area. Optical components shall be coated over their entire clear aperture.

3.3 Coating quality. The coating shall be uniform in quality and condition and shall conform to the following:

3.3.1 Physical. The coating shall have no evidence of flaking, peeling, cracking or blistering.

3.3.2 Cosmetic. When specified the coating shall conform to the cosmetic requirements (stains, smears, discolorations, streaks, cloudiness, etc.) stated on the component drawing or other documents.

3.3.3 Environmental and solubility blemishes. The coated surface shall be free of blemishes such as stains, smears, discolorations, streaks, cloudiness, etc., that would cause non-conformance to the coating's spectral requirements as stated on the component drawing or other documents.

3.3.4 Spatter and holes. Coating spatter and holes in the coating shall be considered as a dig and shall not exceed the allowable dig size and quantity stated on the componet drawing or other documents.

3.3.5 Surface defects (scratch and dig). Coating scratches and digs shall not exceed the values specified for the substrate on the component drawing or other documents. Coating scratches and digs shall be considered separate from the substrate scratch and dig requirements.

3.3.5.1 Transparent coated surfaces. Scratch and dig requirements for transparent coated surface shall be specified by two (2) numbers separated by a hyphen (i.e. 60-40) in accordance with MIL-O-13830.

3.3.5.2 Opaque coated surfaces. Scratch and dig requirements for opaque coated surfaces shall be specified by two letters separated by a hyphen (i.e. F-C). The first letter of the pair is the maximum scratch value; the second letter is the maximum dig value.

3.3.5.2.1 Scratches. The scratch letter defines the width of the scratch in accordance with the following table:

SCRATCH LETTER	SCRATCH WIDTH(MICRONS)	SCRATCH WIDTH (INCHES)
A	5	.00020
B	10	.00039
C	20	.00079
D	40	.00157
E	60	.00236
F	80	.00315
G	120	.00472

3.3.5.2.1.1 Density of maximum size scratches. The accumulated length of all maximum size scratches present shall not exceed 1/4 the average diameter of the coated surface.

3.3.5.2.1.2 Density of all scratches. When a maximum size scratch is present, the sum of the products of the widths designated by the scratch letters times the ratio of their length to the diameter of the coated surface shall not exceed one half the width specified by the scratch letter. When a maximum size scratch is not present, the sum of the products of the widths designated by the scratch letters times the ratio of their length to the diameter of the coated surface shall not exceed the width specified by the scratch letter.

3.3.5.2.2 Digs. The dig letter specifies the average diameter of the dig in accordance with the following table:

DIG LETTER	DIAMETER (MM)	DIAMETER (INCHES)
A	0.05	.0020
B	0.10	.0039
C	0.20	.0079
D	0.30	.0118
E	0.40	.0158
F	0.50	.0197
G	0.70	.0276
H	1.00	0.394

The permissible number of maximum size digs shall not exceed one per each 20 millimeters of (.80 inches) diameter or fraction thereof on any single coated surface. The sum of the diameters of all digs shall not exceed twice the diameter of the maximum size specified by the dig letter per 20 millimeters of diameter.

3.4 Coating durability.

3.4.1 Environmental and pyhsical durability. The coated optical surface shall meet the following service conditions in the order specified:

3.4.1.1 Adhesion. The coated optical surface shall show no evidence of coating removal when cellophane tape is pressed firmly against the coated surface and quickly removed at an angle normal the the coated surface.

3.4.1.2 Humidity. After exposure in an atmosphere of $120^{\circ} \pm 4^{\circ}$F and 95 to 100% relative humidity, the coated optical surface shall meet the requirements of 3.3.1 and 3.3.3.

3.4.1.3 Moderate abrasion. The coated optical surface shall show no signs of deterioration such as streaks or scratches when abraded with a dry, clean cheesecloth pad.

3.4.2 Thermal and cleaning durability. The coated optical surface shall meet the following conditions:

3.4.2.1 Temperature. The coated optical surface shall be exposed to temperatures of -80°F and $+160^{\circ}$F for 2 hours at each temperature. The rate of temperature change shall not exceed 4°F per minute. Subsequent to these exposures, the coated optical surface shall meet the requirements of 3.3.1 and 3.4.1.1.

3.4.2.2 Solubility and cleanability. After immersion in trichloroethylene, acetone and ethyl alcohol and wiping with cheesecloth, the coated optical surface shall show no evidence of coating removal or scratches and shall meet the requirements of 3.3.1 and 3.3.3.

3.4.3 Optical durability requirements. When specified on the component drawing or other documents (See 6.2) the following requirements shall be added or substituted:

3.4.3.1 Severe abrasion. Abrasion by an eraser conforming to MIL-E-12397 shall not cause deterioration such as streaks or scratches on the coated optical surface.

3.4.3.2 Salt solubility. After immersion in a saline solution, the coated optical surafce shall meet the requirements of 3.3.1 and 3.3.3.

3.4.3.3 Water solubility. After immersion in distilled water, the coated surface shall meet the requirements of 3.3.1 and 3.3.3.

3.5 Optical. The coated component shall conform tothe pertinent optical requirements specified on the component drawing or other documents (See 6.2).

4. QUALITY ASSURANCE PROVISIONS

4.1 Responsibility for inspection. Unless otherwise specified in the contract or purchase order, the supplier is responsible for the performance of all inspection requirements as specified herein. Except as otherwise specified, the supplier may utilize his own facilities or any other facilities suitable for the performance of the inspection requirements specified herein, unless disapproved by the Government. The Government reserves the right to perform any of the inspections set forth in the specification where such inspections are deemed necessary to assure that supplies and services conform to prescribed requirements (See 6.2).

4.1.1 General provisions. Definitions of inspections terms shall be as listed in MIL-STD-109.

4.1.2 Witness piece. Unless otherwise specified, witness pieces as defined in 4.1.3 or coated componets may be used to test the optical and durability requirements of the coated component. (See 4.2, 4.3 & 6.2). The witness pieces shall be positioned in the coating chamber such that they represent the optical and durability characteristics of the whole evaporated lot (See 6.4). The Government reserves the right to test the actual coated component with the same test to which the witness pieces were subjected. Should a component fail, even though the witness pieces pass the test, the lot shall be rejected.

4.1.3 Characteristics of the witness piece. When witness pieces are used to test the optical and durability requirements of the coated component they shall exhibit the following characteristics:

a. The witness piece shall have the same refractive index and absorption coeffecient as the componets to be coated.

b. The witness piece shall have a surface finish similar to that of the component to be coated.

c. The witness piece shall be such that it presents no difficulty in measuring and testing the optical and durability requirements of the coating.

d. Where transmission characteristics are required the witness piece shall simulate the axial thickness of the component by either computational methods or using a witness piece that has the same thickness as the coated component.

4.2 First article (initial production) approval. The requirement for first article approval and the responsibility (government or contractors) for first article testing shall be as specified in the contract. (See 6.2). Unless otherwise specified, the sample for first article approval tests shall consists of five (5) coated components plus ten (10) coated witness pieces (See 4.1.2 and 4.1.3). The 5 coated componets shall be tested, as specified herein, for all the requirements of 3.2, 3.3.1, 3.3.2, 3.3.4 and 3.3.5. Five coated witness pieces shall be tested, as specified herein, for all the requirements of 3.4.1 and 3.5. The other 5 coated witness pieces shall be tested as specified herein for all the requirements of 3.4.2 and 3.5. The government reserves the right to subject the coated components to all the tests specified here. The sample shall be coated in the same manner, using the same materials, equipment, processes, and procedures as used in regular production. All parts and materials including packaging shall be obtained from the same source of supply as used in regular production.

4.2.1 Government testing. When the Government is responsible for conducting first article approval tests, the contractor, prior to submitting the sample to the government, shall inspect the sample to insure that it conforms to all the requirements of the contract and submit a record of this inspection with the sample, including certificates of conformance for materials, as applicable.

4.2.2 Contractor testing. When the contactor is responsible for conducting first article approval tests, the sample shall be inspected by the contractor for all the requirements of the contract. The sample and a record of this inspection, including certificates of conformance for materials, shall be submitted to the government for approval. The government reserves the right to witness the contractor's inspection.

4.3 Inspection provisions.

4.3.1 Submission of product.

4.3.1.1 Inspection lot size. The inspection lot size shall consist of all components or subassemblies (unit of product) coated within one (8 hour or 10 hour) work shift.

4.3.1.2 Lot formation and presentation of lots. The unit of product shall be submitted for inspection on a moving inspection lot basis where the components or subassemblies are continuously offered for inspection in the order produced.

NOTE: Each evaporation lot forming a part of a moving inspection lot shall be identified for subsequent evaluation, if required.

4.3.2 Examination and tests.

4.3.2.1 Components and subassemblies. All coated componets and subassemblies shall be inspected in accordance with this specification and the inspection provisions contained in Supplementary Quality Assurance Provisions (SQAP) listed in the Technical Data Package (TDP). Examination and tests related to Section 3 herein shall be performed on a class basis in accordance with MIL-STD-105 and the sampling plans specified in Tables I and II herein. The tabulated classification of defects in Tables I and II shall constitute the minimum inspection to be performed by the supplier after first article approval and prior to Government acceptance or rejection by item or lot.

TABLE I - CLASSIFICATION OF DEFECTS FOR COMPONENTS

CLASS	CHARACTERISTIC	REQUIREMENT	TEST PROCEDURE
CRITICAL:	NONE DEFINED		
MAJOR:	AQL 1.0%		
101.	Coated area	3.2	4.5.1
102.	Coating quality, physical	3.3.1	4.5.2.1
103.	Coating quality, cosemtic	3.3.2	4.5.2.2
104.	Coating quality, spatter & holes	3.3.4	4.5.2.4
105.	Coating quality, scratch & dig	3.3.5	4.5.2.5
MINOR:	NONE DEFINED		

NOTE: The inspection for the characteristics in Table I shall be conducted at a temperature between +60°F and +90°F.

4.3.2.2 Acceptance and rejection. Rejected lots shall be screened for all defective characteristics. Removal of defective units and resubmittal or rejected lots shall be in accordance with "Acceptance and Rejection" as specified in MIL-STD-105.

4.3.2.3 Special sampling.

4.3.2.3.1 Environmental and physical durability. A minimum of one (1) coated witness piece (or one (1) coated component, when required) shall be selected from each evaporation lot to form a minimum of five (5) samples for each inspection lot. The samples shall meet the requirements and tests in Table II.

TABLE II - ENVIRONMENTAL & PHYSICAL DURABILITY

NO.	CHARACTERISTIC	REQUIREMENT	TEST PROCEDURE
301.	Adhesion	3.4.11	4.5.3.1
302.	Coating quality, physical (post humidity)	3.4.1.2, 3.3.1	4.5.3.2, 4.5.2.1
303.	Coating quality, blemishes (post humidity)	3.4.1.2, 3.3.3	4.5.3.2, 4.5.2.3
304.	Moderate abrasion	3.4.1.3	4.5.3.3
305.	Optical	3.5	4.5.6

NOTE: The inspection for the characteristics in Table II shall be conducted at a temperature between +60°F and +90°F.

4.3.2.3.2 Failure of sample. Should any one item of a special sampling fail to meet the specified test requirements, acceptance of the represented inspection lot shall be suspended by the Government until necessary corrections have been made by the contractor and the resubmitted items have been approved. (See 4.3.2.2).

4.4 Inspection equipment. Except as otherwise provided for by the contract, the contractor shall supply and maintain inspection equipment in accordance with the applicable requirements of MIL-I-45607.

4.4.1 Government furnished inspection equipment. Where the contract provides for Government furnished test equipment, supply and maintenance of test equipment shall be in accordance with applicable requirements specified in MIL-I-45607.

4.4.2 Contractor furnished inspection equipment.

4.4.2.1 Government design. All inspection equipment specified by drawing number in specifications or SQAP forming a part of the contract shall be supplied by the contractor in accordance with technical data included in the contract.

4.4.2.2 Contractor equipment. The contractor shall supply inspection equipment compatible with the "Test Methods and Procedures" specified in 4.5 of this specification. Since tolerance of test equipment is normally considered to be within 10% of the product tolerance for which it is intended, this inherent error in the test equipment must be considered as part of the prescribed product tolerance limit. Thus, concept, construction, materials, dimensions and tolerances used in the test equipment shall be so selected and controlled as to insure that the test equipment will reliably indicate acceptability of a product which does not exceed 90% of the prescribed tolerance limit, and permit positive rejection when non-eonforming. Construction shall be such as to facilitate routine calibration of test equipment.

4.5 Test methods and procedures.

4.5.1 Coated area. The coated area of the component or subassembly shall be examined with standard measuring equipment (See 6.4). The coated area shall conform to the requirements of 3.2.

4.5.2 Coating quality.

4.5.2.1 Physical. The coating shall be visually examined by reflection, with the unaided eye, for evidence of flaking, peeling, cracking or blistering. The examination shall be performed using two 15 watt cool white fluorescent light tubes as the light source. The viewing distance from the coated surface to the eye shall not exceed 18 inches. The coated surface shall be viewed against a black matte background. The only illumination in the test area shall be from the light source used for testing. This method of examination is as depicted in Figure 1. The coating shall conform to the requirements of 3.3.1.

4.5.2.2 Cosmetic. The coated component shall be examined using the test method specified in 4.5.2.1 for evidence of discoloration, stains, smears, streaks, cloudiness, etc. The coating shall conform to the requirements of 3.3.2.

4.5.2.3 Environmental and solubility blemishes. The coated witness samples (or coated components, when required) shall be examined using the test method specified in 4.5.2.1 for evidence of discoloration, stains, smears, streaks, cloudiness, etc. The coating shall conform to the requirements of 3.3.3 When evidence of blemishes is found, the blemished area shall be examined (See 4.5.6) for compliance to the spectral requirements specified on the component drawing.

4.5.2.4 Spatter and holes. The coated components shall be examined for dig size and quantity using the applicable test method specified in 4.5.2.5. The coating shall conform to the requirements of 3.3.4.

4.5.2.5 Surface defects (scratch and dig).

4.5.2.5.1 Transparent coated surfaces. Transparent coated surfaces shall be examined utilizing the technique specified in 4.5.2.1, except that (hole) in the coating shall be evaluated by comparison with the Surface Quality Standards for Optical Elements (scratch & dig) Drawing 7641866. The visual appearance of the scratch, and the diameter of the digs in the coating shall conform to the requirements of 3.3.5.1. The lengths and density of all scratches, and the diameters and density of all digs shall conform to the requirements of MIL-O-13830.

4.5.2.5.2 Opaque coated surfaces. Opaque coated surfaces shall be examined utilizing the lumination, reflection and viewing background technique specified in 4.5.2.1. The length and width of scratches, and the dig (hole) diameters shall be determined by use of interferometry, microscopic measuring

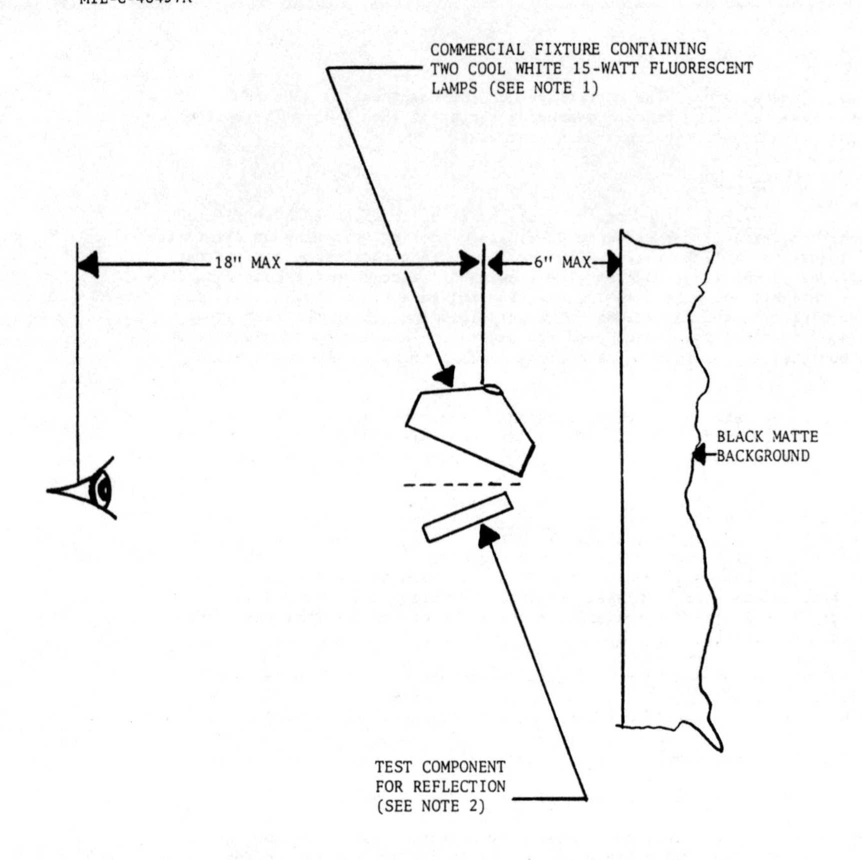

NOTES:

1. THE ONLY ILLUMINATION IN THE TEST AREA SHALL BE FROM THE LIGHT SOURCE USED FOR TEST.
2. TILT AT AN APPROPRIATE ANGLE TO SEE THE COATED SURFACE.

FIGURE 1

devices, calibrated precision comparators, or similar applicable precision measuring devices. The width, length and density of all scratches in the coating shall conform to the requirements of 3.3.5.2. The diameter and density of all digs (holes) in the coating shall conform to the requirements of 3.3.5.2.

4.5.3 Environmental physical durability tests. The coated components or witness samples shall withstand exposure to the following test conditions:

4.5.3.1 Adhesion. The coated component or witness samples shall be subjected to an adhesion test using 1/2" wide cellophane tape conforming to Type I of L-T-90. Press the adhesive surface of the cellophane tape firmly against the coated surface and quickly remove at an angle normal to the coated surface. A visual inspection shall be made for conformance to 3.4.1.1. Subsequent to this test, the coated component or witness sample shall be subjected to the test of 4.5.3.2.

4.5.3.2 Humidity. The coated components or witness samples shall be placed into an environmentally controlled test chamber and exposed to a temperature of 120°F and 95% to 100% relative humidity. The coating shall be exposed for a minimum of 24 hours. Subsequent to this exposure, the coating shall be cleaned and evaluated for conformance to 3.3.1 and 3.3.3. The coated component or witness sample shall then be subjected to the test specified in 4.5.3.3.

4.5.3.3 Moderate abrasion. Within one hour after the humidity test of 4.5.3.2, the coated component or witness sample shall be subjected to a moderate abrasion test. It shall consist of rubbing a minimum of 50 strokes across the surface in straight lines. The abrader shall be a 1/4" thick by 3/8" wide pad of clean dry, laundered cheesecloth conforming to CCC-C-440. The bearing force shall be a minimum of 1 pound and shall be applied approximately normal to the coated surface. The actual test apparatus shall be the Eraser Abrasion Coating Tester of Drawing 7680606 except that the eraser portion shall be completely covered with cheesecloth and the cheesecloth shall be secured to the shaft of the tester with an elastic band. Subsequent to this test, the coating shall be evaluated for conformance to 3.4.1.3. Following the tests of 4.5.3.1, 4.5.3.2 and 4.5.3.3 the coated component or witness sample shall be evaluated for conformance to the provisions of 4.5.6.

4.5.4 Thermal and cleaning durability. The witness samples shall withstand exposure to the following test conditions:

4.5.4.1 Temperature. The witness samples shall be subjected to temperatures of -80°F ± 2°F and +160°F ±2°F for a period of 2 hours at each temperature. After each exposure the coated componets or witness samples shall be stabilized at an ambient temperature between +60° and +90°F and subjected to the test specified in 4.5.3.1. Subsequent to this test, the coating shall be evaluated for conformance to 3.4.1.3 and 3.3.1. The rate of temperature change between any two temperature levels shall not exceed 4°F per minute.

4.5.4.2 Solubility and cleanability. The witness samples shall be immersed, in sequence, in the following solutions maintained at room temperature (+60° to +90°F): trichloroethylene conforming to O-T-634; acetone conforming to O-A-51; and ethyl alcohol conforming to O-E-760. The immersion time in each solution shall be a minimum of ten minutes. Upon removal from each solution, the solvent shall be allowed to evaporate to dryness without wiping or forced drying before proceeding to the next solution. Upon removal from the alcohol solution, and after drying any resultant stains on the coated surface shall be removed by wiping the coating to a clean, stain-free condition with an ethyl-alcohol-mositened cheesecloth. Subsequent to this cleaning, the coating shall be evaluated for conformance to 3.4.2.2, 3.3.1, 3.3.3 and 3.5.

4.5.5 Optional durability tests. The coated componets or witness samples shall withstand exposure to the following test conditions when specified in the contract, purchase order or on the componet drawing. (See 6.2).

4.5.5.1 Severe abrasion. The coated componets or witness samples shall be subjected to a severe abrasion, rubbing the coated surface with a standard eraser conforming to MIL-E-12397 mounted in the Eraser Abrasion Coating Tester of Drawing 7680606. The bearing force shall be between 2 and 2 1/2 lbs. Applied approximately normal to the coated surface. A total of 20 strokes shall be made along a straight line path. Subsequent to this test the coating shall be evaluated for conformance to 3.4.3.1 and then subjected to the test specified in 4.5.3.

4.5.5.2 Salt solubility. The coated components or witness samples shall be immersed for a period of 24 hours in a solution of water and sodium chloride (salt). The mixture shall be 6 ounces of salt per gallon of water at room temperature (+60 to +90°F). Subsequent to this immersion, the coated componets or witness samples shall be removed from the solution, dried with clean, laundered, cheesecloth, and then evaluated for conformance to 3.3.1 and 3.3.3.

4.5.5.3 Water solubility. The coated componets or witness samples shall be immersed for a period of 24 hours in distilled water at room temperature (+60°F to +90°F). Subsequent to this immersion, the coated componets or witness samples shall be removed from the solution, dried with clean laundered cheesecloth, and then evaluated for conformance to 3.3.1 and 3.3.3.

4.5.6 Optical. Appropriate optical instruments such as spectrophotometers, radiometers, etc. shall be used to test conformance to the optical requirements of 3.5.

5. PACKAGING

This section is not applicable to this specification.

6. NOTES

6.1 Intended use. This specification extablishes minimum quality and durability requirements for single or multi-layer interference coatings applied to components in optical systems. It is applicable to coatings utilized on a wide variety of substrate materials for such items as lenses, prisms, mirrors, beam splitters, reticles, laser elements, windows, filters, etc.

Coatings produced to the requirements of this specification may be exposed without damage to those handling and cleaning procedures associated with optical instrument assembly practice and will generally be employed within the protective confines of a sealed instrument.

Certain applications may necessitate the use of additional or expanded quality and durability requirements, some of which are included within the optical durability section of this specification. (See 3.4.3). Other quality and durability requirements with the appropriate test methods must be defined when they are applicable. All additional requirements must be specified on the applicable drawing or other documents. Examples of coating applications where additional requirements must be specified include laser systems where resistance to laser energy damage is necessary; also when coatings are used outside the confines of a sealed instrument, resistance to such environments as salt fog, fungus, extended humidity, etc. may be necessary.

This specification is not intended as a substitute for other coating specifications, e.g., MIL-C-675, MIL-C-14806, etc. It is intended that this specification provide the durability requirements for coating applications not covered by other documents and in some cases to serve as a supplement, by reference, in other coating specifications.

6.2 Ordering data. Purchasers should exercise any desired options offered herein, and procurement documents should specify the following:

a. Title, number and date of this specification.
b. Optional abrasion, salt solubility, and water solubility requirements (see 3.4.3 and 4.5.5)
c. Optical requirements of coating (See 3.5)
d. Responsibility for performance of inspection requirements (See 4.1)
e. Mandatory use of coated componets for tests (See 4.1.2)
f. Requirement for first article approval (See 4.2, 4.2.1 and 4.2.2)

6.3 First article sample. When the government is responsible for conducting first article approval tests the required article sample specified in 4.2 shall be forwarded to:

Commander
US Army Armament Research & Development Command
ATTN: DRDAR-QAF
Dover, NJ 07801

6.4 Definitions.

6.4.1 Evaporation lot. An evaproation lot is defined as the group of parts which has the coating applied at the same time and in the same chamber.

6.4.2 Standard measuring equipment (SME). Standard measuring equipment is defined as the common measuring devices which are usually stocked by commercial supply houses for ready supply (shelf items) and which are normally used by an inspector to perform dimensional inspection of items under procurement. This category also includes commercial testing equipment such as meters, optical comparators, etc.

Custodian:
Army - AR
Air Force - 11

Preparing Activity:
Army-AR

Project No. 6650-0108

MIL-F-48616
SUPPLEMENT 1
29 JULY 1977

MILITARY SPECIFICATION

FILTER (COATINGS), INFRARED INTERFERENCE:
GENERAL SPECIFICATION FOR

This supplement forms a part of Military Specification MIL-F-48616

Specification Sheets

MIL-F-48616/100	Coating, Low Reflectance, Ordinary Requirements
MIL-F-48616/101	Coating, Low Reflectance on Germanium 8.0 to 11.5 Micrometers, High Durability, Special Requirements for
MIL-F-48616/102	Coating, Ultra-Low Reflectance on Germanium 7.5 to 12.3 Micrometers, Special Requirements for
MIL-F-48616/200	Coating, High Reflectance, First Surface, Ordinary Requirements for
MIL-F-48616/300	Filter, Wide Bandpass, 7.7 to 11.7 Micrometers, Ordinary Requirements for (not issued, reserved for future application)
MIL-F-48616/301	Filter, Wide Bandpass, 7.7 to 11.7 Micrometers, Special Requirements for
MIL-F-48616/400	Filter, Narrow Bandpass, Ordinary Requirements for (not issued, reserved for future application)
MIL-F-48616/500	Coating, Long Wavelength Pass Filter, Ordinary Requirements for
MIL-F-48616/600	Coating, Short Wavelength Pass Filter, Ordinary Requirements for

Preparing activity:

Army - MU

FSC-6650

MIL-F-48616/100
29 JULY 1977

MILITARY SPECIFICATION SHEET

COATING, LOW REFLECTANCE, ORDINARY REQUIREMENTS FOR

The complete requirements for procuring the low reflectance coating described herein shall consist of this document and the latest issue of Specification MIL-F-48616.

This specification is approved for use by all Departments and Agencies of the Department of Defense.

Spectral reflectance requirements for low reflectance coatings are given in Table I. Maximum reflectance values for each surface are tabulated for ranges of substrate index of refraction and for several wavelength ranges.

TABLE I

Low Reflectance Coatings - Maximum Percent Reflectance* per Surface

Index of Refraction of Substrate	Wavelength Radio λ_2/λ_1	Reflectance	Dash No.
1.30 to 1.75	1.0	0.3	1
1.30 to 1.75	1.0 to 1.2	0.4	2
1.30 to 1.75	1.2 to 1.5	0.6	3
1.30 to 1.75	1.5 to 2.0	1.5	4
1.75 to 3.00	1.0	0.3	5
1.75 to 3.00	1.0 to 1.2	0.4	6
1.75 to 3.00	1.2 to 1.5	1.0	7
1.75 to 3.00	1.5 to 2.0	4.0	8
3.00 to 4.20	1.0	1.0	9
3.00 to 4.20	1.0 to 1.2	1.5	10
3.00 to 4.20	1.2 to 1.5	3.0	11
3.00 to 4.20	1.5 to 2.0	5.0	12

*For $\lambda_2/\lambda_1 > 1.0$ maximum average reflectance between λ_1 and λ_2 .

NOTES ON TABLE I:

1. The wavelengths λ_1 & λ_2 define the spectral region of interest.

2. The values given in this table apply at an angle of incidence ranging from 0° to 15° and ambient temperature. Reflectance measurements are to be performed at an angle of incidence between 0° and 15° (See 6.4b and 6.4c) of MIL-F-48616. When measurements are made at larger angles, the data shall be corrected to normal incidence.

FSC-6650

MIL-F-48616/100

3. For $\lambda_2/\lambda_1 = 1.0$ spectral resolution required to verify the requirements in Table I shall be 0.5% or less of the specified wavelength. For $\lambda_2/\lambda_1 > 1.0$, the spectral resolution shall be 1.0% or less of $(\lambda_2 + \lambda_1)/2$. Scan speed and time constant shall be set such that the spectral features of interest can be resolved within the accuracy of 4.4.2.3.2.2 of MIL-F-48616.

Ordering Information:

Anti-reflection coatings are used primarily to reduce reflections within optical systems. Performance is strongly dependent on substrate material and wavelength region. For applications for which these coatings are used to enhance transmission throught an optical element by reducing reflection losses at substrate surfaces, the user should specify the required transmission. The user should be aware of the possibility of absorption within these coatings and within the substrate material, and should therefore specify the absorption requirement if applicable.

Ordering Data:

At a minimum, the substrate (See 6.2d) (nominal refractive index value) to be coated and the wavelength region of interest (See 6.2m) (λ_1 & λ_2) and the dash number (See 6.2b) shall be specified by the procuring activity (See 6.2).

Preparing activity:

Army - MU

Project No. 6650-0091-1

MIL-F-48616/101
29 JULY 1977

MILITARY SPECIFICATION SHEET

COATING, LOW REFLECTANCE ON GERMANIUM, 8.0 to 11.5, MICROMETERS, HIGH DURABILITY, SPECIAL REQUIREMENTS FOR

The complete requirements for procuring the low reflectance coating described herein shall consist of this document and the latest issue of Specification MIL-F-48616.

This specification is approved for use by all Departments and Agencies of the Department of Defense.

Spectral requirements exclusive of substrate absorptance for the low reflectance coating on germanium are given the following paragraphs:

1. The average transmittance through two coated surfaces shall be greater than 93% over the 8.0 to 11.5 micrometer region.

2. The minimum transmittance through two coated surfaces at any point between 8.0 and 11.5 micrometers shall not be less than 90%.

3. The average reflectance shall not be greater than 3.0% per surface over the 8.0 to 11.5 micrometer region.

4. The maximum reflectance at any point between 8.0 and 11.5 micrometers shall not be greater than 5.0% per surface.

Additional Durability Requirements:

In addition to the minimum acceptable durability requirements of MIL-F-48616 the coated element shall meet all of the optional durability requirements stated in 3.4.2.3.

NOTE:

Spectral resolution required to verify the above requirements shall be 0.1 micrometer or less. Scan speed and time constant shall be set such that scan time is two (2) minutes or greater for transmittance measurements and ten (10) minutes or greater for reflectance measurements.

Ordering Data:

Not applicable.

Preparing activity:
Army-MU
Project No. 6650-0091-2

FSC-6650

MIL-F-48616/102
29 JULY 1977

MILITARY SPECIFICATION SHEET

COATING, ULTRA-LOW REFLECTANCE ON GERMANIUM,

7.5 TO 12.3 MICROMETERS, SPECIAL REQUIREMENTS FOR

The complete requirements for procuring the ultra-low reflectance coating described herein shall consist of this document and the latest issue of MIL-F-48616.

This specification is approved for use by all Departments and Agencies of the Department of Defense.

Requirements exclusive of substrate absorptance for the ultra-low reflectance coating on germanium are given the following paragraphs:

1. The average transmittance through two coated surfaces shall be equal to or greater than 98% over the 7.5 to 11.5 micrometer region.

2. The average reflectance shall not be greater than 0.3% per surface over the 8.0 to 11.5 micrometer region.

3. The average reflectance shall not be greater than 0.9% 'surface over the 7.5 to 8.0 micrometer region and 1.3% per surface over the 11.5 to 12.3 micrometer region.

NOTE:

Spectral resolution required to verify the above requirements shall be 0.1 micrometer or less. Scan speed and time constant shall be set such that scan time is two (2) minutes or greater for transmittance measurements and ten (10) minutes or greater for reflectance measurements.

Ordering Data:

Not applicable.

Preparing activity:
Army - MU
Project No. 6650-0091-3

MIL-F-48616/200
29 JULY 1977

MILITARY SPECIFICATION SHEET

COATING, HIGH REFLECTANCE, FIRST
SURFACE, ORDINARY REQUIREMENTS FOR

The complete requirements for procuring the first surface high reflectance coating described herein shall consist of this document and the latest issue of Specification MIL-F-48616.

This specification is approved for use by all Departments and Agencies of the Department of Defense.

Requirements for first surface high reflectance coatings are given in Table I. The table provides the minimum allowable reflectance for several wavelength ratios.

TABLE I

First Surface High Reflectance Coatings - Minimum Reflectance per Surface

Wavelength Radio λ_2/λ_1	Minimum Allowable Reflectance*	Dash No.
1 to 1.05	99	1
1.05 to 1.12	98	2
1.12 to 2	97	3
2	96	4

*For λ_2/λ_1, 1.0 maximum average reflectance between λ_1 and λ_2 .

NOTES FOR TABLE I:

1. The wavelengths λ_1 and λ_2 are the extreme values of the wavelength region of interest.

2. The values given in this table apply at an angle of incidence ranging from 0° to 15° and ambient temperature. Reflectance measurements are to be performed at an angle of incidence between 0° and 15° (See 6.4b and 6.4c) of MIL-F-48616. When measurements are made at larger angles, the data shall be corrected to normal incidence.

Ordering Data:

The wavelength region of interest (See 6.2m) (λ_1 & λ_2) and the dash number (See 6.2b) shall be specified by the procuring activity.

Preparing activity:

Army - MU

Project No. 6650-0091-4

FSC-6650

MIL-F-48616/301
29 JULY 1977

MILITARY SPECIFICATION SHEET

FILTER, WIDE BANDPASS, 7.7 to 11.7 MICROMETERS,
SPECIAL REQUIREMENTS FOR

The complete requirements for procuring the wide bandpass filter described herein shall consist of this document and the latest issue of MIL-F-48616.

This specification is approved for use by all Departments and Agencies of the Department of Defense.

Requirements for the wide bandpass filter (coated substrate) are given in the following paragraphs:

1. Average transmittance shall be equal to or greater than 80% from 8.3 to 11.1 micrometers.

2. Absolute 45% transmittance points shall be at wavelengths than 7.7 micrometers and less than 11.7 micrometers.

3. Absolute 5% transmittance points shall be at wavelengths than 7.5 micrometers and less than 12.1 micrometers.

4. Out of band transmittance shall be:

a. Less than 0.1% average from 1.0 to 7.0 micrometers and from 12.6 to 15.0 micrometers.

b. Less than 1.0% absolute from 1.0 to 7.0 micrometers and from 12.6 to 15.0 micrometers.

c. Less than 5% absolute from 7.0 to 7.5 micrometers and from 12.1 to 12.6 micrometers.

NOTES:

1. Spectral resolution required to verify the above requirements shall be 0.03 micrometer or less.

2. Scan speed and time constant shall be set such that scan time is ten (10) minutes or greater fom 7.0 to 12.6 micrometers and twenty (20) minutes or greater total time for the two regions from 1.0 to 7.0 micrometers and 12.6 to 15.0 micrometers.

3. The spectral requirements stated above apply at an angle of incidence ranging from 0° to 5° and at ambient temperature. Measurements will be made at an angle of incidence within this range (See 6.4b and 6.4c).

FSC-6650

MIL-F-48616/301

<u>Ordering Data</u>:

Not applicable.

Preparing activity:

Army-MU

Project No. 6650-0091-5

MIL-F-48616/500
29 JULY 1977

MILITARY SPECIFICATION SHEET

COATING, LONG WAVELENGTH PASS FILTER,
ORDINARY REQUIREMENTS FOR

The complete requirements for procuring the long wavelength pass filters described herein shall consist of this document and the latest issue of MIL-F-48616.

This specification is approved for use by all Departments and Agencies of the Department of Defense.

The following figure illustrates the terms used for describing long wavelength pass filters.

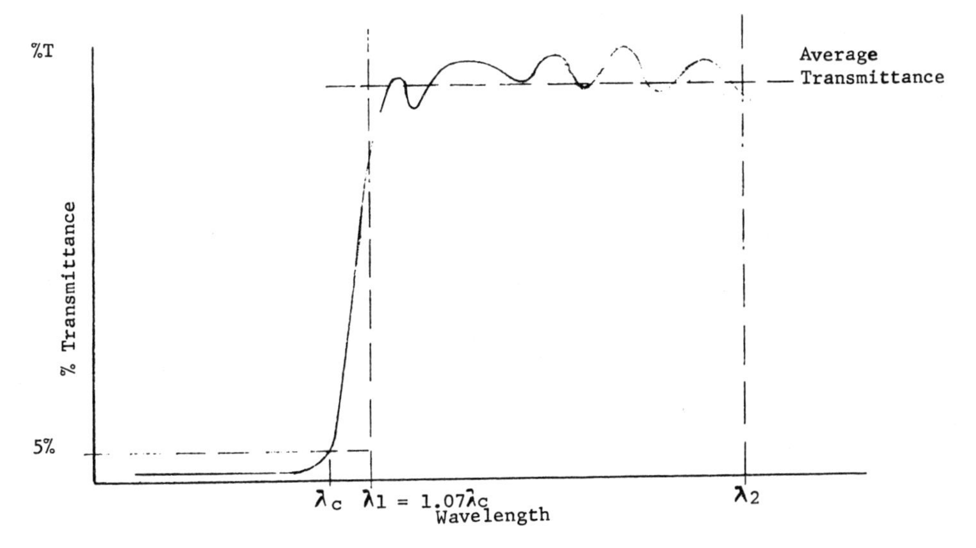

NOTES FOR FIGURE:

1. The cut-on wavelength (λ_c) is the wavelength at which the absolute transmittance is 5%.

2. Average transmittance is calcuated between λ_1, & λ_2 where λ_1 = 1.07 λ_c and λ_2 is specified by the user.

MIL-F-48616/500

Long Wavelength Pass Filter Requirements

TABLE I

Cut-on Wavelength	Longest Wavelength of Transmitting Region	Minimum Average Transmittance	Dash No.
	0.75 to 1.1 um	85%	1
0.7 to 1.1 um	1.1 to 2.2 um	80%	2
	2.2 to 3.5 um	75%	3
	1.18 to 1.8 um	85%	4
1.1 to 1.8 um	1.8 to 3.0 um	80%	5
	3.0 to 5.0 um	75%	6
	1.93 to 3.0 um	80%	7
1.8 to 6.0 um	3.0 to 5.0 um	75%	8
	5.0 to 8.0 um	70%	9
	6.45 to 12.0 um	80%	10
6.0 to 14.0 um	12.0 to 14.0 um	75%	11
	14.0 to 20.0 um	65%	12
	15.0 to 22.0 um	70%	13
14.0 to 25.0 um	22.0 to 30.0 um	60%	14
	30.0 to 50.0 um	50%	15
	26.9 to 30.0 um	60%	16
25.0 to 45.0 um	30.0 to 50.0 um	50%	17

NOTES FOR REQUIREMENTS TABLE:

1. The 5% absolute transmittance point λ_c shall be located to within $\pm$ 2% of the specified wavelength.

2. Transmittance shall be equal to or less than 0.1% absoulte from 0.3 um to 0.93 λ_c.

3. The values in this table apply at an angle of incidence ranging from 0° to 5° and at ambient temperature. Measurements will be made at an angle of incidence within this range (See 6.4b and 6.4c).

4. Spectral resolution to verify the above requirements in the cut-on and transmitting regions shall be 0.5% or less of the cut-on wavelength. Resolution shall be 10% or less of the cut on wavelength in the attenuation region. Scan speed and time constant shall be set such that the spectral features of interest can be resolved within the accuracy of 4.4.2.3.2.2 of MIL-F-48616.

Ordering Data:

At a minimum, the cut-on wavelength (See 6.2m) (λ_c) and the longest wavelength of the transmitting region (See 6.2m) (λ_c) and the dash number (See 6.2b) shall be specified by the procuring activity.

Preparing actitity:

Army - MU

Project No. 6650-0091-6

MIL-F-48616/600
29 JULY 1977

MILITARY SPECIFICATION SHEET

COATING, SHORT WAVELENGTH PASS
FILTER, ORDINARY REQUIREMENTS FOR

The complete requirements for procuring the short wavelength pass filters described herein shall consist of this document and the latest issue of MIL-F-48616.

This specification is approved for use by all Departments and Agencies of the Department of Defense.

The following figure illustrates the terms used for describing short wavelength pass filters.

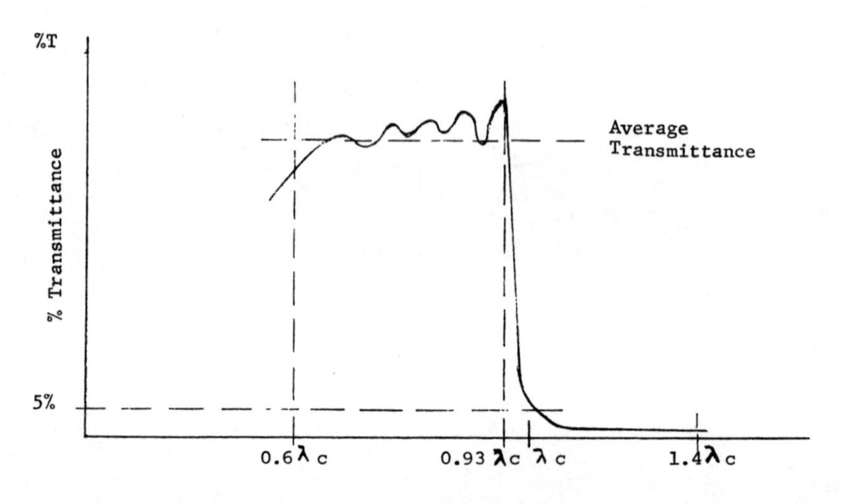

NOTES FOR FIGURE:

1. The cutoff wavelength (λ_c) is the wavelength at which the absolute transmittance is 5%.

2. Average transmittance is determined between 0.6 λ_c and 0.93 λ_c.

FSC-6650

MIL-F-48616/600

Short Wavelength Pass Filter Spectral Requirements

TABLE I

Cut-Off Wavelength (λ_c)	Minimum Average Transmittance	Dash No.
0.9 to 15.0 um	75%	-1
15.0 to 25.0 um	70%	2
25.0 to 35.0 um	60%	3
35.0 to 50.0 um	50%	4

NOTES ON SPECTRAL REQUIREMENTS:

1. The 5% absolute transmittance point shall be located to within ±2% of the specified wavelength.

2. Transmittance shall be equal to or less than 0.1% absolute from 1.07 λ_c to 1.4 λ_c.

3. The values in this table apply at an angle of incidence ranging from 0° to 5° and at ambinet temperature. Measurements will be made at an angle of incidence within this range (See 6.4b and 6.4c).

4. At wavelengths longer than 1.4 λ_c short wavelength pass filters may have high transmittance.

5. Spectral resolution to verify the above requirements in the transmitting and cut off regions shall be 0.5% or less of the cut off wavelength. Resolution shall be 10% or less of the cut off wavelength in the attenuation region. Scan speed and time constant shall be set such that the spectral features of interest can be resolved within the accuracy of 4.4.2.3.2.2 of MIL-F-48616.

Ordering Data:

At a minimum the cut off wavelength (See 6.2m) (λ_c) and the dash number (See 6.2b) shall be specified by the procuring activity.

Preparing activity:

Army - MU

Project No. 6650-0091-7

MIL-F-48616
29 JULY 1977

MILITARY SPECIFICATION

FILTER (COATINGS), INFRARED INTERFERENCE:
GENERAL SPECIFICATION FOR

This specification is approved for use by all Departments and Agencies of the Department of Defense.

1. SCOPE

1.1 Scope.- This specification establishes general performance and durability requirements for thin film coated optical elements which are used in the spectral region from 0.7 um to 50.0 um (See 6.1).

2. APPLICABLE DOCUMENTS

2.1 The following documents, of the issue in effect on date of invitation for bids or requests for proposal, form a part of this specification to the extent specified herein.

SPECIFICATIONS

Federal

L-T-90	Tape, Pressure-Sensitive Adhesive Cellophane and Cellulose Acetate
CCC-C-440	Cloth, Cheesecloth, Cotton Bleached and Unbleached

Military

MIL-E-12397	Eraser, Rubber-Pumice for Testing Coated Optical Elements
MIL-I-45607	Inspection Equipment, Acquisition, Maintenance and Disposition of

STANDARDS

MIL-STD-105	Sampling Procedure and Tables for Inspection by Attributes
MIL-STD-109	Quality Assurance Terms and Definitions
MIL-STD-1241	Optical Terms and Definitions

Beneficial comments (recommendations, additions, deletions) and any pertinent data which may be of use in improving this document should be addressed to: Commander, U. S. Army Armament Research & Development Command, ATTN: DRDAR-TST-S, Dover, N. J. 07801 by using the self-addressed Standardization Document Improvement Proposal (DD Form 1426) appearing at the end of this document or by letter.

FSC 6650

DRAWINGS

U. S. Army, Frankford Arsenal

D7680606 Coating, Eraser Abrasive Tester

2.2 Other publications.- The following documents form a part of this specification to the extent specified herein. Unless otherwise indicated, the issue in effect on date of invitation for bids or request for proposal shall apply:

AMERICAN SOCIETY FOR TESTING AND MATERIALS

B117 - Standard Method of Salt Spray (fog) Testing

(Applications for copies should be addressed to the American Society for Testing and Materials, 1916 Race Street, Philadelphia, Pa. 19103. Technical society and technical association specifications and standards are generally available for reference from libraries. They are also distributed among technical groups and using Federal Agencies.)

3. REQUIREMENTS

3.1 Specification sheets.- Infrared filters and coatings shall comply with the requirements of this specification and the applicable specification sheet except as further defined in applicable drawings or procurement document.

3.1.1 Optical terms and definitions.- Terms and definitions are as used herein and defined in MIL-STD-1241 and 6.3 herein.

3.1.2 Detail specifications.- Specification sheets are categorized as shown in the supplement. As additional specification sheets are generated they will be placed in the appropriate class eg. the first additional specification sheet applicable to the class "Filter, Wide Bandpass, 7.7 to 11.7 Micrometers, Special Requirements for" shall be inserted as MIL-F-48616/301. Additional class designations, when necessary, shall be assigned outside this initial allocation in ascending order i.e. MIL-F-48616 - 700 to 799 would be allocated to the first class addition required.

3.2 Spectral.- The minimum spectral requirements for infrared filters and coatings are specified in the specification sheets. The spectral values for the wavelengths specified herein are given as measured in vacuum. (See 6.4). When specified on the component drawing or the procurement document (See 6.2c) the coated element shall meet the spectral requirements after environmental and durability testing.

3.3 Substrate.- As a minimum, when the substrate material is indicated on the component drawing or procurement document (See 6.2d), the requirements of 3.3.1, 3.3.2 and 3.3.3 shall be specified. When the substrate material and quality (internal and surface defects) are not specified, the substrate shall be chosen so that the optical component shall meet the requirements of 3.2 and 3.4.

3.3.1 Internal defects.- Prior to coating, substrate materials which are transparent or semi-transparent to visible radiation shall be evaluated for internal defects, such as bubbles and other defects which are essentially round in nature. These defects shall not exceed the dig requirements specified for surface defects specified in 3.4.1.3. The substrate materials shall also be evaluated for other internal defects such as striae, fractures, inclusions, etc. The limits on these defects shall be specified by the procurement agency (See 6.2e). For substrate materials which do not transmit visible radiation, the method for evaluation of the internal quality of the materials (See 6.4d) shall be specified by the procurement agency.

3.3.2 Surface defects.- Surface defects on the substrate shall be such that the coated component does not have defects in excess of the surface quality requirements of 3.4.1.2 and 3.4.1.3.

3.3.3 Dimensions.- The substrate shall meet the dimensional requirements of the applicable drawings or procurement document prior to coating (See 6.4e).

3.4 Coated component.- As a minimum, the coated component shall meet the following requirements.

3.4.1 Surface quality.

3.4.1.1 Coating.- The coating shall show no evidence of flaking, peeling, cracking, fingerprints, brush marks, out gassing, blistering, back-coating, crazing, etc. Spatter and holes on or in the coating shall be considered as a dig and shall not exceed the allowable dig size and quantity. (See 3.4.1.3).

3.4.1.2 Scratches.- Surface scratches (coating and substrate) shall not be in excess of the values specified on the component drawing or procurement document (See 6.2f). Scratches are permissible provided the width does not exceed that specified by the scratch letter. The accumulated length of all maximum scratches shall not exceed 1/4 of the average diameter of the element. The scratch letter and corresponding width are shown in Table 1.

3.4.1.2.1 Integrating scratches.- Where the scratches (coating and substrate) do not exceed the requirements of 3.4.1.2, each surface shall be evaluated further by integrating scratches. All scratches of widths less than or equal to the maximum allowable scratch width, and greater than or equal to the minimum scratch width to be considered, (See Table 1) shall be included in the integration. The length of each scratch shall be

MIL-F-48616

multiplied by the scratch width. These products are to be added and the sum divided by the average diameter of the element. If a maximum scratch is present, this resulting value shall not exceed 1/2 the maximum allowed scratch width. If no maximum scratch is present, this value shall not exceed the maximum allowed scratch width.

3.4.1.3 Digs.- Surface digs (coating and substrate) shall not be in excess of the values specified on the component drawing or procurement document (See 6.2g). Digs are permissible on a surface provided the average diameter does not exceed that specified by the dig letter and no more than (1) maximum size dig occurs in any 20mm (0.80") diameter circle on the substrate. The dig letter and corresponding average diamter are shown in Table II.

3.4.1.3.1 Integrating digs.- Where the digs (coating and substrate) do not exceed the requirements of 3.4.1.3, each surface shall be evaluated further by integrating digs. All digs of diameters less than or equal to the maximum allowable dig diameter and greater than or equal to the minimum dig diameter to be considered (See Table II) shall be included in the integration. All digs shall be accumulated such that the sum of the diameters does not exceed twice the diameter of the maximum allowed dig for any 20mm (0.80") diameter circle on the surface. All digs of size B or smaller shall be separated by 1.0mm (0.04") minimum. The measurement of the distance between digs shall not be required for surfaces where digs larger than size B are allowed.

TABLE I

SCRATCH IDENTIFICATION
(See 6.3.1)

Scratch Letter	Scratch Width		Disregard Scratch Widths less than	
	Millimeters	Inches	Millimeters	Inches
A	.005	.00020	.0010	.00004
B	.010	.00039	.0025	.00010
C	.020	.00079	.0050	.00020
D	.040	.00157	.0100	.00039
E	.060	.00236	.0100	.00039
F	.080	.00315	.0200	.00079
G	.120	.00472	.0200	.00079

TABLE II

DIG IDENTIFICATION
(See 6.3.1)

Dig Letter	Average Dig Diameter Millimeters	Average Dig Diameter Inches	Disregard Digs Smaller Than Millimeters	Disregard Digs Smaller Than Inches
A	.05	.0020	.010	.0004
B	.10	.0039	.025	.0010
C	.20	.0079	.050	.0019
D	.30	.0118	.050	.0019
E	.40	.0157	.100	.0039
F	.50	.0197	.100	.0039
G	.70	.0276	.200	.0079
H	1.00	.0394	.250	.0099

3.4.1.4 Cosmetic.- No blemishes (coating and substrate) such as streaks, smears, stains, blotchiness, discoloration, etc. shall be permitted on an optical component lying in a focal plane. Unless otherwise specified on the component drawing or procurement document (See 6.2h), blemishes on a component which lies outside the focal plane in an optical system shall be acceptable when it can be shown that these blemishes do not impair the spectral performance and durability requirements.

3.4.1.5 Coated area.- Optical components shall be coated over their entire clear aperture. In those instances where the clear aperture is not specified on the component or procurement document, the following allowable uncoated area shall apply:

Diameter or Maximum Diagonal Dimension of Optical Element	Maximum Width of Uncoated Area
Up to 2"	0.040"
Greater than 2"	0.040" plus additional width of 0.015" for each inch in diagonal greater than 2"
or	or
Up to 5CM	1MM
Greater than 5CM	1MM plus additional width increasing at the rate of 0.15MM for each CM over 5CM

Tooling marks are measured perpendicular from the edge of the components. In those instances where the clear aperture is specified, areas outside the clear aperture may be coated at the discretion of the contractor.

3.4.2 Surface durability (coating and substrate).

3.4.2.1 Environmental and physical durability.- The coated optical surface shall meet the following service conditions in the order specified:

3.4.2.1.1 Adhesion.- The coated optical surface shall show no evidence of coating removal when cellophane tape is pressed firmly against the coated surface and quickly removed at an angle normal to the coated surface:

3.4.2.1.2 Humidity.- After exposure in an atmosphere at 120° ± 4°F (49°C) and 95 to 100% relative humidity, the coated optical surface shall meet the requirements of 3.4.1.1 and 3.4.1.4.

3.4.2.1.3 Moderate abrasion.- The coated optical surface shall show no signs of deterioration such as streaks or scratches when abraded with a dry, clean cheesecloth pad.

3.4.2.2 Thermal and cleaning durability.- The coated optical surface shall meet the following conditions:

3.4.2.2.1 Temperature.- The coated optical surface shall be exposed to temperatures of -80°F and +160°F (-62°C and 71°C) for 2 hours at each temperature. The rate of temperature change shall not exceed 4°F (2°C) per minute. Subsequent to these exposures, the coated optical surface shall meet the requirements of 3.4.1.1 and 3.4.2.1.1.

3.4.2.2.2 Solubility and cleanability.- After immersion in trichloroethylene, acetone and ethyl alcohol and wiping with cheesecloth, the coated optical surface shall show no evidence of coating removal or scratches and shall meet the requirements of 3.4.1.1 and 3.4.1.4.

3.4.2.3 Optional durability requirements.- When specified on the component drawing or other documents (See 6.2j) the following requirements shall be added or substituted:

3.4.2.3.1 Severe abrasion.- There shall be no visible damage, such as evidence of abrasion or coating removal, to the coated surface when abraded by an eraser conforming to MIL-E-12397.

3.4.2.3.2 Salt solubility.- After immersion in a saline solution, for a period of 24 hours, the coated optical surface shall meet the requirements of 3.4.1.1 and 3.4.1.4.

3.4.2.3.3 Water solubility.- After immersion in distilled water, for a period of 24 hours, the coated surface shall meet the requirements of 3.4.1.1 and 3.4.1.4.

3.4.2.3.4 Salt spray fog.- After exposure to a salt spray fog for a continuous period of 24 hours, the coated surface shall meet the requirements of 3.4.1.1 and the applicable abrasion resistance requirements of 3.4.2.1.3 or 3.4.2.3.1.

4. QUALITY ASSURANCE PROVISIONS

4.1 Responsibility for inspection.- Unless otherwise specified in the contract or purchase order, the supplier is responsible for the performance of all inspection requirements as specified herein. Except as otherwise specified, the supplier may utilize his own facilities or any other facilities suitable for the performance of the inspection requirements specified herein, unless disapproved by the Government. The Government reserves the right to perform any of the inspections set forth in the specification where such inspections are deemed necessary to assure that supplies and services conform to prescribed requirements.

4.1.1 General provisions.- Definitions of inspection terms shall be as listed in MIL-STD-109 and 6.3 herein.

4.1.2 Witness piece.- Unless otherwise specified, witness pieces representing the actual coated component* may be used for spectral and environmental testing. The witness pieces shall be positioned in the coating chamber such that they represent the whole evaporation lot (See 6.3.2). The Government reserves the right to test the actual coated component. Should a component fail, even though the representative witness pieces pass the test, the lot shall be rejected.

4.1.3 Characteristics of the witness piece.- The witness piece shall be of the same material and have a surface finish similar to that of the component to be coated. The witness piece shall be such that it presents no difficulty in measuring and testing the spectral requirements of the coating. Spectral performance of coating applied to small, thick, or curved components may be verified on convenient sized flat, thin (approximately 1.0mm, or 0.40 inch) witness piece.

4.2 First article (initial production) approval.- The requirement for first article approval and the responsibility (Government or contractor) for first article testing shall be as specified in the contract (See 6.2k). Unless otherwise specified, the sample for first article approval tests shall consist of five (5) coated components plus ten (10) coated witness pieces (See 4.1.2 and 4.1.3). The five (5) coated components shall be tested, as specified herein, for all the requirements of 3.4.1. Five (5) coated witness pieces shall be tested, as specified herein, for all the requirements of 3.4.2.1, then 3.2. The other five (5) coated witness pieces shall be tested, as specified herein, for all the requirements of 3.4.2.2, then 3.2. When an optional durability requirement is specified in the contract or order an additional five (5) coated witness pieces shall be tested, as indicated herein, for each optional durability requirement specified. In addition to the foregoing, when the substrate material is identified by component drawings or procurement documents, five (5) uncoated substrates shall be provided and tested as specified herein for all the requirements of 3.3. The Government reserves the right to subject the coated components to all the tests specified here. The sample shall be coated in the same manner, using the same materials, equipment, processes,

* and coated in the same evaporation lot

and procedures as used in regular production. All parts and materials including packaging shall be obtained from the same source of supply as used in regular production.

4.2.1 Government testing.- When the Government is responsible for conducting first article approval tests, the contractor, prior to submitting the sample to the Government, shall inspect the sample to insure that it conforms to all the requirements of the contract and submit a record of this inspection with the sample, including certificates of conformance for materials, as applicable.

4.2.2 Contractor testing.- When the contractor is responsible for conducting first article approval tests, the sample shall be inspected by the contractor for all the requirements of the contract. The sample and a record of this inspection, including certificates of conformance for materials, shall be submitted to the Government for approval. The Government reserves the right to witness the contractor's inspection.

4.3 Inspection provisions.

4.3.1 Submission of product.

4.3.1.1 Inspection lot size.- The inspection lot size shall consist of all components or subassemblies (unit of product) coated within one (8 through 12 hour) work shift.

4.3.1.2 Lot formation and presentation of lots.- The unit of product shall be submitted for inspection on a moving inspection lot basis where the components or subassemblies are continuously offered for inspection in the order produced.

NOTE: Each evaporation lot (coated components and witness pieces) forming a part of a moving inspection lot shall be identified for subsequent evaluation, if required.

4.3.2 Examination and tests.

4.3.2.1 Substrate material.- Uncoated substrate material, when specified on the component drawings or procurement document, shall be inspected in accordance with this specification, the inspection provisions contained in Supplementary Quality Assurance Provisions (SQAP), or other procurements documents listed in the Technical Data Package (TDP). Examination and tests related to Section 3 herein shall be performed on a class basis in accordance with MIL-STD-105 and the sampling plan specified in Table III. The tabulated classification of defects in Table III shall constitute the minimum inspection to be performed by the supplier after first article approval and prior to Government acceptance or rejection by item or lot.

TABLE III - CLASSIFICATION OF DEFECTS

CLASS	CHARACTERISTIC	REQUIREMENT	TEST PROCEDURE
CRITICAL:	NONE DEFINED		
MAJOR:	AQL 0.65% DEFECTIVE		
101.	Internal defects	3.3.1	4.6.4
102.	Surface defects	3.3.2	4.6.5
103.	Dimensions	3.3.3	4.6.6
MINOR:	NONE DEFINED		

NOTE: The inspection for the Characteristics in Table III shall be conducted at a temperature between +60° and +90°F (16°C and 32°C).

4.3.2.2 Components, subassemblies and witness pieces.- Coated components subassemblies and witness pieces shall be inspected in accordance with this specification, the inspection provisions contained in Supplementary Quality Assurance Provisions (SQAP), or other procurement documents listed in the Technical Data Package (TDP). Examination and tests related to Section 3 herein shall be performed on a class basis in accordance with MIL-STD-105 and the sampling plans specified in Tables IV and V herein. The tabulated classification of defects in Tables IV and V shall constitute the minimum inspection to be performed by the supplier after first article approval and prior to Government acceptance or rejection by item or lot.

TABLE IV - CLASSIFICATION OF DEFECTS

CLASS	CHARACTERISTIC	REQUIREMENT	TEST PROCEDURE
CRITICAL:	NONE DEFINED		
MAJOR:	AQL 0.65% DEFECTIVE		
104.	Surface quality, coating	3.4.1.1	4.6.7.1
105.	Surface quality, scratches	3.4.1.2	4.6.7.2
106.	Surface quality, digs	3.4.1.3	4.6.7.2
107.	Surface quality, cosmetic	3.4.1.4	4.6.7.3
108.	Coated area	3.4.1.5	4.6.7.4
MINOR:	NONE DEFINED		

NOTE: The inspection for the Characteristics in Table IV shall be conducted at a temperature between +60°F and +90°F (16°C and 32°C).

4.3.2.3 Acceptance and rejection.- Rejected lots shall be screened for all defective characteristics. Removal of defective units and resubmittal of rejected lots shall be in accordance with "Acceptance and Rejection" as specified in MIL-STD-105.

4.3.2.4 Special sampling.- A minimum of three coated witness pieces, or coated components (when required), shall be selected as samples from each evaporation lot. Each sample shall meet the requirements and tests in Table V. The sequence of test shall be as delineated in Table V.

TABLE V - CLASSIFICATION OF DEFECTS

NO.	CHARACTERISTIC	REQUIREMENT	TEST PROCEDURE
301.	Spectral	3.2	4.6.3
302.	Adhesion	3.4.2.1.1	4.6.8.1
303.	Surface quality, coating (post humidity)	3.4.2.1.2	4.6.8.2, 4.6.7.1
304.	Surface quality, cosmetic (post humidity)	3.4.2.1.2	4.6.7.3
305.	Moderate abrasion	3.4.2.1.3	4.6.8.3

NOTE: The inspection for the Characteristics in Table V shall be conducted at a temperature between +60° and +90°F (16°C and 32°C).

4.3.2.5 Failure of sample.- Should any one item of a special sampling fail to meet the specified test requirements, acceptance of the represented inspection lot shall be suspended by the Government until necessary corrections have been made by the contractor and the resubmitted items have been approved. (See 4.3.2.3).

4.4 Inspection equipment.- Except as otherwise provided for by the contract, the contractor shall supply and maintain inspection equipment in accordance with the applicable requirements of MIL-I-45607.

4.4.1 Government furnished inspection equipment.- Where the contract provides for Government furnished test equipment, supply and maintenance of test equipment shall be in accordance with applicable requirements specified in MIL-I-45607.

4.4.2 Contractor furnished inspection equipment.

4.4.2.1 Government design.- All inspection equipment specified by drawing number in specifications or SQAP forming a part of the contract shall be supplied by the contractor in accordance with technical data included in the contract.

4.4.2.2 Contractor equipment.- The contractor shall supply inspection equipment compatible with the "Test Methods and Procedures" specified in 4.6 of this specification. Since tolerance of test equipment is normally considered to be within 10% of the product tolerance for which it is intended, this inherent error in the test equipment must be considered as part of the prescribed product tolerance limit. Thus, concept, construction, materials, dimensions and tolerances used in the test shall be so selected and controlled as to insure that the test equipment will reliably indicate acceptability of a product which does not exceed 90% of the prescribed tolerance limit, and permit positive rejection when nonconforming. Construction shall be such as to facilitate routine calibration of test equipment.

4.4.2.3 Inspection equipment design requirements.

4.4.2.3.1 Spectrophotometric equipment.- The spectrophotometer utilized for spectral measurements of finished components, or witness pieces, shall have an optical system collimated to f/4.0, or a larger "f/-" number.

4.4.2.3.2 Spectrophotometric measurement accuracies.

4.4.2.3.2.1 Wavelength resolution, scanning speed and time constant.- The wavelength resolution of the spectrophotometer shall be better than that required by the applicable detailed specification sheet. The scanning speed and time constant shall be compatible with the resolution requirement of the item under test.

4.4.2.3.2.2 Wavelength and photometric accuracy.- When not specified in the applicable detailed specification sheet, the spectral parameters shall be measured to the following accuracies in the region from 0.7 um to 50.0 um:

a. Wavelength accuracy.- Plus or minus (±) 0.5% of the wavelength being measured, or in the case of narrow bandpass filters, ± 20.0% of the filter's half-band width.

b. Transmittance (T) accuracy.

TRANSMITTING REGION	ACCURACY
(T > 1.0% absolute)	± 1.5% of full scale
ATTENUATED REGION	ACCURACY
(T ≤ 1.0% absolute)	± 0.1%

c. Reflectance (R) accuracy.

HIGH REFLECTANCE REGION	ACCURACY
($R \geq 80.0\%$)	$\pm$ 1.5% of full scale
MID-REFLECTANCE REGION	ACCURACY
($5.0\% < R < 80.0\%$)	$\pm$ 2.0% of full scale
LOW REFLECTANCE REGION	ACCURACY
($1.5\% < R \leq 5.0\%$)	$\pm$ 0.1%
($R \leq 1.5\%$)	$\pm$ 0.05%

4.5 Inspection documentation.

4.5.1 Spectral.- A copy of the spectral test results shall be supplied with each coating lot and shall contain sufficient data to show compliance with the spectral requirements for the parts being shipped. This data should include as a minimum the following:

a. Type of spectrophotomer used.

b. Properly labelled axes with spectral reference lines (e.g. where appropriate, zero reference and/or full scale reference lines).

c. Test conditions.

d. Actual measured values plus any theoretical or measured corrections to the measured values.

4.5.2 Physical.- Inspection records shall be available which show compliance with all physical requirements.

4.5.3 Environmental.- Inspection records shall be available which show compliance with all environmental durability requirements.

4.6 Test methods and procedures.

4.6.1 Cleaning.- Before and after subjecting a coated sample (component or witness piece) to any inspection or test, the coated sample shall be thoroughly and carefully cleaned to remove dirt, finger marks, smears, etc. The cleaning solution shall be acetone, ethyl alcohol, isopropyl alcohol, or mixtures thereof. Following the cleaning, the coated sample shall be carefully dried with lens tissue or a soft clean cloth. The temperature of the cleaning solution shall not exceed +80°F (27°C).

4.6.2 Temperature conditions.- Unless otherwise specified the testing equipment and coated components, or witness pieces, shall be stabilized at a temperature between +60° and + 90°F (16°C and 32°C), and maintained at that temperature during all spectral testing.

4.6.3 Spectral.- Perform the test for spectral requirements utilizing the test equipment specified in 4.4.2.3. The measured spectral performance of infrared filters and coatings shall conform to the requirements of 3.2. Unless otherwise specified the measured spectral values for the applicable wavelengths shall be recorded as measured in vacuum.

4.6.4 Internal defects.- Substrate materials which are transparent, or semi-transparent to visible radiation, shall be visually examined by transmission with the unaided eye for the requirements of 3.3.1. The examination shall be performed using two 15 watt cool white fluorescent light tubes as the light source. The viewing distance from the substrate surface to the eye shall not exceed 18.0 inches (45.7 cm). The substrate material shall be viewed against a black matte background. The only illumination in the inspection area shall be from the light source used for examination. This method of examination is as depicted in Figure I. The quality of the substrate material shall not exceed the dig tolerance specified in 3.3.1. The method of inspection and the evaluation of transparent and semi-transparent material for internal defects, other than defects which are essentially round in nature, shall be as specified by the procurement agency. The overall internal quality of the substrate material shall conform to the requirements of 3.3.1. The method of inspection to evaluate the internal quality of substrate materials which do not transmit visible radiation shall be as specified on the component drawing, or procurement document (See 6.2e).

4.6.5 Substrate surface defects.- The component substrate shall be examined for scratches and digs by reflection or transmission, where applicable using the inspection technique specified in 4.6.4. Magnification shall be used as needed in these examinations. The length and width of scratches, and the dig diameters shall be determined by use of interferometer, microscopic measuring devices, calibrated precision comparators, or similar applicable precision measuring devices. The width, length and density of all scratches on the substrate surface shall conform to the requirements of 3.4.1.2 and 3.4.1.2.1. The diameter and density of all digs in the substrate shall conform to the requirements of 3.4.1.3 and 3.4.1.3.1. The quality of each substrate surface shall meet the requirements of 3.3.2.

4.6.6 Substrate dimensions.- The substrate shall be inspected by use of standard measuring equipment (See 6.3.3). The substrate dimensions shall conform to the requirements of 3.3.3.

4.6.7 Surface quality of coated components.

4.6.7.1 Coating.- The coating on the component surface shall be examined for evidence of flaking, peeling, cracking, etc by reflection or transmission, where applicable, using the inspection technique specified in 4.6.4. The coating shall conform to the requirements of 3.4.1.1.

4.6.7.2 Scratch and dig.- The coated substrate shall be examined for scratches and digs by reflection or transmission, as applicable, using the inspection technique specified in 4.6.4. Magnification shall be used as needed in these examinations. The length and width of scratches, and the dig (hole) diameters shall be determined by use of interferometer, microscopic measuring devices, calibrated precision comparators, or similar applicable precision measuring devices. The width, length and density of all scratches in the coating or substrate shall conform to the requirements of 3.4.1.2 and 3.4.1.2.1. The diameter and density of all digs (holes) in the coating or substrate shall conform to the requirements of 3.4.1.3 and 3.4.1.3.1.

4.6.7.3 Cosmetic.- The coating and substrate shall be examined by transmission or reflection, where applicable, using the inspection technique specified in 4.6.4. Blemishes on the coating or substrate shall conform to the requirements of 3.4.1.4. Coatings or substrate that exhibit evidence of discoloration shall be inspected for conformance to the applicable spectral requirements (See 4.6.3) specified on drawing, contract or purchase order. The blemished area on the coated substrate shall then be subjected to the moderate abrasion test of 4.6.8.3. The coating and substrate shall conform to the requirements of 3.4.1.4.

4.6.7.4 Coated area.- The coated area of the component shall be inspected by use of standard measuring equipment. The coating of the component shall conform to the requirements of 3.4.1.5.

4.6.8 Environmental and physical durability.- The coating on the component, or witness piece shall withstand exposure to the following conditions in the order specified.

4.6.8.1 Adhesion.- The coated component or witness piece shall be subjected to an adhesion test using 1/2" (12.7mm) wide cellophane tape conforming to Type I of L-T-90. Press the adhesive surface of the cellophane tape firmly against the coated surface so as to cover the stained area, then quickly remove it at an angle which is normal to the coated surface. Immediately following the removal of the adhesive tape, the coated surface of the component, or witness piece, shall be examined by reflection using the inspection technique specified in 4.6.4. The coating shall conform to the requirements of 3.4.2.1.1. Subsequent to this test the coated component, or witness piece, shall be subjected to the test in 4.6.8.2.

4.6.8.2 Humidity.- The coated component, or witness piece, shall be placed into an environmental controlled test chamber and exposed to a temperature of + 120° ± 4°F (49° ± 2°C) and 95% to 100% relative humidity for a minimum of 24 hours. Subsequent to this exposure the coated component, or witness piece, shall be removed from the test chamber, cleaned (See 4.6.1), dried, and then subjected to the examinations specified in 4.6.7.1 and 4.6.7.3. The coating shall conform to the requirements of 3.4.2.1.2. The coated component, or witness piece, shall then be subjected to the test specified in 4.6.8.3.

4.6.8.3 Moderate abrasion.- Within one hour after the humidity test of 4.6.8.2 the coated component, or witness piece shall be subjected to a moderate abrasion by rubbing the coated surface with a 1/4 inch (6.4mm) thick by 3/8 inch (9.5mm) wide pad of clean dry, laundred cheesecloth conforming to CCC-C-440 affixed to an abrasion tester that conforms to Drawing D7680606. The cheesecloth pad shall completely cover the eraser portion of the tester and be secured to the shaft with an elastic band. The cheesecloth pad shall be rubbed across the coated surface from one point to another over the same path for 25 complete cycles (50 strokes) with a minimum force of 1.0 pound (0.45kg) continuously applied. The length of the stroke shall be approximately equal to two widths of the cheesecloth pad when the diameter or area of the component permits. The abrasion tester shall be held approximately normal to the surface under test during the rubbing operation. Subsequent to the rubbing operation the component, or witness piece, shall be cleaned (See 4.6.1), dried and then subjected to an examination by reflection using the inspection technique specified in 4.6.4 for evidence of physical damage to the coating. The coating on the component, or witness piece, shall meet the requirements of 3.4.2.1.3.

4.6.9 Thermal and cleaning durability.- The coated component, or witness piece, shall withstand exposure to the following test conditions:

4.6.9.1 Temperature.- The coated component, or witness piece, shall be subjected to temperatures of $-80^{o}F \pm 2^{o}F$ ($-62^{o} \pm 1^{o}C$) and $+ 160^{o}F \pm 2^{o}F$ ($71^{o}C \pm 1^{o}C$) for a period of 2 hours at each temperature. After each exposure the coated components or witness pieces shall be stabilized at an ambient temperature between $+60^{o}$ and $+90^{o}F$ ($16^{o}C$ and $32^{o}C$) and subjected to the examination and test specified in 4.6.7.1 and only the adhesion test of 4.6.8.1. The coating shall conform to the requirements of 3.4.2.2.1.

4.6.9.2 Solubility and cleanability.- The coated component, or witness piece, shall be immersed, in sequence, in the following solutions maintained at room temperature ($+60^{o}$ to $+90^{o}F$): trichloroethylene; acetone; and ethyl alcohol. The immersion time in each solution shall be a minimum of ten minutes. Upon removal from each solution, the solvent shall be allowed to evaporate to dryness without wiping or forced drying before proceeding to the next solution. Upon removal from the alcohol solution, and after drying any resultant stains on the coated surface shall be removed by wiping the coating to a clean, stain-free condition with an ethyl-alcohol-moistened cheesecloth. Subsequent to this cleaning, the coating shall be subjected to the examinations specified in 4.6.7.1 and

4.6.7.3. The coating shall conform to the requirements of 3.4.2.2.2.

4.6.10 Optional durability tests.- The coated components or witness samples shall withstand exposure to the following test conditions when specified in the contract, purchase order or on the component drawing (See 6.2j).

4.6.10.1 Severe abrasion.- The coated components, or witness piece, shall be subjected to a severe abrasion by rubbing the coated surface with a standard eraser conforming to MIL-E-12397 mounted in a eraser abrasion coating tester that conforms to the requirements of Drawing D7680606. The eraser shall be rubbed across the surface of the component, or witness piece, from one point to another over the same path for 10 complete cycles (20 strokes) with a force of 2.0 to 2.5 pounds continuously applied. The length of the stroke shall be approximately equal to 3 diameters of the eraser when the diameter or area of the component, or witness piece, permits. The eraser abrasion tester shall be held approximately normal to the surface under test during the rubbing operation. Subsequent to the rubbing operation the component or witness piece, shall be cleaned (See 4.6.1), dried and then subjected to an examination by reflection using the inspection technique specified in 4.6.4 for evidence of physical damage to the coating. The film coating on the component, or witness piece, shall meet the requirements of 3.4.2.3.1.

4.6.10.2 Salt solubility.- The coated component, witness piece, shall be immersed for a period of 24 hours in a solution of distilled water and sodium chloride (salt). The mixture shall be 6 ounces (170 grams) of salt per gallon (3.8 liters) of water at room temperature (16° to $32^{o}C$). Subsequent to this immersion the coated component, witness piece, shall be removed from the solution and gently washed, or dipped in clean running water not warmer than $100^{o}F$ ($38^{o}C$) to remove salt deposits. The coated component, witness piece, shall then be cleaned (See 4.6.1), dried, and then subjected to the examination specified in 4.6.7.1 and 4.6.7.3. The coated component, or witness piece, shall conform to the requirements of 3.4.2.3.2.

4.6.10.3 Water solubility.- The coated components, or witness piece, shall be immersed for a period of 24 hours in distilled water at room temperature ($+60^{o}F$ to $+90^{o}F$). Subsequent to this immersion, the coated components, or witness piece, shall be removed from the solution, cleaned (See 4.6.1), dried, then subjected to the examination specified in 4.6.7.1 and 4.6.7.3. The coating on the component, or witness piece, shall meet the requirements of 3.4.2.3.3.

4.6.10.4 Salt spray (fog).- The coated component, witness piece, shall be subjected to a salt spray fog test in accordance with ASTMB-117-73, for a continuous period of 24 hours. Subsequent to the salt spray fog exposure the coated component, or witness piece, shall be washed and dried as specified. The coated component or witness piece, shall then be cleaned (See 4.6.1), dried, and then subjected to the examination specified in 4.6.7.1. Subsequent to the visual examination for physical defects the coated component or witness piece, shall be subjected to the test in 4.6.8.3 or 4.6.10.1, as specified in the contract or order. The coated component or witness piece, shall meet the requirements of 3.4.2.3.4.

5. PACKAGING

5.1 Packaging and packing.- In the absence of specified packaging data in the contract, or purchase order, each thin film coated optical element shall be individually wrapped in a chemically inert paper that will not scratch, leave a residue or corrode the coated element (See 6.4g). The wrapped coated element shall be immobilized in a unit container that provides adequate protection during handling and shipment.

5.2 Marking.- Each packaging and shipping container shall be marked to show the following:

a. Lot or batch number.
b. Contract or purchase order number.
c. Part or drawing number.
d. Quantity of parts in container.
e. Manufacturer's name or trademark.

5.3 Shipping container.- Shipping containers shall be marked with words "DELICATE OPTICAL COMPONENTS REQUIRING SPECIAL HANDLING". The appropriate side of the container shall be clearly marked to indicate "TOP" or "OPEN THIS SIDE".

6. NOTES

6.1 Intended use.- This document is intended to be used for all types of infrared interference filters and coatings. It is anticipated that most of the coatings/filters covered by this document will not include specialized characteristics such as fluorescence, high energy handling properties, conduction, etc. However, specialized characteristics as those mentioned, or those in which more exacting performance capabilities are needed can be imposed by the generation of additional specification sheets that include the desired specific properties. The primary requirement on the specification sheet is the spectral characteristic of the filter or coating. However, requirements in addition to spectral requirements may be added, if needed. The specification sheet in conjunction with the general document will then adequately specify the required filter and/or coating.

6.2 Ordering data.- Purchasers should exercise any desired options offered herein, and procurement documents should specify the following:

a. Title, number and date of this specification.

b. Number and revision of the specific detailed specification sheet and dash number, if applicable.

c. Post environmental spectral evaluation (See 3.2).

d. Substrate material (See 3.3).

e. Internal defect requirement (See 3.3.1).

f. Scratch value (See 3.4.1.2).

g. Dig value (See 3.4.1.3).

h. Cosmetic requirement (See 3.4.1.4).

j. Optional durability requirements (See 3.4.2.3).

k. Requirement for first article approval (See 4.2, 4.2.1 and 4.2.2).

m. Supplemental spectral requirements (See specification sheets).

6.3 Definitions.

6.3.1 Scratch and dig identification.- This is specified by two letters separated by a hyphen (i.e. F-F). The first letter of the pair is the maximum scratch letter. The second letter is the maximum dig letter.

6.3.2 Evaporation lot.- An evaporation lot is defined as the group of parts which has the coating applied at the same time and in the same chamber.

6.3.3 Standard measuring equipment (SME).- Standard measuring equipment is defined as the common measuring devices which are usually stocked by commercial supply houses for ready supply (shelf items) normally used by an inspector to perform dimensional inspection of items under procurement. This category also includes commercial testing equipment such as meters, optical comparators, etc.

6.4 Informational notes:

a. Spectrophotometer measurements: Spectrophotometers are calibrated for wavelength using* standard air (pressure =760mm Hg and temperature = 15°C). The purpose is to clearly identify the index of refraction of the medium in which the calibration is to apply and eliminate possible fluctuations in poorly defined conditions. For most applications, such details in calibration are inconsequential since the change in operating conditions affects measuring only slightly. For example, the difference between wavelengths measured in vacuum and the same wavelengths measured in standard air amount to about 0.025%. (Reference: Commission on Molecular Structure and Spectroscopy. International Union of Pure and Applied Chemistry, Table of Wavenumbers for a Calibration of Infrared Spectrometers. (Butterworths, 1961).

* standards evaluated either in a vacuum or in

b. Temperature effects on coatings:

If the application for these coatings is for other than normal room temperature, the user should request from the manufacturer specific performance information. Spectral performance is temperature dependent and allowance for temperature effects can frequently be included in filter design. Coating durability is also temperature dependent.

c. Angle of incidence effects on coatings:

Normally, as the angle of incidence of the radiation increases from 0° (normal incidence to the optical element) the spectral features of the thin film coating shift to shorter wavelengths with changes in transmittance and reflectance also occurring. The same holds true as the radiation changes from parallel to convergent beam. Allowance for non-normal angles of incidence and convergency angles can be included in filter design. For angles of incidence greater than approximately 20°, spectral performance is influenced by the polarization state of the incident radiation. For a particular application, the user should request pertinent information from the manufacturer.

d. Internal defects of substrates:

Internal defects of spectral filter substrates can seriously deteriorate the related system performance. Techniques of various degrees of precision exist for evaulation of visually opaque substrates for internal defects. Some of these are image spoiling property test, snooper scope evaluation, infrared microscope test, laser scanning, X-ray, etc.

The higher the resolution and lower the tolerance of the system to scattered radiation, the more exacting the tests that should be selected.

Improperly oriented birefringent substrates can seriously deteriorate higher resolution performance. A favorable orientation should be specified.

e. Physical thickness of coatings:

Coatings in the infrared spectral region can be several thousandths of an inch thick. Consequently, allowance must occasionally be made for coating thicknesses as well as substrate thickness when calculating focal length and designing filter holders.

f. Test conditions:

Should in-use conditions differ from these test conditions, the customer should request from the manufacturer the pertinent adjusted parameters either by theoretical calculation or change in the test conditions. Adjustments that may be required shall include the effects of temperature, angle of incidence, convergence angle of the incident radiation, and polarization of the incident radiation. The optical component or witness may absorb energy during spectrophotometric tests thus causing a temperature change. Controlled temperature during testing should be specified by the customer.

g. Wrapping tissue:

Papers which contain certain chemicals, or are fabricated of bleached wood pulp, may cause deterioration of optical coatings. Care must be exercised to insure that chemically inert paper made from unbleached wood pulp is used to wrap optics.

Custodians:
Army - AR
Navy - SH
Air Force - 99

Preparing activity:
Army - AR
Project No. 6650-0091

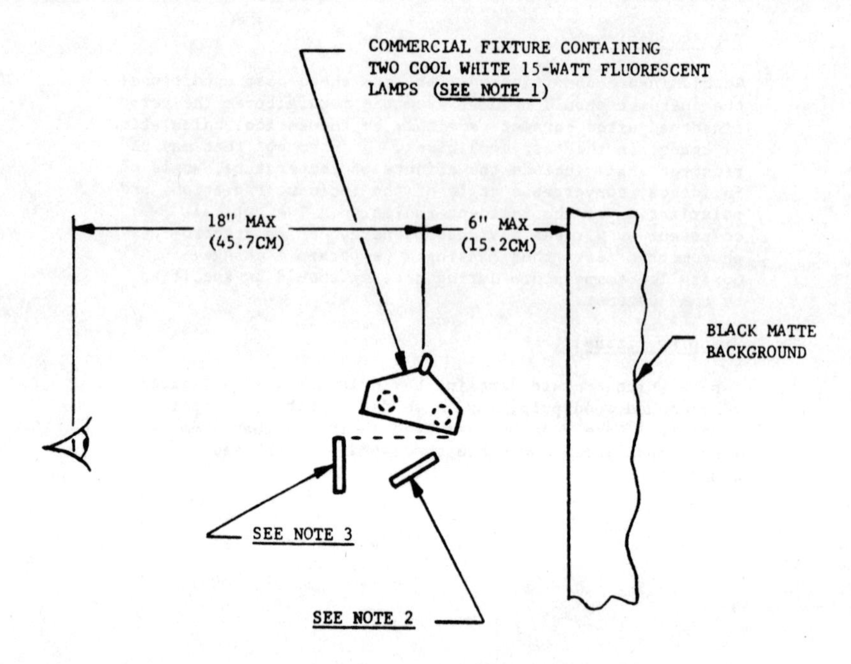

NOTES

1. THE ONLY ILLUMINATION IN THE TEST AREA SHALL BE FROM THE LIGHT SOURCE USED FOR TEST.

2. RELATIVE POSITION OF COMPONENT FOR INSPECTION BY REFLECTION.

3. RELATIVE POSITION OF COMPONENT FOR INSPECTION BY TRANSMISSION.

FIGURE 1

Appendix C

BASIC Program to Calculate Thin Film Filter Performance

Since Berning's landmark work and review (1963) on the mathematics of thin film filter performance calculations, there have been many computer programs written to carry out the numerical drudgery. Baumeister (1986) has recently published an overview of currently commercially available software packages that perform these calculations. DeBell published a review of calculational algorithms in 1978. The majority of the modern programs use the matrix formulation as their mathematical basis.

Liddell (1981) published a monograph on computer-aided thin film design calculations. She includes a listing of a FORTRAN program that evaluates the performance of a given design.

This Appendix includes a listing of a computer program written in the BASIC language for calculating the transmission, reflection, and absorption of a thin film design. It is compatible with both the IBM PC and the Apple Macintosh classes of computers. The comments in the listing explain the functioning of the program.

```
REM Program name  THIN FILMS by S.P. Fisher

REM   This program calculates the following properties of a stack of thin
REM film coatings at a given wavelength:
REM                         reflection (R),
REM                         transmission (T),
REM                         phase shift upon reflection (rho),
REM                     and phase shift upon transmission (tau).
REM It uses the matrix computational method outlined by Peter H. Berning
REM in "Physics of Thin Films" Vol. 1, p.87-92, Academic Press, 1966.
REM
REM This program is intended to be a starting point for more generalized
REM cases of thin film coatings that would add absorption in each film and
REM oblique angles of incidence.
REM
REM Below is an outline of the THIN FILMS program and the variables used in each step
REM
REM  1.  Data entry
REM      1.1  Set constant(s), (pi)
REM      1.2  Enter the index of the incident medium, (No)
REM      1.3  Enter the index and physical thickness of each layer of thin film, (Nx,Hx)
REM      1.4  Enter the complex index of substrate, (index=Ns + jKs)
REM      1.5  Enter the wavelength range and increment, (wvl1,wvl2,inc)
REM      1.6  Print all the input data to system printer for verification.
REM
REM Repeat steps 2 through 6 for each incremental wavelength throughout
REM the wavelength range of interest.
REM
```

```
REM  2.  Calculate the characteristic matrix for the thin film stack
REM      2.1  Initialize the characteristic matrix.
REM      2.2  Find the characteristic matrix for each layer, (Y)
              Note: Matrix notation and definition is as follows:
REM
REM                      |Y1+jY2   Y3+jY4|   |    cosθ     (j/Nx)sinθ|
REM                  Y = |               | = |                       |
REM                      |Y5+jY6   Y7+jY8|   |(j*Nx)sinθ    cosθ     |

REM
REM                  where θ = 2*pi*Nx*Hx/wavelength
REM
REM      2.3  Multiply the individual layer matrices to
REM           obtain the characteristic matrix for the stack, (M)
REM
REM            M = [Y(layer1)]*[Y(layer2)]*...*[Y(layer L)]
REM
REM
REM
REM  3.  Calculate the reflected and transmitted electric field amplitudes, (Eo and Ho)

REM          |Eo|             |1 |
REM          |  | = [ M ] *   |  |
REM          |Ho|             |Ns|
REM
REM
```

```
REM  4.  Calculate the reflection (R), the transmission (T), the phase shift upon
REM      reflection (rho), and the phase shift upon transmission (tau) for the thin film
REM      stack
REM             | Eo  -  (Ho/No) |**2                    4 Ns
REM    R =      |    ---------   |          , T = ----------------
REM             | Eo  +  (Ho/No) |                No*|Eo + (Ho/No)|**2
REM
REM                | Eo  -  (Ho/No) |               |       1       |
REM   rho = arg    | -------------- |  , tau = arg  |  -----------  |
REM                | Eo  +  (Ho/No) |               | Eo + (Ho/No)  |
REM
REM
REM  5.  Determine the correct quadrant for rho and tau.
REM
REM  6.  Print out R,T,A,rho, and tau.
REM

REM 1. Data entry
DIM N(99),h(99),label(70)
   CLS
    pi=3.141593
    PRINT "Enter comments, up to 70 characters"
    INPUT "comments=?",comments$
    PRINT "Enter index of incident medium, e.g. air=1.0"
    INPUT "No=?",No
    L=1
    Layers:
    PRINT "Enter the index of layer";L;". Enter N=0 to stop input"
```

```
INPUT "N=?",N(L)
IF N(L)=0 THEN GOTO Substrate
PRINT "Enter the physical thickness of layer";L
INPUT "h=?",h(L)
L=L+1
GOTO Layers
Substrate:
L=L-1
IF L=0 THEN PRINT "No film entered. The calculation will be for the uncoated
surface."

PRINT "Enter the substrate index, e.g. glass=1.52"
INPUT "NS=?",Ns
PRINT "Enter the substrate absorption, e.g. glass=0.0"
INPUT "Ks=?",Ks
PRINT "Enter the beginning wavelength, the ending wavelength, and the wavelength
increment."
INPUT "wvl1=?,wvl2=?,wvlinc=?",wvl1,wvl2,wvlinc
LPRINT "COMMENTS"
LPRINT" ";comments$
LPRINT
LPRINT "INPUT PARAMETERS"
LPRINT USING" Incident index: No=###.###";No
FOR i=1 TO L
 LPRINT USING" Layer###, index: N =###.### Phys.Thk:h=###.####";i,N(i),
NEXTi
LPRINT USING" Substrate index: Ns=###.### Ks=###.####";Ns,Ks
LPRINT
Ks=-Ks
```

```
FOR wvl=wvl1 TO wvl2 STEP wvlinc

REM 2.  Calculate the characteristic matrix for the thin film stack.
REM Initialize the characteristic matrix.
 M1=1
 M2=0
 M3=0
 M4=0
 M5=0
 M6=0
 M7=1
 M8=0
FORi=1 TO L
REM X is set equal to M before finding the matrix for each layer, then X is used
REM  to find the new product of M times Y for i number of layers.
 X1=M1
 X2=M2
 X3=M3
 X4=M4
 X5=M5
 X6=M6
 X7=M7
 X8=M8
REM Calculate the characteristic matrix for each film.
  Phi=2*pi*N(i)*h(i)/wvl
  Y1=COS(Phi)
  Y7=Y1
  temp=SIN(Phi)
```

```
Y4=temp/N(i)
Y6=temp*N(i)
REM These are zero because the films are not absorbing.
Y2=0
Y3=0
Y5=0
Y8=0
 REM Calculate the product of the characteristic matrices for each layer
   M1=X1*Y1-X4*Y6
   M2=X2*Y1+X3*Y6
   M3=X3*Y7-X2*Y4
   M4=X1*Y4+X4*Y7
   M5=X5*Y1-X8*Y6
   M6=X6*Y1+X7*Y6
   M7=X7*Y7-X6*Y4
   M8=X5*Y4+X8*Y7
 NEXTi

REM 3.  Calculate the reflected and transmitted electric field amplitudes, Eo and Ho
  Eo1=M1+M3*Ns-M4*Ks
  Eo2=M2+M3*Ks+M4*Ns
  Ho1=M5+M7*Ns-M8*Ks
  Ho2=M6+M7*Ks+M8*Ns
REM 4.  Calculate R,T,A,rho and tau
REM       n is a temporary complex numerator,
REM       d is a temporary complex denominator,
REM       dsqr is the magnitude of d,
```

```
REM       cr1,cr2,ct1, and ct2 are complex components resulting from complex division
    n1=Eo1-Ho1/No
    n2=Eo2-Ho2/No
    d1=Eo1+ Ho1/No
    d2=Eo2+ Ho2/No
    dsqr=d1^2+d2^2
    cr1=(n1*d1+ n2*d2)/dsqr
    cr2=(n2*d1- n1*d2)/dsqr
    ct1=d1 /dsqr
    ct2=-d2/dsqr
REM  Results for R,T,rho,and tau
   R=(cr1^2+cr2^2)*100
   T=(4*Ns/(No*dsqr))*100
   IF Ks<>0 THEN T=0
     rho=ATN(cr2/cr1)*180/pi
     tau=ATN(ct2/ct1)*180/pi
REM 5.  Determine the correct quadrant for rho and tau
REM  re-calculate rho as necessary
  quad=1
  IF crl<0 THEN IF cr2<0 THEN quad=3 ELSE quad=2 ELSE IF cr2<0 THEN quad=4
  IF quad=2 OR quad=3 THEN rho=rho+180
  IF quad = 4 THEN rho=360+rho
REM  recalculate tau as necessary
  quad=1
  IF ctl<0 THEN IF ct2<0 THEN quad=3 ELSE quad=2 ELSE IF ct2<0 THEN quad=4
  IF quad=2 OR quad=3 THEN tau=180+tau
  IF quad=4 THEN tau=360+tau
  IF Ks<>0 THEN tau=0
```

```
REM 6.  Print out R,T,rho, and Tau.
  IF wvl=wvl1 THEN LPRINT "PERFORMANCE RESULTS"
  IF wvl=wvl1 AND L=0 THEN LPRINT" Note: Performance is for the uncoated substrate
    IF wvl=wvl1 THEN LPRINT"  wvl   Refl   rho    Trans   tau"
    LPRINT USING"#####.### ###.#### ###.## ###.#### ###.##";wvl,R,rho,T,tau
    PRINT

NEXT wvl
LPRINT
LPRINT
LPRINT
PRINT "Program ended."
END
```

REFERENCES

An excellent bibliography on thin film technology is given at the end of the chapter written by Dobrowalski on optical filters in the Optical Society of America's *Optics Handbook* (Driscoll and Vaughan, 1978).

Abelès, F. (1950). La détèrmination de l'indice et de l'épaisseur des couches minces transparentes. *J. Phys. Rad.*, 11, 310.

Agranoff, Joan (1985). *Modern Plastics Encyclopedia 1985–1986,* vol. 62, no. 10A, McGraw-Hill, New York.

Allen, Thomas H., Joseph H. Apfel, and C. K. Carniglia (1978). A 1.06 micrometer laser absorption calorimeter for optical coatings. In Glass, Alexander J., and Arthur H. Guenther, eds.: *Laser Induced Damage in Optical Materials: 1978*. Washington, D.C., National Bureau of Standards Special Publication #541.

Anthon, Erik (1982). Spatially uniform detector assemblies. In Seddon, R. I., ed.: *Optical Thin Films*. Proceedings of the SPIE, volume 325, Bellingham, Wash., 98227.

Apfel, Joseph H. (1965). Circularly wedged optical coatings. II. Experimental. *Appl. Opt.,* 4(8):983.

Apfel, Joseph H. (1977). Optical coating design with reduced electric field intensity. *Appl. Opt.,* 16(7):1880.

Apfel, Joseph H. (1982). Phase retardance of periodic multilayer mirrors. *Appl. Opt.*, 21(4):733.

Apfel, Joseph H. (1984). Graphical method to design internal reflection phase retarders. *Appl. Opt.,* 23(8):1178.

Austin, R. (1970). Narrow band interference light filter. United States Patent #3,528,726, September 15.

Baker, Martin L., and Victor L. Yen (1967). Effects of the variation of angle of incidence and temperature on infrared filter characteristics. *Appl. Opt.,* 6(8):1343.

Ballik, E. A. (1971). Area and wavelength sensitivity of a photomultiplier. *Appl. Opt.,* 10(3):689.

Bartell, F. O., E. L. Dereniak, and W. L. Wolfe (1980). The theory and measurement of bidirectional reflectance distribution function (BRDF) and bidirectional transmittance distribution function (BTDF). In *Radiation Scattering in Optical Systems*. Proceedings of the SPIE, vol. 257, SPIE, Bellingham, Wash.

Bartolomei, Leroy A. (1976). Striped dichroic filter with butted stripes and dual lift-off method for making same. U.S. Patent #3,981,568, September 21.

Baumeister, Philip (1981). Theory of rejection filters with ultranarrow bandwidths. *J. Opt. Soc. Am.,* 71(5):604.

Baumeister, Philip (1986). Computer software for optical coatings. Photonics Spectra, p. 76, July.

Beauchamp, W. T. (1986). Critical issues for coating on large high power laser components. In *Optical Component Specifications for Laser-Based Systems and Other Modern Optical Systems*. Proceedings of the SPIE, vol. 607, SPIE, Bellingham, Wash.

Bell, Alan E., and Fred W. Spong (1978). Antireflection structures for optical recording. *IEEE Journal of Quantum Electronics,* QE14(7):488.

Benjamin, P., and C. Weaver (1959). Adhesion of metal films to glass. *Proc. Roy. Soc. A,* 254:177.

Benjamin, P., and C. Weaver (1961). The adhesion of evaporated metal films on glass. *Proc. Roy. Soc. A,* 261:516.

Bennett, H. E. (1978). Scattering characteristics of optical materials. *Opt. Eng.* 17(5):480.

Berning, P. H. (1963). Theory and calculations of optical thin films. In Hass, G., ed.: *Physics of Thin Films,* vol. 1. Academic Press, New York and London, pp. 69–121.

Birth, G. S., and D. P. DeWitt (1971). Further comments on the areal sensitivity of end-on photomultipliers. *Appl. Opt.,* 10(3):687.

Born, Max, and Emil Wolf (1970). *Principles of Optics,* 4 ed. Pergamon Press, Oxford.

Buckmelter, J. R., T. T. Saito, R. Esposito, L. P. Mott, and R. Strandlund (1975). Dielectric coated diamond turned mirrors. In *Laser Induced Damage in Optical Materials,* National Bureau of Standards Special Publication 435.

C. I. E. (1978). Supplement No. 2 to C. I. E. Publication No. 15 (1971), Bureau Central de la C. I. E., Paris.

Carniglia, C. K. (1979). Scalar scattering theory for multilayer optical coatings. *Opt. Eng.,* 18(2):104.

Carniglia, C. K., and Joseph H. Apfel (1980). Maximum reflectance of multilayer dielectric mirrors in the presence of slight absorption. *J. Opt. Soc. Am.,* 70(5):523.

Clarke, D., and J. F. Grainger (1971). *Polarized Light and Optical Measurement*. Pergamon Press, Oxford.

Cox, J. T., and G. Hass (1958). Anti-reflection coatings for germanium and silicon in the infrared. *J. Opt. Soc. Am.,* 48(10):677.

Cox, J. T., J. E. Waylonis, and W. R. Hunter (1959). Optical properties of zinc sulfide in the vacuum ultraviolet. *J. Opt. Soc. Am.,* 49(8):807.

Crabb, R. L. (1972). Evaluation of cerium stablized microsheet coverslips for higher solar cell outputs. Conference Records of the Ninth IEEE Photovoltaic Specialists Conference.

DeBell, G. W. (1978). Computational methods for optical thin films. In *Optical Coatings Applications and Utilization II*. Proceedings of the SPIE, vol. 140, SPIE, Bellingham, Wash.

Demichelis, F., E. Mezzetti-Minetti, L. Tallone, and E. Tresso (1984). Optimization of optical parameters and electric field distribution in multilayers. *Appl. Opt.,* 23(1):165.

Dept. of Defense (1962). *Mil-HNDBK-141*. Defense Supply Agency, Washington, D. C.

Dirks, A. G., and H. J. Leamy (1977). Columnar microstructure in vapor-deposited thin films. *Thin Solid Films,* 47:219.

Dobrowolski, J. A. (1986). Comparison of the Fourier transform and flip-flop thin-film synthesis methods. *Appl. Opt.,* 25(12):1966.

Donovan, T. M., P. A. Temple, Shiu-Chin Wu, and T. A. Tombrello (1980). The relative importance of interface and volume absorption by water in evaporated films. In Bennett, H. E., A. J. Glass, A. H. Guenther, and B. E. Newnam, eds.: *Laser Induced Damage in Optical Materials, 1979*. National Bureau of Standards Special Publication no. 568, Washington, D. C.

Driscoll, Walter G., and William Vaughan, eds. (1978). *Handbook of Optics*. McGraw-Hill, New York.

Edwards, D. K., J. T. Gier, K. E. Nelson, and R. D. Roddick (1961). Integrating sphere for imperfectly diffuse samples. *J. Opt. Soc. Am.,* 51(11):1279.

Ennos, Anthony, E. (1966). Stresses developed in optical film coatings. *Appl. Opt.,* 5(1):51.

Epstein, L. Ivan (1952). The design of optical filters. *J. Opt. Soc. Am.,* 42(11):806.

Glang, R., R. A. Holmwood, and R. L. Rosenfeld (1965). Determination of stress in films on single crystalline silicon substrates. *Rev. Sci. Instr.* 36, no. 1, p. 7 (Jan.).

Gee, John R., Ian J. Hodghinson, and H. Angus Macleod (1985). Moisture-dependent anisotropic effects in optical coatings. *Appl. Opt.,* 24(19):3188.

Guenther, K. H. (1981). Nodular defects in dielectric multilayer and thick single layers. *Appl. Opt.,* 20(6):1034.

Guenther, K. H., and H. Enssle (1986). Ultrasonic precision cleaning of optical components prior to and after vacuum coating. In *Conference on Thin Film Technologies*. Proceedings of the SPIE, vol. 652, SPIE, Bellingham, Wash.

Hahn, R. E., and B. O. Seraphin (1978). Spectrally selective surfaces for photothermal solar energy conversion. In Hass, Georg, and Maurice H. Francombe, eds.: *Physics of Thin Films*. Academic Press, New York.

Haisma, Jan, Jan H. T. Pasman, Johan M. M. Pasmans, and Pieter van der Werf (1985). Wide-spectrum tint-free reflection reduction of viewing screens. *Appl. Opt.,* 24(16):2679.

Harding, G. L., B. Window, D. R. McKenzie, A. R. Collins, and C. M. Horwitz (1979). Cylindrical magnetron sputtering system for coating solar selective surfaces onto batches of tubes. *J. Vac. Sci. Technol.,* 16(6):2105.

Harrick, N. J. (1967). *Internal Reflection Spectroscopy*. Interscience-Wiley, New York.

Hass, G., and J. B. Ramsey (1969). Vacuum deposition of dielectric and semiconductor films by a CO_2 laser. *Appl. Opt.,* 8(6):1115.

Heavens, O. S. (1950). Some factors influencing the adhesion of films produced by vacuum evaporation. *J. Phys. Rad.,* 11:355.

Heavens, O. S. (1965). *Optical Properties of Thin Solid Films*. Dover, New York.

Herpin, André (1947). Calcul du pouvoir réflecteur d'un système stratifié quelconque. *Comptes Rendus,* 225:182.

Hill, Russell J., ed. (1976). *Physical Vapor Deposition*. Temescal, Berkeley, Calif.

Holland, L. (1956). *Vacuum Deposition of Thin Films*. Chapman & Hall, London.

Illsley, Rolf F., Alfred J. Thelen, and Joseph H. Apfel (1969). Circular variable filter. United States Patent #3,442,572, May 6.

Jackson, John David (1962). *Classical Electrodynamics*. John Wiley & Sons, New York.

Jacobs, Carol (1981). Dielectric square bandpass design. *Appl. Opt.*, 20(6):1039.

Jacobson, M. (1986). *Deposition and Characterization of Optical Thin Films*. Macmillan, New York.

Jenkins, F. A., and H. E. White (1957). *Fundamentals of Optics*. 3 ed., McGraw-Hill, New York.

Jet Propulsion Laboratory (1976). *Solar Cell Array Design Handbook*. Report SP 43-38, Table 7.14-1 in volume 2, JPL, 4800 Oak Grove Drive, Pasadena, Calif., 91103.

Johnston, S. C., and S. F. Jacobs (1986). Some problems caused by birefringence in dielectric mirrors. *Appl. Opt.*, 25(12):1878.

Keller, Peter (1983). 1976 CIE-UCS chromaticity diagram with color boundaries. Proceedings of the Society for Information Display, 24(4):317.

Kelley, K. L. (1943). Color designations for lights. National Bureau of Standards Research Paper RP1565, *Journal of Research of NBS,* 31:271; also see the EIA Publication of the JEDEC IGC, Optical characteristics of cathode ray tubes, 1955.

Kittel, Charles (1968). *Introduction to Solid State Physics,* 3 ed., John Wiley, New York, p. 150 et seq.

Klinger, R. E., and C. K. Carniglia (1985). Optical and crystalline inhomogeneity in evaporated zirconia films. *Appl. Opt.*, 24(19):3184.

LaBudde, Edward V., Robert A. LaBudde, and Craig M. Shevlin (1983). Theoretical modeling, calculations, and experiments characterizing the laser-induced hole-formation mechanism of an in-contact optical disk medium. In *Optical Data Storage*. Proceedings of the SPIE, vol. 382, SPIE, Bellingham, Wash.

Leybold-Heraeus Vacuum Products, Inc. (1984). *Product and Vacuum Technology Reference Book,* p. 77.

Li, H. H. (1976). Refractive index of alkali halides and its wavelength and temperature derivatives. *J. Physical and Chemical Reference Data,* 5(2):329.

Li, H. H. (1980). Refractive index of silicon and germanium and its wavelength and temperature derivatives. *J. Physical and Chemical Reference Data* 9(3):561.

Liddell, Heather M. (1981). *Computer-aided Techniques for the Design of Multilayer Filters*. Adam Hilger Ltd., Bristol.

Lyon, D. A. (1946). Method for coating optical elements. United States Patent #2,398,382, April 16.

Macleod, H. A. (1969). *Thin-Film Optical Filters*. American Elsevier, New York.

Macleod, H. A. (1983). Design of switching filters. *J. Opt. Soc. Am.*, 73(12):1879.

Maissel, Leon I., and Reinhard Glang, eds. (1970). *Handbook of Thin Film Technology*. McGraw-Hill, New York.

Martinu, L., H. Biderman, and L. Holland (1985). Thin films prepared by sputtering magnesium fluoride in an RF planar magnetron. *Vacuum,* 35(12):531.

Maxwell-Garnett, J. C. (1904). *Phil. Trans. Roy. Soc.,* A203:385 [also see A205:237 (1906)].

McCarthy, Donald E. (1967). Black polyethylene as a far infrared filter. *J. Opt. Soc. Am.*, 57(5):699.

McNally, James J., Ghanim A. Al-Jumaily, John R. McNeil, and B. Bendow (1986). Ion assisted deposition of optical and protective coatings for heavy metal fluoride glass. *Appl. Opt.*, 25(12):1973.

Merck (1976). *The Merck Index,* 9 ed. Martha Windholz, ed., Merck and Co., Inc., Rahway, N.J.

Minot, Michael Jay (1976). Single layer, gradient refractive index antireflection films effective from 0.35 to 2.5 micrometers. *J. Opt. Soc. Am.*, 66(6):515.

Mittal, K. L. (1978). *Adhesion Measurement of Thin Films, Thick Films, and Bulk Coatings*. American Society for Testing and Materials, Philadelphia.

O'Hanlon, John F. (1980). *A User's Guide to Vacuum Technology*. Wiley-Interscience, John Wiley & Sons, New York.

Ordal, M. A., L. L. Long, R. J. Bell, S. E. Bell, R. R. Bell, R. W. Alexander, Jr., and C. A. Ward (1983). Optical properties of the metals Al, Co, Cu, Au, Fe, Pb, Ni, Pd, Pt, Ag, Ti, and W. *Appl. Opt.,* 22(7):1099.

Presland, A. E. B., G. L. Price, and D. L. Trimm (1972). Hillock formation on thin silver films. *Surface Science,* 29:424; and The role of microstructure and surface energy in hole growth and island formation in thin silver films. *Surface Science,* 29:435.

Pulker, H. K. (1984). *Coatings on Glass*. Elsevier, Amsterdam.

Rancourt, J. (1981). Design and production of tellurium optical data disks. In *Advances in Laser Scanning Technology*. Proceedings of the SPIE, vol. 299, SPIE, Bellingham, Wash.

Rancourt, J. D. (1984). Anti-halo coatings for cathode ray tube faceplates. In Proceedings of the Society for Information Display, vol. 25, no. 1, p. 43.

Rancourt, J., W. T. Beauchamp, V. Foster, and I. Sachs (1984). Emissivity enhancement of fused silica for space applications. Proceedings of the Seventeenth IEEE Photovoltaic Specialists Conference, p. 201.

Rancourt, James D., and Robert L. Martin, Jr. (1986). High temperature lamp coatings. In *Optical Thin Films II: New Developments*. Proceedings of the SPIE, vol. 678, SPIE, Bellingham, Wash.

Samsonov, G. V., ed. (1978). *The Oxide Handbook,* 2 ed. IFI/Plenum Press, New York, p. 429.

Sankur, Haluk, and R. Hall (1985). Thin-film deposition by laser-assisted evaporation. *Appl. Opt.,* 24(20):3335, 3343.

Savage, J. A. (1985). *Infrared Optical Materials and their Antireflection Coatings*. Adam Hilger, Ltd., Bristol and Boston, p. 144.

Sax, N. Irving (1984). *Dangerous Properties of Industrial Materials,* 6 ed. Van Nostrand Reinhold, New York.

Scheele, S. R. (1977). Scattering from infrared transparent materials. In *Stray Light Problems in Optical Systems*. Proceedings of the SPIE, vol. 107, p. 48, SPIE, Bellingham, Wash.

Schott Optical Glass, Inc., Schott Glass Catalog, Duryea, Penn.

Seddon, R. I. (1977). Opportunities in thin films to meet energy needs. In *The Business Side of the Optical Industry III*. Proceedings of the SPIE, vol. 111, SPIE, Bellingham, Wash.

Seeley, J. S., R. Hunneman, and A. Whatley (1980). Temperature-invariant and other narrow-band IR filters containing PbTe, 4–20 micrometers. In *Symposium on Contemporary Infrared Sensors and Instruments*. Proceedings of the SPIE, vol. 246, p. 83, SPIE, Bellingham, Wash.

Shurcliff, William A. (1962). *Polarized Light Production and Use*. Harvard University Press, Cambridge.

Smith, Howard M., and A. F. Turner (1962). Vacuum-deposited thin films using an optical maser. *J. Opt. Soc. Am.,* 52(11):1319.

Sommer, A. H. (1973). Stability of photocathodes. *Appl. Opt.,* 12(1):90.

Song, Dar-Yuan, R. W. Sprague, H. Angus Macleod, and, Michael Jacobson (1985). Progress in the development of a durable silver-based high-reflectance coating for astronomical telescopes. *Appl. Opt.,* 24(8):1164.

Southwell, William H. (1980). Multilayer coating design achieving a broadband 90° phase shift. *Appl. Opt.*, 19(16):2688.

Southwell, William H. (1982). Using deposited thin-film Fabry-Perot cavities to measure thin-film absorption. *J. Opt. Soc. Am.*, 72(12):1731.

Southwell, William H. (1985). Coating design using very thin high- and low-index layers. *Appl. Opt.*, 24(3):457.

Spiller, Eberhard (1981). In Attwood, D. T., and B. L. Henke, eds.: *Topical Conference on Low-Energy X-ray Diagnostics*. AIP Conference Proceedings 75, p. 124, New York.

Stone, John M. (1963). *Radiation and Optics*. McGraw-Hill, New York.

Stowers, Irving F. (1978). Advances in cleaning metal and glass surfaces to micron level cleanliness. *Journal of Vacuum Science and Technology*, 15(2):751.

Stowers, Irving F., and Howard G. Patton (1978). Cleaning optical surfaces. In *Optical Coatings Applications and Utilization II*. Proceedings of the SPIE, vol. 140, SPIE, Bellingham, Wash., p. 16.

Strong, John (1938). *Procedures in Experimental Physics*. 1 ed. Prentice-Hall, New York, p. 376.

Thelen, Alfred (1963). Multilayer filters with wide transmittance bands. *J. Opt. Soc. Am.*, 53(11):1266.

Thelen, Alfred (1965). Circularly wedged optical coatings. I. Theory. *Appl. Opt.*, 4(8):977.

Thelen, Alfred (1971). Design of optical minus filters. *J. Opt. Soc. Am.*, 61(3):365.

Thelen, Alfred (1973). Multilayer filters with wide transmittance bands II. *J. Opt. Soc. Am.*, 63(1):65.

Thelen, Alfred (1981). Non-polarizing edge filters. *J. Opt. Soc. Am.*, 71(3):309.

Thelen, Alfred (1987). *Design of Optical Interference Coatings*. Macmillan, New York.

Turner, A. F., and P. W. Baumeister (1966). Multilayer mirrors with high reflectance over an extended spectral range. *Appl. Opt.*, 5(1):69.

Underwood, James H. (1986). Multilayer mirrors for X-rays and the extreme UV. *Optics News*, 3:20.

Vasicek, A. (1960). *Optics of Thin Films*. North Holland, Amsterdam.

Walsh, John W. T. (1958). *Photometry*. 3 ed. Dover, New York.

Weast, Robert C. (1974). *CRC Handbook of Chemistry and Physics*. 55 ed. CRC Press, Cleveland.

Wirtenson, G. Richard, and Douglas Flint (1976). Large aperture wedge filter for contrast enhancement. *J. Opt. Soc. Am.*, 66(2):161.

Wolfe, W. (1965). *Handbook of Military Infrared Technology*. Office of Naval Research, Department of the Navy, Washington, D.C.

Yoldas, B. E., and T. W. O'Keeffe (1979). Antireflective coatings applied from metal–organic derived liquid precursor. *Appl. Opt.*, 18(18):3133.

Young, Leo (1967). Multilayer interference filters with narrow stop bands. *Appl. Opt.*, 6(2):297.

Young, Matt (1985). The scratch standard is only a cosmetic standard. *Laser Focus/Electro Optics*, 21(11):138.

Young, Matt (1986). Scratch-and-dig standard revisited. *Appl. Opt.*, 25(12):1922.

Young, M., E. G. Johnson, and R. Goldgraben (1985). In *Measurement and Effects of Surface Defects and Quality of Polish*. Proceedings of the SPIE, vol. 525, SPIE, Bellingham, Wash., p. 70.

Young, P. A. (1971). Thermal runaway in germanium. *Appl. Opt.*, 10(3):638.

Index